CONSTRUCTING A SENSE OF PLACE

Constructing a Sense of Place

Architecture and the Zionist Discourse

Edited by
HAIM YACOBI

Routledge
Taylor & Francis Group

LONDON AND NEW YORK

First published 2004 by Ashgate Publishing

2 Park Square, Milton Park, Abingdon, Oxon OX14 4RN
711 Third Avenue, New York, NY 10017, USA

Routledge is an imprint of the Taylor & Francis Group, an informa business

First issued in paperback 2017

British Library Cataloguing in Publication Data
Constructing a sense of place: architecture and the
   Zionist discourse. - (Design and the built environment
   series)
   1. Architecture - Israel - History - 20th century 2. City
   planning - Israel - History - 20th century 3. Zionism
   4. Israel - Buildings, structures, etc.
   I. Yacobi, Haim
   720.9'5694

ISBN 978-0-7546-3427-0 (hbk)
ISBN 978-1-138-26427-4 (pbk)

Library of Congress Cataloging-in-Publication Data
Constructing a sense of place: architecture and the Zionist discourse / edited
by Haim Yacobi
      p. cm. -- (Design and the built environment)
   Includes bibliographical references and index.
   ISBN 0-7546-3427-2
   City planning--Israel. 2. Architecture--Israel. 3. Zionism--Philosophy.
   4. Jews--Israel--Identity. 5. National characteristics, Israeli. I. Title:
   Architecture and the Zionist discourse. II. Yacobi, Haim. III. Design and
   the built environment series.

HT169.I784C65 2003
307.1'216'095694--dc22

                                                                2003056293

Contents

List of Figures and Tables   *vii*
Information about Contributors   *x*

INTRODUCTION

Whose Order, Whose Planning?   3
Haim Yacobi

PART I: RESHAPING *TERRA NULLIUS*

1   Contested Zionism – Alternative Modernism: Erich Mendelsohn
    and the Tel Aviv Chug in Mandate Palestine   17
    Alona Nitzan-Shiftan

2   The Flight of the Camel: The Levant Fair of 1934 and the Creation
    of a Situated Modernism   52
    Sigal Davidi Kunda and Robert Oxman

3   Mold   76
    Zvi Efrat

4   Horizontal Ideology, Vertical Vision: Oscar Niemeyer and Israel's
    Height Dilemma   89
    Zvi Elhyani

PART II: FRONTIERS

5   Trapped Sense of Peripheral Place in Frontier Space   119
    Erez Tzfadia

6   The Political Construct of the 'Everyday': The Role of Housing in
    Making Place and Identity   136
    Rachel Kallus

PART III: MIXED SPACES – SEPARATED PLACES

7   Urban Iconoclasm: The Case of the 'Mixed City' of Lod   165
     Haim Yacobi

8   Planning to Conquer: Modernity and its Antinomies in the
     'New-Old Jaffa'   192
     Mark LeVine

PART IV: LANDMARKS OF IDENTITY

9   Academia and Spatial Control: The Case of the Hebrew
     University Campus on Mount Scopus, Jerusalem   227
     Diana Dolev

10   Re-Placing Memory   247
     Yael Padan

11   Geometrical Wounds: The Work of Zvi Hecker in Israel, 1990
     and 2000   264
     Timothy Brittain-Catlin

PART V: PLACE/KNOWLEDGE

12   On Belonging and Spatial Planning in Israel   285
     Tovi Fenster

13   Fragile Guardians: Nature Reserves and Forests Facing Arab
     Villages   303
     Naama Meishar

EPILOGUE

14   A Moment of Change? Transformations in Israeli Architectural
     Consciousness Following the 'Israeli Pavilion' Exhibition   329
     Shelly Cohen

Index   *352*

List of Figures and Tables

Figure 1.1   The logo of the 1994 'Bauhaus in Tel Aviv' celebrations   18
Figure 1.2   Zeev Rechter, Engle House, Tel Aviv, 1933   20
Figure 1.3   Erich Mendelsohn, Schocken library, Jerusalem, 1937,
             a detail of the exterior   20
Figure 1.4   Tel Aviv in the 1930s   24
Figure 1.5   Illustrations from Arieh Sharon, 'Planning of
             Cooperative Houses', *Habinyan* (August 1937)   28
Figure 1.6   Weizmann House, exterior view   38
Figure 1.7   Weizmann House, view of the courtyard   39
Figure 1.8   Hadassah Hospital, aerial view of the courtyards   40
Figure 2.1   General view of the fairgrounds, 1934   53
Figure 2.2   Galina Restaurant, 1934. One of the most outstanding
             pavilions in the fairgrounds. Architects Averbuch,
             Gidoni, Ginzburg.   63
Figure 2.3   Cover of catalogue for the Levant Fair 1934.
             Designer unknown.   66
Figure 2.4   Poster for the 1936 Levant Fair (left). Designers, the
             Samir brothers. Poster for the 1934 Levant Fair (right).
             Designer unknown.   71
Figure 4.1   View of Mount Carmel and Haifa Lower Town, 1964.
             On hilltop: Dan Carmel Hotel. Photograph: Fritz
             Cohen, courtesy Government Press Office,
             Jerusalem, Photography Department.   93
Figure 4.2   Detail of Brise-Soleil of the Administration Building,
             Hebrew University, Givat Ram Campus (D. Carmi,
             Z. Meltzer, R. Carmi, Jerusalem, Israel, 1954–56) on
             cover of *Handasah V'adrichalut* (IAEA Journal),
             April 1956   95
Figure 4.3   Y. Perlstein, G. Ziv Arch., S. Ben-Avraham Eng.: Shalom
             Mayer Tower, Herzl Square, Tel Aviv, 1959–66.
             Developers: Wolfson-Clore-Mayer. Aerial
             photo-montage. From the promotional brochure 'The
             Shalom Mayer Tower', issued by the developers.   99
Figure 4.4   O. Niemeyer (with A. El-Hanani and I. Lotan),
             Kikar Hamedinah Project, Tel Aviv, 1964. Developer:

	Tel Aviv Municipality. Early version with three towers. Photograph: Y. Lior, courtesy Y. Lior and *Ha'aretz* daily photographs archive.	102
Figure 4.5	O. Niemeyer, Nordia Plan, Tel Aviv, early version, 1964, model photographed against the background of Tel Aviv beach. Photographer unknown, from the promotional brochure 'NP' (Nordia Plan), issued by the developer Y. Federman.	103
Figure 4.6	O. Niemeyer, H.G. Muller, S. Rawett, G. Dimanche, Haifa University, 1964. Developer: Haifa municipality. Photo montage of the plan model on the site of the Carmel National Park.	107
Figure 4.7	O.Niemeyer, H.G. Muller, S. Rawett, G. Dimanche, Negev Plan, a proposal for a new city in the Negev Desert (southern Israel), 1964. Developer: Ministry of Housing and Development. Photograph of a model buried in the sands of Tel Aviv sea shores. Photographer unknown.	109
Table 5.1	'Describe your sense of the town in which you reside': mean scores for the semantic differential measurement	128
Table 5.2	'Describe your sense of the town that you reside in, before and after the arrival of the "Russian" (FSU) immigrants': mean scores for the semantic differential measurement	131
Figure 6.1	A view of Gilo housing (Gilo housing cluster number 11, architect Salo Harshman)	137
Figure 6.2	Decorated defensive wall on Anafa Street, Gilo, 2001	138
Figure 6.3	View of Beit Jalla from a living room window of an apartment on Anafa Street, Gilo, 2002	138
Figure 6.4	Residential neighborhoods constructed around Jerusalem 1967–90	146
Figure 6.5	Plan of Gilo	148
Figure 6.6	A study of Gilo as viewed from Bethlehem	150
Figure 7.1	An aerial photograph of Lydda, 1918. Source: The Hebrew University Aerial Photographs Collection.	171
Figure 7.2	The new western design for Lydda Junction housing district. Source: The Israeli Rail Company Archive.	172
Figure 7.3	Holliday's Town Planning Scheme. Source: The Israeli Antiquities Authority Archive. In Hymann, 1994.	173

Figure 7.4   An aerial photograph of Lydda, 1944. Source: The
                Hebrew University Aerial Photographs Collection.   174
Figure 7.5   Polcheck's Town Planning Scheme. Source: Polcheck's
                private archive.   176
Figure 7.6   Bar's Town Planning Scheme. Source: Bar, 1953.   181
Figure 9.1   The Dome of the Rock viewed through the Hebrew
                University Synagogue glass wall, Mount Scopus,
                Jerusalem   228
Figure 9.2   Drawing of the proposed west front of the Hebrew
                University, Frank Mears, 1919   233
Figure 9.3   An aerial view of the Hebrew University, 1974,
                Jerusalem   236
Figure 10.1   The Valley of the Communities by architects Dan Zur
                and Lippa Yahalom   255
Figure 10.2   Yad Layeled, The Children's Memorial Museum by Ram
                Karmi and Partners (on the left)   258
Figure 11.1   The Spiral House   266
Figure 11.2   Adi Nes, *Untitled*, 1996 (courtesy Dvir Gallery,
                Tel-Aviv)   276
Figure 11.3   The Palmach House   277
Figure 12.1   The grave of Iz A-din el Kassam in Balad ash-Sheikh
                graveyard, Nesher, Haifa   287
Figure 12.2   The disputed bridge crossing over some graveyards,
                Balad ash-Sheikh graveyard, Nesher, Haifa   288
Figure 14.1   R. Segal and E. Weitzman, *A Civilian Occupation: The
                Politics of Israeli Architecture* (2002), cover of the
                catalog. Designed by Tartakover Design.   337
Figure 14.2   Z. Efrat, 'Borderline Disorder' (2002), detail of the
                Israeli installation at the 8th Architecture Biennale   346
Figure 14.3   S. Cohen, Z. Tuvi, A. Tsruya and N. Meishar,
                'Under Construction' (2002), detail of the Israeli
                installation at the 8th Architecture Biennale   347

Information about Contributors

Timothy Brittain-Catlin (Boaz ben Manasseh) is an architect and taught architectural history at the Bezalel Academy of Art and Design, Jerusalem, from 1993–2000. During the same period he was project architect, responsible for town and master planning, at Thomas M. Leitersdorf, Architects, in Tel Aviv. He is now a general studies tutor at the Architectural Association in London, and is a regular contributor to the *World of Interiors* and the *Architectural Review*; and he is currently completing his doctorate on the domestic architecture of A.W.N. Pugin at Cambridge University.

Shelly Cohen is a chief curator of the Architects' House Gallery, BArch, Technion, Israel Institute of Technology, BA Philosophy and Art History Tel Aviv University, Currently an MA student at the Cohn Institute for the History and Philosophy of Science and Ideas, Tel Aviv University. Winner of the State Rechter Architecture Prize, 2003.

Sigal Davidi Kunda received her BA in Architecture in 1993 and her Master of Science in Architecture in 2001 at the Technion Israel Institute of Technology, both with distinction. She is a practising architect and lectures on the history of Israeli Architecture at the Holon Academic Institute of Technology and the College of Management Academic Studies Division.

Diana Dolev has completed a PhD at the Bartlett School, University College London. Her thesis deals with the connections between the various master plans for the Hebrew University in Jerusalem and Zionist national identity.

Zvi Efrat is a partner in Efrat-Kowalsky Architects, Tel Aviv. He is the head of the department of Architecture at the Bezalel Academy of Art and Design Jerusalem. He is also a PhD candidate at the School of Architecture, Princeton University and a co-editor of *Architecture: In Fashion* (Princeton Architectural Press, 1994). His forthcoming book, *The Israeli Project: Buildings and*

Architecture 1948–1973 (Tel Aviv Museum of Art and Orion Press) is due in April 2005.

Zvi Elhyani earned a BArch with honors at the Bezalel academy of Art and Design, Jerusalem, and recently his MscArch Magna Cum Laude at the Faculty of Architecture and Town Planning, the Technion, Haifa, where he is currently studying towards a PhD in architectural history. His master's thesis research covers the work of Oscar Niemeyer in 1960s Israel. He is a regular contributor for Israeli and international publications on art and architecture.

Tovi Fenster is a Senior Lecturer at the Department of Geography and Human Environment, Tel Aviv University, Israel. She has published articles and book chapters on ethnicity, citizenship and gender in planning and development. She is the editor of *Gender, Planning and Human Rights* (1999, Routledge). She is the writer of *The Global City and the Holy City: Narratives on Knowledge, Planning and Diversity* (2004, Pearson). She is one of the founders and has been a chairperson of Bimkom – Planners for Planning Rights in Israel.

Rachel Kallus is a Senior Lecturer at the Technion, where she teaches architecture, urban design and town planning. She is also a practising architect, and has worked in the United States, The Netherlands and Israel. She received her Master of Architecture from the Massachusetts Institute of Technology and her Doctorate from the Technion, Israel Institute of Technology. Her research in the field of housing and urbanism focuses on the interplay between policy and design and examines the relationships between policy measures and their physical outcomes, especially in relation to equity, equality and social justice.

Mark LeVine is Assistant Professor of History at the University of California, Irvine. He received his PhD in Middle Eastern Studies at New York University in 1999 and has been a fellow at the NYU's International Center for Advanced Studies, Cornell's Society for the Humanities, and a Jean Monnet Fellow at the Robert Schuman Centre for Advanced Studies at the European University Institute in Florence. He is also a contributing editor at *Tikkun* magazine and writes frequently for academic, journalistic and web publications. His first book, *Overthrowing Geography: Jaffa, Tel Aviv and the Struggle for Palestine*, is forthcoming from the University of California Press.

Naama Meishar is a landscape architect, an artist and an art curator. She is a graduate of the Technion-Israel Institute of Technology. She is currently researching a MA thesis on aspects of gender, ethnicity and class in marginal environments in Israel at the Cultural Studies department at the Hebrew University in Jerusalem. Her art concerns aspects of nature production. She lives in Tel Aviv.

Alona Nitzan-Shiftan is an architect and a doctoral student at the Program in History, Theory and Criticism of Art and Architecture at the Massachusetts Institute of Technology. She received the Lady Davis Fellow 2000–2002 at the Center for Advanced Study in the Visual Arts at the National Gallery, Washington DC. She has taught at the Technion between 1997 and 2000, and has written on issues concerning architectural modernism, post-war architectural culture, nationalism, Orientalism and historiography, mostly in the Israeli context.

Robert Oxman is Professor Emeritus of Architecture and Design and the former Dean of the Faculty of Architecture and Town Planning, Technion Israel Institute of Technology. He has published widely in the fields of contemporary architectural theory and the history of Israeli modernism.

Yael Padan is a practising architect, a partner in 'Architype' Architects, Jerusalem. She also studied at the Bartlett School of Architecture, University College London, obtaining an MA in History of Modern Architecture. Her articles that deal with Israeli contemporary and architectural history and theory are published in the *Architectural Design Journal*. Also, she is one of the founders of Bimkom – Planners for Planning Rights in Israel.

Erez Tzfadia is Lady Davis postdoctoral fellow at the Department of Geography at the Hebrew University, Jerusalem. He studied previously at Ben-Gurion University, Beer-Sheva, where he has written a PhD dissertation on 'Immigrants in Peripheral Towns in the Israeli Settler Society: Mizrahim in Development Towns Face Russian Migration' (2002). His research and teaching are in social geography, focusing especially on settler societies facing persisting waves of immigration, accelerating urbanization, and intensifying ethnic conflicts. His approach is critical, attempting to dissect official policies, ideologies and legal structures, and ascertain their influence on the structural stratification and tension within peripheral urban localities.

Haim Yacobi is an architect and planner, graduated from the Bezalel academy of Art and Design, Jerusalem. He received his Master from the Bartlett School of Architecture, University College London, obtaining an MA with overall Distinction in International Housing Studies. He also studied at Ben-Gurion University, Beer-Sheva, where he has written a PhD dissertation on Israeli Jewish-Arab mixed cities. His main field of research deals with architecture, planning and power relations, and his article were published in several international journals. He is also one of the establishers of Bimkom – Planners for Planning Rights in Israel.

INTRODUCTION

Whose Order, Whose Planning?

Haim Yacobi

> ... [N]either cities nor places in them are unordered, unplanned; the question is only whose order, whose planning, for what purpose ... (Marcuse, 1995, p. 244).

Introduction

In the last two decades scholars from different disciplines such as political sciences, sociology, history and geography have produced a vast and rich body of knowledge that examines critically the Zionist national project.[1] Beyond the differences and often disagreements among the writers, they have all contributed to an understanding of the Israeli national space as a product of power relations resulting from social characteristics that can be located within the wider definition of a settler society. This discourse acknowledges the spatial meaning of the Zionist project and the central role of the Land of Israel in its national-territorial and conflictual level. The analysis of the Zionist project has raised two interrelated attitudes: the first presents it as the *return* of the ancient Jewish people from the Diaspora to their homeland and the second highlights the *colonial* nature of the European society, established in Israel by the European Jews.[2]

The two attitudes schematically presented here are indeed central for understanding the territorial disputes, as noted by Oren Yiftachel (1999, p. 372):

> ... [T]he 'return' of the Jews to their ancestors' mythical land, and the perception of this land as a safe haven after generations of persecution had a powerful liberating meaning. Yet, the darker sides of this project were nearly totally absent from the construction of an unproblematic 'return' of Jews to their biblical Promised Land.

This discussion, beyond its socio-historical interest, is significantly important for the analysis of the processes in which the Israeli built environment and open landscape were approached, physically shaped and culturally constructed. Yet, despite the extensive academic discourse mentioned above, the significance of architecture and planning to the very act of constructing national space in both tangible and symbolic dimensions has been generally neglected.

This book aims to examine critically the inherent nexus between ideology and the construction of *a sense of place* and to explore the role of architecture and planning as efficient yet polemic practices that serve the hegemonic agenda. Indeed, this book which includes chapters by an emerging generation of architects, planners and geographers proposes some critical viewpoints on the role of architecture and the Zionist discourse. In addition, the different themes that will be discussed in the book will undoubtedly be of some relevance to other cases of settler societies.

However, this book aims to go beyond the sociological debates and interpretations. It will explore the process of constructioning a sense of place in its dual meaning: the physical as well as the discursive. Two interrelated questions will be discussed:

a) *What is the role of architects, planners, architecture and planning as mediators between national ideology and the politization of space?*

This question focuses on the epistemological dimension of the spatial practices, and suggests investigating the 'unquestionable' veracities that the professional disciplines of architecture and planning use in order to 'rationalize' power relations in the name of modernity, order, hygiene and efficiency.

b) *What is the contribution of the very act of shaping the landscape to the construction of a sense of place?*

This question refers to the phenomenological significance of constructing a sense of place, and examines not just what a sense of place is, but how it is produced, transformed and reproduced within a given social context.

The Politization of Sense of Place – a Theoretical Outline

In order to deal with the questions presented above, it will be important to outline some of the theoretical debates about the terms 'place' and 'sense of place' which are used in the fields of architecture, geography and the social sciences. My intention is not to review the literature and the different definitions of these terms, but rather to try and link their main insights to the context of this book.

The distinction between 'place' and 'space' emerged during the 1970s, when a qualitative shift in the field of geography paved the road to the development of social and cultural geography. Tuan (1971) for instance, located 'space' as a general term in opposition to 'place' that was defined as material. This distinction also appeared in the definition of absolute space as a container of material objects, in opposition to relational space that was defined as perceived and socially produced (Madanipour, 1996). Relph (1976, p. 141) emphasized the phenomenological dimension, claiming that 'place' is not an abstract but an experienced phenomenon linked to a process which involves the perception of objects and activities that are used as sources of personal and collective identities.

Space and place became fundamental terms in the field of architectural theory and criticism. Christian Norberg Schultz in his well-known book *Genius Loci* (1979) followed the line of arguments described above and claimed, similar to the definition of absolute space, that 'space' is nothing but the relationships between objects. On the other hand he argued that 'place' is a defined built or natural space that has *meaning,* which stem from personal and collective memories as well as from identity. Indeed, 'space' will transform into 'place' only when we are identified with and define ourselves through it. This work was viewed by many as a critic of the modernist movement in architecture, claiming that it had produced 'spaces' but not 'places'. Schultz's phenomenological approach opened a re-understanding of space and its symbolic attributes and urged architects and planners to produce 'places'.[3]

Simultaneous with the emergence of the phenomenological perspective, a new generation of geographers and urban sociologists, including among others Manuell Castels and David Harvey, presented a structuralist-Marxist approach. They saw the capitalist system as a social structure that can be a key for understanding the organization of space. Furthermore, spatial practices such as of planning and architecture that were seen by the phenomenologists as agents for the production of 'places', were viewed by the Marxist critics as tools in the service of capitalism, which aims to balance between private and collective capital, and thus contains a potential for social oppression (Castells, 1978). Indeed, this school of thought was significant in revealing interrelations between society, space, culture and economy.

Yet, the Marxist point of view lacks an understanding of the everyday practices of the users and their struggle to transform capitalist space into place. In this context it is important to mention the work of Henri Lefebvre who aspired to integrate theories and abstract thought with practices and the tangible daily urban experiences. For Lefebvre (1991) space is a social product and thus a 'sense of place' – though he does not use this term – cannot be seen solely as a reflection of either experience or knowledge. Rather, it is the juxtaposition of three interrelated dimensions: perceived space, conceived space and lived space.

Perceived space, following Lefebvre, relates to physical space and the way in which it is organized. Perceived space contains the very functional uses of space, including infrastructures and the built environment that shape our spatial experiences. Conceived space relates to the way in which professionals such as scientists, mathematicians, planners and architects represent space. This conceptualized space is the result of epistemological processes and developments that cannot be seen as autonomous from the socio-political context in which they are produced. Finally, lived space embodies images, symbols and associative ideas of the 'users' that give meaning to space.

I would like suggest that the construction of a sense of place is based on the three dimensions suggested by Lefebvre; through these lenses the act of shaping the landscape using planning and architecture cannot be discussed aesthetically or technologically but demands refrences to the ideological aspect and the power relations that stand behind it and may siggest answer to Marcuse's questions cited above.

In this context I would like to mention Said's (1993) claim that no one is completely free from the struggle over space. This struggle is both complex and interesting since it is not only about soldiers and cannons but also about ideas, forms, images and imaginings. This struggle is exacerbated while dealing with national space that represents geopolitical and social order, which aims to correlate between the homogeneity of population and its collective identity and the territorial borders. However, in reality there are many cases in which different and contested groups share the built or the natural landscapes, claiming symbolic and often tangible sovereignty on it. The appropriation of space and its connection to those in power that are motivated to reinforce the hegemonic narrative, is expressed in the use and implementation of specific settlement structures, as well as in the use of certain architectural styles. Indeed, the relations between power mechanisms and architecture are not a modern phenomenon. Yet, as noted by Foucault (1982) only since the beginning of the eighteenth century, in correspondence with the development of European nationalism, had architecture and town planning become disciplines with an additional dimension that expresses the state's power as an institution which conducts territorial, social and political order.

An example for the link between architectural style and national identity can be found in eighteenth-century Germany, with the rise of romanticism that represented an essentialist attitude to German nationalism. These ideas were supported and expressed by the architectural discourse at that time which claimed that architecture is a plastic expression to the human spirit, so a specific style – and in the German case the Gothic style – represents the authenticity of German nationalism. Another example, given by Forty (1996), relates to the perception of Gothic revival in the nineteenth century as representative of the English national spirit. This linkage was supported at the time by 'scientific' facts, defined by Collins (1967) as the 'obsession for archeology' that provided proofs for the sentiments to the primordial past.

It can be argued that similar to other cultural representations, architecture is first and foremost a statement of an ideological program. Vale (1992) claims that government buildings concisely symbolize the political power of the state which imposes a certain collective identity and not another. Supported by Bourdieu's conceptualization of the *habitus*, Kim Dovey (1999) follows Vale's line of argument suggesting the term 'framing of place'. According to him 'the key linkage to built form ... is that authority becomes stabilized and legitimized through its symbols' (Dovey, 1999, p. 12). This produces and frames meaning that is transformed into an 'unquestionable' daily reality.

Beyond the general discussion presented above, the very particularity of the Israeli spatial reality calls for a localization of these theories and attitudes towards the meaning of the built environment. The Israeli 'place' is a product of a contested socio-historical process, characterized by the motivation for controlling national space and framing it in a total manner. Such a decisive approach generates counter-products which are also spatially expressed.

The Structure of the Book

The first part of this book is called 'Reshaping *Terra Nullius*'. This section aims to present and critically discuss some of the concepts that have shaped the architectural and planning discourse within the Zionist context, namely: modernism, orientalism and colonialism.

The first chapter, 'Contested Zionism – Alternative Modernism: Erich Mendelsohn and the Tel Aviv Chug in Mandate Palestine' by Alona Nitzan-Shiftan, discusses in details the history and ideology of modern architecture during the 1930s, presenting the different Zionist ideological streams. Beyond the focus on Erich Mendelsohn, which was one of the leading modern architects, Nitzan-Shiftan argues that there was a dialogue between modernist form and the social, political and ethical convictions of the Zionist project. She highlights the intense debates that took place during the 1930s about architectural forms vis-à-vis the revaluation of the self: Jewish culture, nation, society and religion and the other: the Orient and the 'Arab'.

The second chapter is 'The Flight of the Camel: The Levant Fair of 1934 and the Creation of a Situated Modernism' written by Sigal Davidi Kunda and Robert Oxman. In this chapter the writers explore how the plan of the Levant Fair site had aimed to introduce modern and innovative architecture into the urban context of Tel Aviv. Through a detailed study they examine the rhetoric of design in the Levant Fair, arguing that its goal was to establish an identity of the Jewish society in Palestine. Modernist progress was to be manifested – 'a new form for new social order' – while establishing the notion of 'situated Modernism'. The writers' argument is contextualized within the shift that appeared in the Zionist attitude towards conquesting land, in which the urban as a cultural strategy had emerged from the situation of being secondary to the significance of agricultural settlements.

The dispute concerning the Zionist spatial characteristics also appears in Zvi Efrat's chapter, 'Mold'. Efrat deals with the dramatic change, which took place after the establishment of the state of Israel in 1948. This chapter analyses the ideological shift that aimed to change the 'Jewish inclination' from being a 'parasite bourgeoisie' to the formation of a 'landed workers' community'. More specifically, this chapter analyses the first national plan, known as Sharon's plan, which was intended to disperse the incoming Jewish immigrant population and had shaped the spatial characteristics of the modern State of Israel. The plan included constructing 30 new towns and about 400 new agrarian settlements that have replaced about 400 erased Palestinian villages. Indeed, Efrat's main argument is that ideologically Sharon's plan was the expression of Arcadian utopianism, presenting the profound anti-urban sentiments and socialist radicalism that had characterized the hegemonic parties of the Zionist Movement.

Indeed, the post-1948 Zionist ideology was pro-agrarian, and the 'vertical option' was quite alien to it. The spatial strategy of 'horizontal' low-density expansion over the new national territory had dominated the planning and architecture discourse. In the fourth chapter – 'Horizontal Ideology, Vertical Vision: Oscar Niemeyer and Israel's Height Dilemma' – Zvi Elhyani analyzes the critic of Oscar Niemeyer. The well-known planner of Brasilia was invited to Israel in 1964. Elhyani investigates Niemeyer's challenging vertical high-density proposals that included large business and residential projects in the Tel-Aviv area, a mega-skyscraper project, and a visionary proposal for a compact-vertical city for 40,000 inhabitants in the Negev desert.

While the first part of the book focuses on the 'perceived space' – the way in which professionals and decision-makers have framed the Israeli sense of place – the second part, entitled 'Frontiers' presents the lived space; the images and meanings of the sense of place as defined by the 'users'. More specifically, in spatial terms Part II will explore one of the very characteristic phenomenon of Israeli context – the frontier; an essential spatial attribute, which constructs the Israeli sense of place on both social and physical aspects.

The chapter by Erez Tzfadia – 'Trapped Sense of Peripheral Place in Frontier Space' – deals with the development towns in Israel, established as part of the first national plan discussed by Zvi Efrat in Part I. Tzfadia's chapter explores the links between social and spatial peripheriality, evident not only within the Palestinian indigenous society, but has also excluded in the 1950s the oriental Jewish immigrants from Israel's centers of authority and economic wealth. This pattern reappeared 40 years later when a large wave of Jewish immigrants from the former Soviet Union arrived and settled in the development towns. According to Tzfadia this event reframed the nature of the towns and changed them into 'Russian' places, while it is evident that the former immigrants' sense of place is different from what it used to be. For the earlier residents the towns have become unfamiliar both due to massive construction that was changed the architectural appearance and due to the change of population that has modified the nature of the public spaces.

Rachel Kallus analyses the reproduction of a different typology of frontier. In her chapter – 'The Political Construct of the 'Everyday': The Role of Housing in Making Place and Identity' – she deals with Gilo, a Jewish neighborhood built as part of the Israelization process of Jerusalem following the 1967 war. This neighborhood has been under fire in the beginning of the second wave of the Palestinian Intifada. Following a detailed study she points out the dual political and social role of public housing in Israel. The first is shaping land, through the Zionist proclamation of a new (Jewish) territory, and the second is shaping identity, by determining a new (Israeli) citizen. Indeed, public housing's space has never been private, claims Kallus, but has always been perceived as a national asset. According to Kallus's argument the private homes, the base of everyday civilian life, have become the guardians of national territory.

Part III, 'Mixed Spaces – Separated Places', discusses the role of architecture, town planning and urban design in two Jewish-Arab 'mixed cities'. The first chapter – 'Urban Iconoclasm: The Case of the "Mixed City" of

Lod' – critically analyzes the dynamics in which the contested urban landscape of the city of Lod is produced, transformed, and reproduced. Following a detailed study which goes back to the British Mandate period and covers the different plans that were designed for the city until the 1990s, this chapter argues that similar to other studies of colonial urbanism, the production of physical and social division in the city of Lod has re-ordered perceptions of reality. This is expressed in an epistemological antinomy of 'here' and 'there'; 'we' and 'they'; 'enlightened new' and 'backward old' respectively.

Mark LeVine in his chapter 'Planning to Conquer: Modernity and its Antinomies in the "New-Old Jaffa"' explores the role of architecture and town-planning in the struggles of territory and identity in the city turned neighborhood of Jaffa. He argues that Zionism, as a modern nationalist movement, is an inherently colonial discourse, and Tel Aviv, the 'modern capital' of Zionist Palestine and now globalized Israel, cannot be understood or examined other than as a colonial city. Through a detailed discussion of architectural discourse he suggests that once we begin to expose the cracks of the four-fold matrix of modernity-colonialism-capitalism-nationalism, the rich history of Jaffa and its struggle with modernity—long buried under the debris of Tel Aviv's perpetual 'growth' and (today) gentrification—begin to emerge.

Part IV, entitled 'Landmarks of Identity', presents to the reader particular cases, which are articulated within the theoretical discussion concerning architectural representations of power relations. Diana Dolev's chapter – 'Academia and Spatial Control: The Case of the Hebrew University Campus on Mount Scopus, Jerusalem' – relates to the significance of the Mount Scopus campus of the Hebrew University, which exceeds its function as a research and higher education institute. Dolev suggests that since Zionist activists had resolved, at the beginning of the twentieth century, to construct the Hebrew University buildings on Mount Scopus, the combination of the institution and its site has created a powerful symbol of Jewish domination. This symbolism has gained extra weight after the 1967 war, accentuating the meaning of the university as a national emblem. The campus architecture became a visual manifestation of Jewish/Israeli sovereignty over the occupied territories, including east Jerusalem.

Yael Padan's chapter 'Re-Placing Memory' deals with one of the fundamental national narratives of the Zionist movement; that of the memory of the holocaust. The chapter argues that preserving the memory of the Holocaust is becoming an increasingly important issue as temporal distance grows from the event itself. Memory, which was passed on by the survivors, is now becoming an organized collective experience. She discusses the current increase in buildings and monuments devoted to the memory of Holocaust commissioned by the Holocaust Martyrs' and Heroes' Remembrance Authority, and by survivors' organizations. Padan claims that in Israel, physical distance from the sites of the Holocaust leaves memory in the realm of narrative, and the museum or memorial is intended to provide a new setting for the projection of memory onto built form. Also, it is suggested that these buildings serve as places of pilgrimage for the Israeli civil religion. The buildings express the importance of memory in constructing national identity, but they also demonstrate changes in the national attitude towards its past.

'Geometrical Wounds: The Work of Zvi Hecker in Israel, 1990 and 2000' written by Timothy Brittain-Catlin deals with the work of the Israeli architect Zvi Hecker. Contextualizing architecture within the Israeli cultural sphere, Brittain-Catlin argues that Hecker's experiments – with constructional techniques explicitly derived from or recalling the practical expediencies of battle, the bunker, the sandbags, the life-or-death improvisations, the pile-caps cropping up as windows, and the denying of heroic forms in favor of an almost pessimistic lack of boundary or mass – form a shift which coincides with the political history of the mid-1990s.

In the last section, 'Place/Knowledge', the epistemological dimension of planning is critically examined. Tovi Fenster's chapter – 'On Belonging and Spatial Planning in Israel' – explores the discursive relationships between notions of sense of place, belonging and spatial planning in Israel. More specifically this chapter is focussed on the conflict that had occurred between Muslim organizations and communities and the local Jewish council of Nesher around the preservation of an old Muslim graveyard in which one of the Arab leaders Iz A-din el Kassam is buried. Through this case study, Fenster explores the dilemmas as to how to define belonging, and who defines which land and its symbolism belongs to whom. She argues that this debate is accentuated especially in the situation of the Palestinian in Israel whose 'sense of place and belonging' contradicts or is perceived as threatening by the 'sense of place and belonging' of the majority hegemonic Jews.

Naama Meishar's chapter entitled 'Fragile Guardians: Nature Reserves and Forests Facing Arab Villages' is a detailed study which examines the way in which post-colonial Palestinian environmental culture at the Arab village of Ein Houd is produced. Meishar criticizes the mode of institutive Israeli open space culture from the theoretical, physical and planning points of view, arguing that this culture is characterized by nostalgia for an imagined and caged-in space of primal nationhood and nature. As a result, she suggests, nature preservation in Israel does not relate to nature on the basis of commitment through biocentric ecology, nor through social ecology, but rather manifests anthropocentric-ethnocentric ecology interrelations.

Finally, at the 'Epilogue' section of this volume Shelly Cohen points out the transformations that had appeared in architectural discourse in Israel in the last few years. Her chapter, 'A Moment of Change? Transformations in Israeli Architectural Consciousness', analyses the Israeli Pavilion and the debates that surrounded the Berlin Congress, as a test case for the coming of age of the architectural discourse in Israel. Cohen suggests that new directions are currently being formed in the Israeli architectural discourse by a new generation of architects, curators and researchers. This shift is expressed in the re-definition of the relations between architecture, politics and the state, which produces a new local discourse that replaces the regional discourse in Israeli architecture and attempts to outline the theoretical context of its transition into a political discourse.

Notes

1   See for example: Greenstein (1995); Kedar (1998); Lustick (1999); Morris (1993); Shafir and Peled (1998); Yiftachel (1997).
2   For a rich overview on this debate see the introduction part in Shafir and Peled (2002).
3   See for example: Alexander, Ishikawa and Silverstein (1977); Gehl (1987).

References

Alexander, C., Ishikawa, S. and Silverstein, M., *A Pattern Language* (New York: Oxford University Press, 1977).
Castells, M., *City, Class and Power* (London: Macmillan, 1978).
Collins, P., *Changing Ideals in Modern Architecture* (Montreal: McGill University Press, 1967).
Dovey, K., *Framing Places – Mediating Power in Built Form* (Routledge, London and New York, 1999).
Forty, A., '"Europe is No More Than a Nation made up of Several Others ..." Thoughts on Architecture and Nationality, Promoted by the Taylor Institute and the Martyrs' Memorial in Oxford', *AA Files* (32) (1996).
Foucault, M., 'Space, Knowledge, Power', interview with Paul Rabinow, in N. Leach (ed.), *Rethinking Architecture – A Reader in Cultural Theory* (London and New York: Routledge, 1982 [1997]), pp. 367–80.
Gehl, Y., *Life Between Buildings – The Use of Public Space* (New York: Van Nostrand Reinhold, 1987).

Greenstein, R., *Genealogies of Conflict* (Hanover and London: Wesleyan University Press, 1995).

Kedar, S., 'Majority Time, Minority Time: Land, Nation and the Law of Adverse Possession in Israel', *Iyyuney Mishpat*, 21 (3) (1998), pp. 665–746 (Hebrew).

Lefebvre, H., *The Production of Space* (Oxford: Blackwell, 1991).

Lustick, I., 'Israel as a Non-Arab State: The Political Implications of Mass Immigration of Non-Jews', *Middle East Journal*, 53 (3) (1999), pp. 416–33.

Madanipour, A., *Design of Urban Space – An Inquiry into a Socio-spatial Process* (Chichester: John Wiley and Sons, 1996).

Marcuse, P., 'Not Chaos, but Walls: Postmodernism and the Partitioned City', in S. Watson and K. Gibson (eds), *Postmodern Cities and Space* (Oxford: Blackwell, 1995), pp. 187–98.

Morris, B., *The Birth of the Palestinian Refugee Problem, 1947–1949* (Tel Aviv: Am Oved, 1993) (Hebrew).

Norberg-Schultz, C., *Genius Loci – Towards a Phenomenology of Architecture* (New York: Rizzoli, 1979).

Relph, E., *Place and Placelessness* (London: Pion, 1976).

Said, E., *Culture and Imperialism* (New York: Vintage Books, 1993).

Shafir, G. and Peled, Y., 'Citizenship and Stratification in an Ethnic Democracy', *Ethnic and Racial Studies*, 21 (3) (1998), pp. 408–27.

Shafir, G. and Peled ,Y., *Being Israeli – The Dynamics of Multiple Citizenship* (Cambridge: Cambridge University Press, 2002).

Tuan, Y., *Space and Place: The Perspective of Existence* (Minneapolis: Minneapolis University Press, 1977).

Vale, L.J., *Architecture, Power, and National Identity* (New Haven and London: Yale University Press, 1992).

Yiftachel, O., 'The Political Geography of Ethnic Protest: Nationalism, Deprivation and Regionalism among Arabs in Israel', *Transactions: Institute of British Geographers*, 22 (1) (1997), pp. 91–100.

Yiftachel, O. '"Ethnocracy": The Politics of Judaizing Israel/Palestine', *Constellations*, 6 (3) (1999), pp. 364–90.

PART I
RESHAPING *TERRA NULLIUS*

Chapter 1

Contested Zionism – Alternative Modernism: Erich Mendelsohn and the Tel Aviv Chug in Mandate Palestine[1]

Alona Nitzan-Shiftan

In memory of Royston Landau[2]

Introduction

In preparation for the 1994 celebrations of the 'Bauhaus in Tel Aviv', the newly renovated city was painted white. To its numerous visitors, Tel Aviv boasted the largest concentration worldwide of 1930s modernist buildings. The impact of this 'live museum' was compounded by numerous exhibitions and street-festivities that promoted the Mandate period 'International Style Architecture' as a national heritage (Figure 1.1). Scholarship was crucial to this campaign: a large number of new publications focused attention on architects who shaped the built landscape of the Jewish population in Mandate Palestine known as the Yishuv.[3] The pattern emerging from these studies is of a perfect fit between modern architecture and Zionism.[4] Both Le Corbusier and the leaders of the Zionist movement, the argument goes, were simultaneously 'creating something out of nothing'.[5]

This chapter challenges such depictions of a neat juncture between modern architecture and Zionism by exposing the profound ideological tensions embedded in the architectural production of the 1930s. I illustrate this tension by pointing to the difference between the work of the Tel Aviv architectural circle – known by the Hebrew word for 'circle', Chug – and the architecture of Erich Mendelsohn in Palestine.[6] I propose that the architecture of the Chug reflected the ideology of the socialist leadership, which was inspired by Herzl's *political* Zionism. By contrast, Mendelsohn's architecture gave form to Martin Buber's interpretation of Ahad Ha'am's *cultural* Zionism.[7] These affiliations corresponded to different strands of architectural modernism: the Chug espoused an international 'new architecture', while Mendelsohn

Figure 1.1   The logo of the 1994 'Bauhaus in Tel Aviv' celebrations

developed a localist modernism with Orientalist touches. They both expressed very different positions regarding the desired national identity of Jewish settlers in Mandate Palestine, and the modern architecture that could best express such identity.

During the 1920s and 1930s modernism and Zionism were plural movements still debating their eventual forms. Both were ultimately reduced to a type of official story, which rendered them indispensable to the myth of statehood. In order to unify an official narrative about the origin of Israeli architecture historians advocated a visual reading of modern architecture as a formal style. This approach, which was established in the famous 1932 'International Style' exhibition at the Museum of Modern Art in New York,[8] underlined the 1984 pivotal exhibition 'White City: International Style Architecture in Israel, a Portrait of an Era'.[9] The exhibition's new emphasis on the formal properties of the country's existing modernist buildings presented to the bourgeois Israeli of the 1980s and 1990s a tangible past of physical objects rather than an ideology no longer in vogue. Once popularized as a style, this modernist vernacular was re-politicized as a visual emblem of 'the modest' Zionist spirit which produced it.[10] The 'portrait of an era' thus confirmed the hegemonic ideology that had formed the State of Israel. Moreover, it suggested to Israelis alternative architectural roots in Tel Aviv. After six years of Palestinian uprising and a growing internal conflict between secular and religious factions, the 1994 'Bauhaus' or 'International Style' celebrations provided the secular bourgeoisie with a reassuring architectural heritage.

The popularity of the Bauhaus in Israeli consciousness echoes an active agreement between a leading architectural trend, a national ideology, and a historiography which binds the two.[11] The Chug is an example of such a bond. It was formed in 1932 by young members of the Yishuv, who returned to Palestine after acquiring architectural education and apprenticeship in Europe. Being integrated into and reflecting the socialist leadership of the Yishuv helped the Chug to institutionalize modern architecture in Tel Aviv and beyond. In order to gain such a powerful position they needed a clear message, which eventually subsumed the initial diversity of the group's members. The architecture promoted by the Chug became that of the newly founded state by the end of the British Mandate in 1948, with Arieh Sharon, a Bauhaus graduate and a Chug founding member, as its head architect.

Unlike the Chug, Erich Mendelsohn openly disapproved of the mainstream socialist ideology. His practice in Palestine between 1934 and 1941 was sponsored by the circles of the World Zionist Organization on the one hand, and the government of the British Mandate on the other. The volume of his built work in Palestine was outstanding in comparison to other modern masters in those years. While Le Corbusier was drawing visionary plans for Algiers and Mies van der Rohe was designing hypothetical 'Court Houses', Mendelsohn was constructing villas, hospitals and colleges, which vastly affected the Zionist landscape.

Mendelsohn shared with the Yishuv the Zionist conviction of establishing a home for the Jewish people in Palestine, and with the Chug the will to build this home in the modern mode. Both parties built prolifically and made significant contributions to the built environment and to Israeli architectural discourse, an influence that extends well into the present. Yet, the complexion of Mendelsohn's firmly grounded plain volumes, courtyards, and carefully punched blank walls opposes the white architecture of the Chug, with its playful strip windows, curves and *piloti* (Figures 1.2 and 1.3). This difference illustrates a tension predicted in 1933 by Yohanan Ratner, the head of the Technion School of Architecture and a major military figure. Ratner envisioned a contention in 'the future development of Palestine architecture' between 'the straightforward modern style' and what he saw as a growing infiltration of 'Oriental influences' into the New Architecture.[12] Mendelsohn's architecture was indeed rooted in Oriental cultural analysis that the Chug rejected. The force of the difference between his architecture and that of the Chug unsettles the neatly constructed juncture between Modern architecture and Zionism.

In the 1960s architectural texts saw modern architecture as the expression of a teleology which began in the 1920s and culminated in the present.[13] Scholars

Figure 1.2   Zeev Rechter, Engle House, Tel Aviv, 1933

**Figure 1.3   Erich Mendelsohn, Schocken library, Jerusalem, 1937, a
            detail of the exterior**

then felt that Mendelsohn's work could neither be ignored nor be followed by mainstream Israeli architects. They were therefore convinced that his influence on Israeli architecture had reached a dead end.[14] The emergence of an art-historical discourse that packaged the modern architecture of the Mandate period as an historical style drastically changed the reception of Mendelsohn's work during the 1980s and the 1990s. Paradoxically, this new official reading of Israel's architectural heritage included Mendelsohn's architecture in Palestine as a major contribution to the International Style legacy despite his many efforts to denounce 'internationalism' in architecture.

As these examples show, any discussion of modern architecture in the 1930s tangles with history and ideology. This chapter, by arguing for a dialogue between modern form and the social, political and ethical convictions of the Zionist project, draws attention to the variety of opinions and intense debates in the 1930s about architectural forms vis-à-vis the revaluation of self (Jewish culture, nation, society and religion) and other (the Orient and the 'Arab'). Such exposure divests the International Style of its all-encompassing claim over modern architecture in Mandate Palestine.

A National Movement and an Architectural Movement

Modern architecture matured during the 1920s as a multifaceted avant-garde. Its advocates promoted technological progress, clean and bare aesthetics, functionalism and efficient means of production. The Modern Movement in Architecture, which was founded in 1928, saw in these merits a jumping board toward a universal architectural expression, an attractive position to socialist movements from Weimar to Moscow. These same characteristics triggered the rejection of modern architecture by most nationalistic regimes. Indeed, this kind of architecture denied the existence of a history describing a territorially bounded ethnic and linguistic community, a necessary prerequisite for the cultivation of national ideologies. It is not surprising that hardly any of the post-World War I national regimes of Europe, the birthplace of modern architecture, embraced this architecture as their national expression.

Yet, in Palestine it was precisely the national instinct, which provoked Jewish immigrants to gather as an ethnic and linguistic community that led architects and ideologists alike to embrace modern architecture as the appropriate expression of Zionism. They clung to its attributes of progress as well as to its lack of identity with forms associated with European nations. Moreover, Jewish architects found a potential for the new vernacular of the

East in the association with Oriental forms, thus transforming the object of Nazi contempt into a virtue. Standard texts on *Israeli* architecture establish its origins in the early experiments with modernism.[15] During the 1920s, the story goes, the battle for national expression opposed Orientalists (Baerwarld) and Modernists (Kaufmann). By the 1930s, 'the second decade of Israeli architecture',[16] modernism had become the visual mould for the Zionist project.

Zionism

The early Zionist movement consisted of two dominant ideological strands: the political and the cultural. Inspired by Leon Pinsker and Theodor Herzl, political Zionism saw the pervasive phenomenon of xenophobia manifest in anti-Semitism. Jews, they thought, provoked a 'reasonable fear' of the unfamiliar. The inevitable failure of emancipation was their premise; they believed a political solution was essential for the acceptance of Jews on an equal footing with other nations. Concerned with the problem of the *Jews* rather than with *Judaism*, they did not insist on Palestine as the sole territorial possibility.

Cultural Zionism, on the other hand, led by Ahad Ha-am, took on the identity crisis of modern Judaism. An emancipated and secular Jew, Ahad Ha-am attempted to reconcile his loss of religious faith with a community traditionally crystallized around religious notions. He overcame this apparent incongruity by replacing the cohesive element of religion with the modern construct of nation: religion was now assumed to have an instrumental value in enforcing the essential being of the Jews as nation. Secularized Judaism became the seed-bed for Zionist identity. If the original nationhood was formed in the land of Israel, then its revitalization must take place on that same land.

Reacting to anti-Semitic discrimination, political Zionism minimized differentiating features: the goal was to become a nation like any other, within a community of nations. Ahad Ha-am did not believe this total political solution to be feasible, however. His formulation of cultural Zionism is indebted to his anticipation of potential Arab-Jewish conflict under political Zionism's regime. Rather than laying the foundation for a nation-state, pioneering Jews in Palestine were assigned in his mind the role of a spiritual core for the world's Jewry.

The political achievement of the 1917 Balfour Declaration, in which Britain openly approved of the Palestinian Zionist enterprise, boosted the position of synthetic Zionism, led by Chaim Weizmann. Weizmann (whose formal villa was Mendelsohn's first building in Palestine), was Ahad Ha-am's disciple. He engaged political as well as cultural aspects of the Zionist enterprise: on the one hand he worked toward a political charter from the community of nations; on the other, he advocated Jewish settlement in Palestine and the foundation of cultural institutions for the Jewish renaissance, the archetype of which was the Hebrew University. This symbolic project of spiritual revival was the nucleus of Mendelsohn's architectural effort in Mandate Palestine.

The settlement of Palestine became the main objective of the Zionist movement after its seventh congress in 1905. Meanwhile workers' parties obtained a growing influence over the Jewish settlers. For Weizmann and the World Zionist Organization, the socialist sector in Palestine was but one of the Organization's multiple bodies. However, with the growing anti-Semitism in Europe, the Labour division became predominant in Zionist consciousness and action. Its ideology placed the working class at the centre of the Zionist project. Labour leaders intended to infuse Herzl's political Zionism with activism: immigration, acquisition of lands and settlement. Redeeming the soil and building the land were means for engineering a new society for the New Jew, the ultimate creative task of Labour Zionism. A revolution that started at the level of the individual required a drastic change of life-style. Accordingly, the socialist sector launched mostly housing projects as well as cultural and health-care institutions, many of which were designed by Chug members.

The Chug

The Tel Aviv Chug stood at the forefront of the building boom that shaped Tel Aviv's city-scape during the 1930s (Figure 1.4). The founding members of the Chug – Yoseph Neufeld, Zeev Rechter and Arieh Sharon – like many of its later affiliates, first arrived in Palestine circa 1920 from Eastern Europe.[17] Upon their arrival, they were integrated into the workers' circles, the stronghold of Labour Zionism. They left Mandate Palestine a few years later to study architecture in Europe. At that point, their political affiliation with socialist Zionism as well as their identity as part of the Jewish society in the Land of Israel – *Eretz Israel* – were already established. This identity inspired their pursuit of new trends in the architectural schools of Vienna, Rome, Dessau and Paris during the later 1920s. Once they returned to Palestine in the early 1930s, it was a return 'home'. They were already well-connected in the Yishuv and were practically natives in comparison to those architects who first arrived in the 1930s after the rise of Fascism in Europe.[18]

Figure 1.4   Tel Aviv in the 1930s

The fruitful collaboration between Sharon, Neufeld and Rechter upon their return was based on a shared belief in modern architecture, no less than on the professional interests of young individual professionals who were trying to make an impact on the local market. They owed the non-dogmatic character of their architecture to their diverse European training. Neufeld worked with Mendelsohn and Taut, Rechter worked in France under the 'Corbusian spell', while others worked with Le Corbusier himself. The affinity of Sharon's Kibbutz background with Hannes Meyer's Bauhaus pedagogy is a particularly good example of this active compliance between a Zionist ideology and modernism as still a pluralist and formally undetermined architectural concept.

The Chug launched its first publication, *Habinyan Bamizrah Hakarov* (*Construction* or *Building in the Near East*) in 1934. An opening article in favour of organic architecture by Neufeld[19] set the tone of this pluralistically modern publication, which was intended to represent the entire community. A dispute over this goal resulted in an editorial split, however, after which the Chug established its mature and more exclusive publication, *Habinyan* (*Construction* or *Building*). The editorial rhetoric of *Habinyan's* three issues[20] supported the Zeitgeist portrayal of the Modern Movement's rational principles on the one hand, and its grounding in programme, function, economy and construction method on the other. Both were thoroughly compatible with the Zionist spirit of renewal. This position was most profoundly articulated in the theme articles by Arieh Sharon and Julius Posener, which were in accord with the Yishuv's mainstream ideology. Such a clear, unified message was a powerful tool for institutionalizing modern architecture as the national expression of the Jewish population in Palestine.

The omission of the locally specific from the Chug's publication title, which indicated a narrower message of a functional, international 'new architecture', followed larger cultural and political processes in the Yishuv. As part of the Jewish revival in Palestine, the Chug's activity was inextricably linked to the Yishuv's quest for revolution during this historical moment: the negation of Diaspora life in favour of the construction of 'a national home' (eventually in the form of a nation state); the negation of the bourgeoisie, in favour of an agrarian working society; and the negation of Orientalism (in the face of emerging Arab nationalism) in favour of a new collective image, which would generate the 'sabra' myth – the stereotype of the Israeli born, the native of the Land. By extension, this triple negation shaped the physical collective image of the Jewish settlement in Palestine.

The Negation of the Immediate Past of the Diaspora

The modern movement's rupture with the past was amplified in the Zionist context. For those architects who fled from Europe (the vast majority of Jewish architects practicing in the 1930s),[21] modernism signified a break with the anti-Semitism of their subordinated life in exile. It symbolized the transition from centuries of being a stereotyped minority in Europe to the promise of constructing an autonomous Self. In Palestine, those architects were faced with the urgent task of anchoring this quest for national identity in the built environment, that is, of endowing the 'national home' of the Jewish people with a physical form.

For Julius Posener, who was part of the Chug's intellectual force in the late 1930s, the making of a house particular to the 'Land of Israel' required the effacement of all former identities. In an editorial for the Chug's publication, *Habinyan,* he insisted on the 'lack of self-habits and admiration for building in the modern style'[22] as the necessary starting point for the Jewish settler in Palestine. Posener dissociated the Jew in Palestine from the European nationalistic discourse that identified a people's character with its traditional dwellings. If habit and habitation originated in a common linguistic root, Posener explains, then the 'lack of national-habits' eliminated the options of a familiar national home. He thus pointed at the immanent connection between the erasure of past memories and the creation of a strictly *new,* 'modern, clean character of building'[23] for the Land of Israel.

There were inherent contradictions in the quest for a housing type for a nation of immigrants. Those conflicts underlined Posener's argument regarding the process of rendering the essentially foreign as familiar:

> The land of Israel is not a foreign one in the eyes of the Jew. When he builds, he does not bring from the Diaspora the house of his fathers; on the contrary, the Jew wants to construct here, for the first time in his life, a house of his own, and moreover, this house should be the house of the country in which he is settling.[24]

The request that the house of the Jew 'be the house of the [wishfully familiar] country in which he is settling' did not merely express the ambition to shelter members of the emerging Zionist entity, but also expressed the aspiration to instill a new national consciousness in the inhabitants by means of modern dwelling. A popular song of this period – 'We have come to the homeland to build [it] and be rebuilt [in it]' – reflects an agenda similar to that of the socialist leadership: the new society is not only the *creation* of the New Jew, but the *creator* of that Jew as well. This view anticipated the subsequent 'moulding' of Jews coming from Arab countries to fit the new national 'creation', based as it was on Western models and built according to the vision of a European avant-garde.

The Negation of the Bourgeois Life Style

For the Zionist pioneers of the 1920s (among whom were many of the Chug's members), life in exile was identified with bourgeois culture. Against this culture in Europe, which had traditionally subordinated Jewish life, they aspired to be liberated individuals, healthy in body and spirit, rejuvenated by the direct connexion to the soil of *Eretz Israel.* Accordingly, the fundamental values of the early Yishuv were to inhabit the country and to work its land – a joint mission of agriculture and construction. Notions such as 'the culture of the land' granted the worker a moral significance. The construct of the pioneer represented health, strength, enthusiasm, morality and love of life: everything that seemed out of reach for the landless Diaspora Jew. The community of the *kibbutz* enabled a new collective consciousness.[25] For architects the kibbutz also provided prestigious commissions. The alignment of construction with agriculture extended the aura of the pioneer to the architect.

Arieh Sharon, for example, capitalized on this connection throughout his career, as the motto of his autobiography *Kibbutz + Bauhaus* affirms.[26] Owing to his kibbutz experience with beehives, Hannes Meyer exempted Sharon from the preparation of Gropius's famous workshops.[27] Back in Palestine he was the architect closest to the socialist political circles, a connection that eventually granted him the position of head architect of the nascent Jewish state.[28]

The Negation of the Enthusiasm with the East

The firm decision to ground the Jewish national revival in the biblical land of Palestine, was based on a bipolar Zionist reading of history which saw in the Yishuv a revival of the ancient community of the Hebrew people.[29] The identity between the Oriental and the biblical, which underlined European Orientalist thinking, suggested to the architects of the 1910s and 1920s a potential for national expression in the wealth of Palestine's indigenous forms. By the early 1930s, this proposition was no longer tenable owing to tension with local communities.

The British Mandate and the Balfour Declaration provoked national sentiment among the Arab population of Palestine. Sporadic violent confrontations between the Arab and the Jewish populations accelerated in 1929, and later between 1936 and 1939. Labour leaders then planted the seeds of what would be debated as 'the Arab Question'. Berl Katzanelson, *the* Labour ideologist, saw in 'the Arab' the hostile Gentile he had feared in Europe.[30] For Ben-Gurion, his partner in the leadership of Labour Zionism, the role of 'the Arab' in the history of Palestine was one of destruction. Under Arab hands, he believed, the grandeur of antiquity fell into ruins.[31]

The gap, which this conflict opened between the Zionist project and the local Oriental culture, obliged Zionists to de-Orientalize the attributes of the Biblical Land. In architectural circles, as well in socialist culture at large, the weight shifted in the early 1930s from a fascination with the Orient toward an appeal to the natural qualities of the land, with which architects were now preoccupied.[32] *Batim min Ha'hol* (*Houses from the Sand*), the title of the publication that accompanied the International Style conference festivities in 1994, continues to celebrate this legacy of Tel Aviv, the first Jewish city, growing from the dunes. For socialists, architecture, as well as agriculture, were not an extension of pre-existing culture, but a product of deliberate acts of labour.

The natural conditions of the land gradually became the primary determinant in architectural production. Architects began to think of themselves as 'researchers' who based their design strategy on culturally neutral factors such as climate. 'Serious arguments between the followers of the western direction and those of the south-west direction'[33] decided how to site a building along the axis that best served its climate control. Such arguments now constituted the functional discourse endorsed by *Habinyan.* Arieh Sharon combined a Bauhaus education with his kibbutz experience to devise an architectural regionalism thoroughly determined by climate, whether it was the Mediterranean sun angles or the sea breeze of Tel Aviv (Figure 1.5). Further functional considerations to generate the appropriate life-style for the worker, or alternatively, the employment of modern building technique, led him to build in what became known as the 'Bauhaus vernacular' – plain cubical, white housing blocks, adorned only by the characteristic balcony, *brise-soleil* and *piloti.* Such were many of the cooperative housing projects, which crowned the architectural efforts of Joseph Neufeld, Israel Dicker, and Carl Rubin, among others. This 'Bauhaus vernacular', together with the kibbutz, became the shibboleth of the Zionist landscape.

For Posener, a latecomer less committed to Zionism than Sharon, the negation of the East was further complicated. From his previous work with Mendelsohn, he retained a belief in the Orient's formal potential for a new

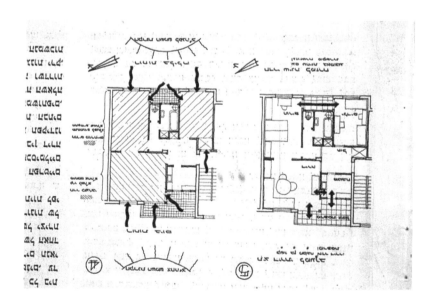

Figure 1.5   Illustrations from Arieh Sharon, 'Planning of Cooperative Houses', *Habinyan* **(August 1937)**

architecture. His approach was the reverse of the one Lutyens employed in India, the one that envisioned New Delhi's architecture as an 'Englishman dressed for the weather'. Unlike the colonizer's objective of preserving British culture, Posener's purpose was precisely the dissolution of such pre-existing identities. The 1920s Orientalist approach of 'clothing a typical English or German villa with an Oriental costume'[34] was a double sin: it maintained European identity and alluded to a rejected Arab culture. Rather, Posener aspired to an architecture that was Oriental enough for a Westerner to live in – a superior alternative to Western architecture. To that end, the Orient needed to be abstracted from its cultural contingencies. Posener was quick to disqualify the Arab village as a model for the Jewish settler in Palestine.[35] Stripping the Palestinian Arab from claims to authoritative architectural knowledge, he directed his reader toward Mediterranean building types ranging from Granada to Teheran. Adhering to formal types of high architecture, rather than to a tangible tradition, liberated the Jewish builder from the necessity to comply with any immediate architectural history. It typified the Yishuv's insecure possession of the land and culture of Palestine.

A Tabula Rasa

As we have seen, the emergence of the Zionist movement in Europe stood remote from its realization in Palestine. The quest of Jewish people for a national identity in Mandate Palestine could not rely upon an immediate past or local culture. The absence of a shared visual heritage allowed the region to be construed as a *tabula rasa*. The ground was thus prepared for the positivist Zionist project whose visual mould was set by modern architecture, the declared epitome of universal rationality.

The Chug adhered to the legacy of the enlightenment. The Zionist project, they believed, was akin to an experiment in which architecture was part of the historical revival on the Promised Land. 'The new village', Posener proclaimed, 'is built ... on the ground of scientific suppositions, in a modern way, or more correctly, it is based on hypothesis'.[36] This line of thinking is similar to Hannes Meyer's re-definition of architecture as 'functional, biological interpretation' which 'logically leads to pure construction'.[37] Indeed, Meyer's disqualification of 'architecture as "an emotional act of the artist"'[38] suited the socialist underpinning of the Chug's members. Likewise, they chose the word *binyan,* meaning building or construction, rather than 'Architecture', as the title of their publications. Meyer's influence on Sharon, his student and employee, is less surprising than the sudden positivism of Posener, who later became a protagonist of expressionist architecture. In the context of the Yishuv, the stark white house was, for Posener, the proper traceless home for the uprooted Jew, 'an apartment free from past memories'.[39]

Buildings

Inspiration was drawn from the abundance of European magazines to which the Chug subscribed. The architectural vocabulary of the Bauhaus, Le Corbusier and Mendelsohn, among others, was popularized and used. There is, however, an obvious gap between the model and its fulfillment. The façades of Tel Aviv, for example, seemingly manifested Le Corbusier's five points: the flat roof, the strip windows and the *piloti,* in particular. But behind this well-composed advertisement for an updated architecture there was neither flowing space nor a liberated façade; the buildings were not entirely lifted by *piloti* and the roof-space was rarely utilized. The strip windows were converted into balconies and, together with the actual wall, constituted a double screen. Stylistically this climatic innovation adhered to the horizontal ocean liner, but it disregarded the Corbusian liberation of façade from construction.

As in many Near and Middle-Eastern countries, Mendelsohn's European architecture became an essential part of the localized modern architectural vocabulary. Mendelsohn was furious to see his curved forms pasted all over Tel Aviv's residential quarters upon his arrival there in 1934. He was not only oblivious to the challenge of housing, but he saw in his famed curvilinear forms a specific response to the lifestyle of the European metropolis. He mocked the efforts of 'imitators' to grasp 'new signs visible on the architectural horizon'. While acknowledging the need for quick, cheap housing, he had little tolerance toward architects' ambition to 'typify the new world' by employing Western materials and architectural methods indiscriminately.[40] In Palestine he searched for an alternative to the Europe from which he had escaped – a new beginning. 'The infallible sign of an original beginning is the birth of an original style' he insisted.[41] But in Palestine, the search for an indigenous style had already been largely abandoned by that time.

Mendelsohn

> I love Eretz Israel and call myself its true child. Whatever work I did, especially my non-realistic outbursts in sketches and conceptions, got its strength from the biblical simplicity, this fulfils itself and embraces the whole world at the same

time. I know that the inimitable quality of my first constructions is of Jewish origins. Early in my youth, I was conscious of it, and that early consciousness made me see the necessity of Zionism. I saw in Zionism the only chance of finding myself and being really creative.[42]

Upon his second arrival in Palestine in 1934 Mendelsohn was already considered one of the luminaries of modern architecture.[43] His way to Palestine was paved by the efforts of some important dignitaries: Zalman Schocken, an industrialist and benefactor of Jewish art, who had previously commissioned Mendelsohn to design his department stores in Germany; Kurt Blumenfeld, the leader of the Zionist movement in Germany, who was Mendelsohn's friend from his youth; above all, Chaim Weizmann, the head of the World Zionist Organization and the first president of Israel. It was no accident that Mendelsohn built a house and library for Schocken in Jerusalem and that Weizmann's representative villa in Rehovot was his first commission in Palestine. Not everybody, however, was quite so excited about Mendelsohn's arrival. The writer S.Y. Agnon, for example, a future winner of the Nobel Prize for literature, who knew Mendelsohn in Berlin before the Great War, was cold if not cynical. Against the Yishuv's enthusiastic reception of 'the great architect', Agnon reveals Mendelsohn's derisive comments on the socialist ward heelers' understanding of architecture, art and beauty.[44]

Mendelsohn's anti-socialist leaning exemplified a long-standing tension between German and East European Jewries, or more accurately, between the Zionist movement in Germany and the predominantly Eastern European leadership of the Yishuv in Palestine. In Europe, the former had been confident of its cultural superiority. In Palestine, meanwhile, the latter was presenting its socialist project as a *fait accompli*. The opposition between the two (by no means unified) groups was based on their different conceptions of Zionism. This section of the chapter will focus on the ties between the particular traits of Mendelsohn's Zionism and his architecture in Palestine between 1934 and 1941. In what follows, I elaborate on Mendelsohn's views about nationalism, Orientalism and the construction of his Semitic identity vis-à-vis 'the Arab' as underlying the Orientalist complexion of his modern architecture in Palestine.

Mendelsohn, Buber and Supranationalism

Mendelsohn's early commitment to Zionism was opposed by the vast majority of German Jews.[45] Recently emancipated, they laboured with limited success to realize their equal status. Flaunting their Jewishness publicly interfered with their hope for cultural assimilation into German society. For the few who chose to manifest their Judaism in a national form well before the advent of Fascism, the commitment to Zionism was not primarily a question of political urgency or menacing anti-Semitism. It was above all a question of an identity crisis in the face of a deceptive emancipation.

Mendelsohn, born in 1887, belonged to an already post-assimilated generation. Breaking with the thin-blooded Judaism of their parents, Mendelsohn's contemporaries found a calling in Martin Buber's definition of Judaism and its national praxis. Like most German Zionists around the 1910s, Mendelsohn was deeply moved by Buber's *Drei Reden uber das Judentum* ('Three Speeches on Jewishness').[46] Buber's mixture of religion (reconstructing Judaism for the post-assimilated Jew), art (Expressionism and an emphasis on the primacy of music) and politics (Zionism and supra-nationalism) described Mendelsohn's relationship to his Jewishness.[47] As a member of early twentieth-century German avant-garde circles,[48] Buber strove to mesh his early teachings regarding the creativity of the Jewish persona with the objective of national revival.

Buber had initially followed the cultural Zionism of Ahad Ha-am. However, whereas Ahad Ha-am had embraced the loss of religious faith in the enlightenment, it was this enlightenment with which Buber was disillusioned. Like Ahad Ha-am, he realized that the emancipation of the Jews in Europe constituted a final break with the tradition of Jewish religious practice. Seeking continuity with the spiritual tradition, he condemned the Jewish law and practice of the Diaspora as a rigid authoritative religion. His reconstruction of Jewish culture as a unique spiritual and aesthetic sensibility was grounded in what he called the *religiosity* of the original biblical Jew. Palestine was the locus of this renewal. From there Judaism emerged as a creative religious sentiment, which made a moral contribution to world history.[49]

Buber insisted on the subjugation of a global national-system to an elevated system of values, the adherence to which provides an antidote for the malaise of nationalism. A 'people' for Buber was an organic impulse, a 'nation' was a construct which fulfils that impulse, and 'nationalism' was no more than a programme of cannons, flags, and military decoration that tended to repeat from nation to nation. Instead of identical, carefully bordered entities clashing with one another, Buber called for nations to give form to the distinct 'impulses' of different peoples. In his 'social utopia' of 'supranationalism', nations will complement each other and exemplify human cultural diversity to the full extent.[50]

Mendelsohn's lecture of 1919 subscribes to this analysis. He declares 'Internationalism' to be the 'nationless anaesthetization of a decaying world'. Alternatively, 'supranationalism maintains national borders ... yet frees humanity'.[51] His political convictions extended into architecture. In 1923, he lectured against the effacement of the culturally and geographically specific. On the Bauhaus exhibition of that year he said: 'to call this apparent conformity [of the new modern architectural concept] simply "international" is more verbal indolence than an expression of conviction'.[52] The dissemination of 'Internationalism' within architectural discourse was similar in his mind to the globally uniform structure of nationalism in political discourse. His opposition to both deepened once he was expelled from Germany. Against European nationalism, which led to 'the new war of today', he saw the gift of 'a national home in Palestine' as double-edged, 'because it induced the Jewish people to think of a State of their own, and deviated it from the major goal, to become an equal member of the Semitic commonwealth of nations'.[53]

Reversed Orientalism and the Mediterranean

Besides criticizing nineteenth-century nationalism, Mendelsohn's notion of a Semitic commonwealth reveals the debt of his Zionism to an enthusiasm for the 'mystical Orient', fashionable at the turn of the century in Germany.[54] Operating in this ambience, Buber's early addresses (the bible of German Zionists[55] which Mendelsohn endorsed in 1915) polarized East and West. This method is the reverse of Edward Said's now famous definition of Orientalism; that is, rather than a confident West asserting itself against a degenerate East, each was constructed as distinct alternatives – but only for the sake of dismissing the former. In this age termed the 'Asiatic Crisis', Buber professed that 'the great spiritual traditions' of the Orient would balance out Western excesses of materiality. Buber suggested that the Jews were the mediating agents for this mission, having 'already acquired all the wisdom and all the skills of the Occident without losing its original Oriental character'.[56] The condemned 'German Oriental' was reinstated as virtuous.

In his pamphlet of 1940, *Palestine and the World of Tomorrow*, Mendelsohn returned to this messianic view. Palestine, he said, is 'where intellect and vision – matter and spirit [genuinely] meet' en route to becoming 'a part of the New-World which is going to replace the world that has gone'.[57] The Zionist enterprise is thus entrusted with a consecrated responsibility: 'Palestine is a country full of magic. It still remains the Holy Land, and that is why there devolves upon us the sacred obligation to take care of it'.[58] As

Palestine fosters the spiritual genesis, Mendelsohn strives for an architectural one as well, equivalent to the originating moment symbolized in the Athenian Acropolis:

> The opponents of Zionism constantly refer to the smallness of the country. They seem to forget that size has nothing to do with significant effort. The Athenians were a small group in a small country but the Acropolis still remains to remind us of the glory that was Greece.[59]

The comparison to Greece is not only one of degree, but also one of kind. Both Judaea and Greece, Mendelsohn argued, were stripped of their spirituality by Western culture. What was, then, 'spirituality' for Mendelsohn, and how did this notion evolve between Europe and Palestine? Why did Mendelsohn locate the hub of his aspiration for an architectural revival in the Mediterranean rather than in the Middle East or Asia, as Buber did? And finally, how did his search for Jewish roots mesh with these architectural concerns?

During his pre-World War I Munich years among fellow expressionists, Mendelsohn nurtured an image of the artist as possessing spiritual strength that emanated from the inner-self.[60] Such was the initial outburst of creativity, which shaped his Einstein Tower. Suspecting, however, that this design hardly touched reality,[61] he directed his energy to grapple with modernity. More specifically he sought a way of generating trust between the individual and the reality of the modern metropolis. His aspiration 'to break through the will to reality'[62] underlined the horizontality of his European buildings with which he intended to pacify the fast pace of modern life. By 1933 Mendelsohn's will to penetrate that technological reality had vanished together with his belief in the capacity of artists to counter the 'accidental market of technical civilization'[63] in Europe with spiritual values. 'Technology', he lamented, 'established the predominance of the materialistic conception of life, that puts the question of life's purpose above life itself. A utilitarianism which destroys the essence of life based upon the unity of mind and matter.'[64]

Mendelsohn's disillusion with technology on the one hand, and the German nationalization of the Biedermeier style on the other, was compounded by his perception of the regression of modernism in the Soviet Union and its reduction to a style in the US.[65] As a result, Mendelsohn attempted to break the ties that avant-garde architects forged between modern architecture and modernity.[66] That is, he looked for alternative architectural laws with which to substantiate modern architecture. Those laws should be as universally valid as 'the International Architecture', yet be rooted in *culture* rather than in progress and technology. He sought for such 'eternal' laws in the architecture of antiquity and the vernacular of the Mediterranean basin. This architecture, he believed, embodied primeval principles rather than academic rules. In the 'villages untouched by civilization', Mendelsohn believed he could find 'the instinct of true building'.[67] When he attempted to authenticate his architecture in Palestine with organicism and spirituality, he referred to a cultural substance and a presence of history in the midst of a fleeting technological reality.

Mendelsohn's focus on the Mediterranean rather than on an Asiatic Orient reveals his debt to the larger architectural discourse of his time. Hoffmann and Olbrich, for example – both highly regarded by Mendelsohn – expressed active interest in the vernacular architecture of the Mediterranean as early as 1895.[68] Hoffmann exhibited sketches from the Adriatic shore and Capri, while Olbrich drew similar cubic white architecture in Tunisia.[69] Mendelsohn's own sketches from his 1923 visit to Palestine reveal comparable inspirations. His enthusiastic encounter with the Greek architect Demetrios Pikionis in 1931 suggested to him a non-Eurocentric venue for admiring the Parthenon and the architecture of antiquity.[70] Through his acquaintance with Marcello Piacentini he could learn about the emerging *mediterraneità* discourse in Italy to which he contributed a 1932 piece in *Architettura*.[71] In the early 1930s Mendelsohn already entertained the idea of a Mediterranean academy on the shore of the French Riviera with the Dutch architect Wijdeveld and the French painter Amedée Ozenfant.[72] He attempted to relocate this academy in Palestine, for the purpose of which he recruited Julius Posener, who later became a Chug member.

Mendelsohn's architectural concerns coincided with his political convictions. He contended that 'the decline of the creative power of the Mediterranean and the loss of its political importance [were] in close and continuous relation to each other'.[73] The West, he claimed, had treated the Mediterranean as a nostalgic but useless gem. Through a *tour de force* in history, Mendelsohn reclaimed the region's former 'place as master of the world'.[74] Toward that end, he separated the virtues of the 'cradles of civilization' from its materialistic consumption by the West:

> The Mediterranean peoples get no profit from the exploitation of their enigmal splendour for the sake of the Europe-American pseudo styles. They sell the copyright of their genuine creations for the tips left behind by romantic artists, snobs and eager archaeologists. They get no royalties for their unique creations, being listed into a pattern book of architectural details and decorations.[75]

Mendelsohn's description of an Orient exploited by a materialistic Occident resorted to the strategies often used by European colonizers. The strategy was to imagine a geographic region, in this case the Mediterranean basin, as possessing a unified cultural heritage. Italian colonizers, for example, used this strategy to bind North Africa to the cultural domain of Europe, thus justifying their colonial enterprise.[76] Mendelsohn employed this strategy inversely. The architecture of Mediterranean culture, he claimed, granted civilization with inventions ranging from the alphabet, science, and trade to the institution of moral and civic law. To the rest of Europe he attributed the invention, and the consequent advancement, of technology. Technology, he claimed, was not the crown of human achievement. On the contrary, a blind faith in the 'glaring clarity of the intellect' would result in a loss of 'the creative power in man'.[77]

For Mendelsohn, sparking the cultural and political revival of the Mediterranean was the gist of the Zionist project. Mendelsohn's plans for the Mediterranean academy reflected his cultural Zionism. This position was clearly articulated by his 1940 biographer: 'Palestine can never adequately be the National Home for the whole of Jewry; it is too small. But it will always be the centre of Jewish culture; it will always be the spiritual home of the Jews wherever they may live'.[78] Mendelsohn's Zionism was not concerned with an overall political solution for the Jews nor with the architectural predicament of housing the gathered exiles. For him the Zionist project was a springboard for a global reconstruction. If a successful implementation of the Zionist project would inaugurate a new world order, then architecture would be its expressive medium.

Semitism and the Arabs

If Semites were doomed by the Nazis, Mendelsohn's answer was to strike back with a Semitic revival. The 'process of the world's remodelling' would start with 'the Semitic world as a signal of national renaissance of the Mediterranean'.[79] Such architectural genesis rested upon a collaboration between Arabs and Jews whose kinship dated back to the biblical times on which Mendelsohn based his Jewish identity. A prerequisite for this process was a successful reconstruction of a *Semitic* Zionist identity in Palestine – distinct from that of the European colonizer.

Mendelsohn was careful to emphasize this distinction. 'Jews', he claimed, 'return to Palestine neither as conquerors nor as refugees'.[80] In order to dispute the 'misconception' of Zionism as 'colonialism in the sense of the imperialist colonial enterprises of the 19th century',[81] Mendelsohn had to secure a mutual Arab-Jewish ownership over the Palestinian territory and its past. 'The age-old Semitic culture', he maintained, 'was based on historically and geographically fixed conditions'.[82] Those fixed conditions were now projected into the future; arrested history was the means with which Mendelsohn could substantiate his ambition 'to become a cell of a future Semitic commonwealth to which [Jews] belong by their race, tongue and character'.[83] With this ethnic bond to Arabs, and in turn the bond of this ethnicity to the Palestinian territory, Mendelsohn could argue for the necessity of Jewish settlement in Palestine. Again, Mendelsohn used a colonial strategy, ethnic essentialism, in order to achieve opposite ends. Colonial powers essentialized the ethnic difference between them and their colonial subjects in order to justify their 'civilizing mission'. Mendelsohn, who was expelled from Germany, had to essentialize his ethnic distinction from Europe, in order to restore his rejected identity.

This aspiration for a unified Semitic identity contained inherent conflicts. Mendelsohn depended on the heritage of the Arab builder to whom he attributed unique aesthetic and spiritual qualities. 'To build in Palestine means to ... include the organic consciousness of the fellah, his age-old experience with regard to the climatic conditions, into our own sense of building ... Our own sense was, in turn, Western and progressive.'[84] For Mendelsohn it was 'the return of the Jews to Palestine [that would] bring new life into that country dormant since centunes'.[85] Such a portrayal of a stagnant Palestine laden with uncontaminated spirituality and waiting to be activated by Western knowledge, was not foreign to colonial discourse. Indeed, among recent architectural achievements in Palestine, Mendelsohn listed only British and Zionist ones.[86] Paradoxically, however, Mendelsohn insisted that the Jew in Palestine had to be 'upgraded' in order 'to reach equal rank among his neighbours'.[87] Likewise, his patronizing attitude towards the Arab fellah was that of a local elite which nationalizes indigenous cultural production. Mendelsohn the elitist stratified society by class rather than nation. He associated with Arab intellectuals, whose opinion in 'matters of taste' he highly appreciated.[88] When needed, he would conspire with the British colonial elite: 'I have called in the Romans – for the sake of an ideal'[89] the ideal of the Hebrew University.

Such anecdotes contribute to an understanding of the way Mendelsohn was suspended between different worlds, belonging fully to none. His architectural reputation relied on his European/German/Berliner persona, an identity he irretrievably lost to Fascism. The British citizenship he was granted as a German émigré in 1938, as well as his dual practice in Britain and Palestine until that year, created an unease about his commitment to the Zionist cause among his peers. To his great disappointment the British were reluctant to accommodate his ambitions as well: his failed efforts to become the chief planner of the British administration in Palestine instigated his final departure to the US in 1941.

Architecture

In Palestine, Mendelsohn tried to locate himself between East and West, where load bearing walls and ground-hugging buildings contrasted with the buoyant image and light materials of modern architecture. Mendelsohn's first commission in Palestine, the villa for Chaim Weizmann, the head of the World Zionist Organization (Figures 1.6 and 1.7), is an excellent example of the difficulty of locating Zionist identity in the gap between the Orient and Europe. 'To the first', Mendelsohn admitted, Jews 'do not yet belong; to the second they do not belong any more'.[90] As a result, various architectural languages are juxtaposed rather than synthesized in this symbolic design. Zionist identity is thus suspended *between* East and West, located in that unutterable interval between the two.

The villa thrives on the tension between its opaque box-like shell, which grants it a quality of introvertedness, and its light, open modern core. It is a tension between different construction methods – masonry walls versus

Figure 1.6   Weizmann House, exterior view

Figure 1.7   Weizmann House, view of the courtyard

columns and beams; a tension between different textures – continuous wall punctured by a few, well-selected openings versus a peristyle courtyard surrounded by generous fenestration; finally, a tension between different typologies – an object around which a procession takes place versus a central open space, a void that controls the entire scheme with the help of the over-monumentalized staircase. Here, Mendelsohn's European depiction of a building as a grand sweeping gesture made room for a sequence of 'static' images composed into an architectural procession.[91] This inspiration from the Parthenon further illustrates aspects of Mendelsohn's Mediterranean vision.

Mendelsohn's most prestigious commissions were the linked institutional complexes of the Hebrew University and the Hadassah Hospital (1936–39). Located on the edge of the Judaean mountains and overlooking the desert with its Arab villages, this site embodied everything 'the romanticism of the East' stood for: 'Everywhere is present that organic unity with the soil whether it is the primitive Arab hut, or the whole formation of a village hanging on a mountainous slope, or built round a hill with the sheik's house as its crown.'[92] Mendelsohn's sketches for those projects betray conflicting Orientalist inspirations: on the one hand he wanted to enclose Islamic-style inner courtyards, and on the other to disperse the mass of the building into the landscape.

Mendelsohn's biographer explains that the analysis of the Judaean organic oneness, and 'the instinct of true building' found there, 'is the adaptation of the plan of a village to the swell of the hill and the erection of the corpus from the stone or clay available on the spot.'[93] Likewise, the long horizontal blocks of Hadassah emerge out of the hill they embrace (Figure 1.8). Mendelsohn wrapped the technologically up-to-date medical structure within a thin shell, a vertical stone cladding, which responded to British regulations for a stone façade. Unlike the British revival of indigenous building techniques and crafts, the pastiche of Hadassah's stone veneer reveals the building's modern character while concealing its actual construction. As a continuous envelope it defines and bounds the plain rectangular masses, which compose the site. Mendelsohn achieved the effect of building blocks emerging out of the hill by stretching the uniform stone cladding toward that continuous seam between building and earth. The façade's rhythmic windows and balconies are mere cut-outs from this veneer, which allow glimpses into the thoroughly modern interior. Their size and density were designed according to the climatic conditions of the site.

Figure 1.8   Hadassah Hospital, aerial view of the courtyards

These horizontal buildings lack the sweeping lines of Mendelsohn's European architecture. The complex expresses serenity and is well integrated into the site. Mendelsohn, however, maintained a tension between the land-hugging presence of the complex and the spacious modern interior, where he exposed the construction method and modern materials of the building. He allowed the two to integrate in the formation of a series of courtyards between the two main horizontal blocks. The visitors' entrance hall, for example, is the dividing zone between the two main courtyards. It is constructed as an opaque horizontal volume, wrapped by stone cladding, which hovers over the glazed entrance hall.

Hadassah's rows of punched windows and the three domes of its monumental entrance echo the Orientalist depiction of the desert's rhythm. Such sublime depiction can be contrasted with the socialist myth of *hfrachat ha'shmama* (the cultivation of the wasteland) or *kibush ha'Negev*, that is, conquering the desert with the most advanced (agricultural) expression of civilization.

Confrontation with the Chug

Mendelsohn was solicited to write for the Chug's journal upon his arrival to Palestine in 1934. In a paragraph of 12 brief sentences, he undermined not only their programme, but that of the mainstream Zionism it represented.[94] Starting from the consensus that 'the hope of the Hebrew people is the construction of their national 'home', he agreed that 'this construction to great extent is of an economical character.' 'However', Mendelsohn was quick to qualify, 'the world will not judge us according to the quantity of citrus export.' Rather, he says, we will be evaluated 'according to the spiritual value of our spiritual production'. Mendelsohn expressed here the Buberian critique of nationalism, which argues against similar national-economical entities competing for materialistic superiority. The aspiration to be a nation like all others was completely foreign to Mendelsohn's cultural Zionism.

Mendelsohn furthered his belief that spiritual production is primarily expressed in the 'public image of our cities', where 'technique and form embody the will'. The debt of his thought to his close friend Lewis Mumford, on the one hand, and to the notion of *Gesamtkunstwerk* on the other, is beyond the scope of this chapter. It is apparent, however, that his emphasis on the architect's will to create the city as a total work of art adheres to a belief in the spiritual power of art to redeem society. The centre of gravity had shifted from these notions, which were prominent immediately after World War I in European architectural discourse, toward the architecture of *Neue Sachlichkeit*.

The shift suited the Zionist aspirations of the Tel Aviv architects. 'Here', Posener's editorial tells us, 'there is no past or experience. Instead of those: demands. The demands of agriculture, of security, of hygiene, of society, of poverty.'[95] Mendelsohn thought such material 'arrogance' necessarily 'leads to the sterilization of human endeavour'.[96] For the Chug, however, Mendelsohn's attempt to impose the public image of 'our cities' from above, regardless of its coherence or beauty, was divorced from the social practice they laboured to establish.

The social aspect was not high on Mendelsohn's list of priorities. His immigration to Palestine post-dated the codification of Labour Zionism, which emerged from an Eastern European ambience. For Mendelsohn the emergence of a 'world paragon' relied exclusively on the courage and responsibility of the architect's vision and his patron. Whereas Mendelsohn's patrons were wealthy individuals like Weizmann and Schocken, most of the Chug's members worked for socialist organizations or the building owners of the rapidly growing Tel Aviv. Carl Rubin, a former Mendelsohn employee and Chug member, was quick to disqualify Mendelsohn's architect-patron exclusivity on the pages of the very same publication. Instead, he called for an encompassing 'publicness', rooting 'the profession' in societal responsibility.[97] At the core of the dispute was Mendelsohn's claim for the spiritual superiority of the artist (for whom the work of art was a religious outlet), a notion that was thoroughly incompatible with the secular socialism of the Chug.

Concluding Remarks

In this chapter I have attempted to divest the International Style of its all-encompassing claim to Modern Architecture in Palestine in order to seek the variety of voices debating architectural forms there in the 1930s. These voices reflected the social, ethical and political convictions of multiple strands within Zionism, which informed architectural production in Mandate Palestine. Such acknowledgement provides a clearer picture of the active compliance between modern architecture and the society in which it functioned. As an example, I have contrasted the architectural position of Erich Mendelsohn to that of the Tel Aviv Chug. The Chug succeeded in synthesizing its initial plurality into a clear architectural message, which complied with mainstream socialist leadership. Mendelsohn's determination to inform his architecture with Orientalist inspirations was part of the challenge posed to the Yishuv's leaders by intellectuals of German origin.[98] I have suggested that these opposing positions stem from the early conceptual split between cultural and political Zionism.

Mendelsohn's task of bridging East and West as a message to the world at large granted a solid form to the cultural Zionism of Ahad Ha-am, as mediated by Martin Buber. By contrast, the 'International Style' architecture of the Chug reflected the Herzlian will to become a nation like any other. The socialist reading of Herzl in the form of Labour Zionism supported an architecture, which did not address any particular country, but rather was a 'civilized' manifesto of the age.

By the end of the 1930s there was an overwhelming consensus in the Yishuv about the goal of the Zionist movement: a sovereign nation-state for Jews. Mendelsohn's unequivocal support of a Jewish-Arab bi-national solution for Palestine strictly contradicted the position of the Yishuv's leadership. It divulged the Semitic core of Mendelsohn's Zionist persona. Only together could the Semites – Arabs and Jews – inaugurate the spiritual revival of the East to which he aspired: the dawn of a Mediterranean Renaissance.

In blatant opposition to Mendelsohn's reliance on Arab collaboration, for the Chug the success of the Zionist project was independent of, and consciously indifferent to, Arab culture. Putting the 'New Hebrew' rather than 'the Semite' at the core of Zionist identity, their Zionist enterprise focused inward – working, building and populating the Land of Israel, as well as reviving its (secular) Hebrew culture.[99] As the notion of an embattled nation in the midst of Arab hostility was gaining momentum, the affinity to a larger Orient gradually diminished.

These ideological paths were allied to different classes and forms of patronage. Mendelsohn's patrons belonged to the upper classes: Zionist statesmen, industrialists, and the intelligencia, as well as the government of the British mandate. Consequently, his commissions were of a representative or symbolic nature. The Chug catered to the working classes – the 'elite' of the Yishuv – as well as to the growing middle class of Tel Aviv. Accordingly, their commissions consisted mostly of housing projects and Socialist institutions.

The role of the artist in these two settings was differently conceived. Mendelsohn saw the artist as a spiritual leader. His aspiration to be the head architect of the British Mandate in Palestine, or alternatively the architect of the Hebrew University, emerged from his firm belief in the need to direct the visual and spiritual image of the Zionist project from above. Only the creative artist, he concluded, could devise a total work of art, which would elevate the Zionist project to a world paragon, capable of countering Western materialism. Conversely, the architects of the Chug saw themselves as pioneers whose contribution to the building of the homeland was equal to that of the agricultural enterprise. If the success of the latter relied on scientific progress, so did the architectural enterprise depend on technological innovations. The Chug's *sachlich* architecture was often a communal enterprise. Years later, in 1949, the first act of Arieh Sharon once he became the chief architect of Israel was to gather his friends and colleagues into a collaborative working team.

The opposition between Mendelsohn and the Chug did not question the general framework of an alliance between modern architecture and Zionism. Consequently, for both parties the national objective of settling the land invested its soil with uppermost importance. Its interpretation, however, differed greatly. In another context of criticizing the modern movement, Sibyl Moholy-Nagy presented an opposition between the modernist approach and that of the settler: 'The speculator lines up bulldozers and cement mixers and then poses the challenge: "What can I do to the land?" The settler asked: "What can the land do for me?"'[100] The former approach was the pride of Labour Zionism while the latter approximates Mendelsohn's Orientalist aspirations.

Labour Zionism's views on building are reflected in the words of a popular song which promises to dress the homeland with 'a gown of concrete and cement'. Ploughing the soil and building the country were means to redeem the land. The labour invested in them would redeem the workers from their subordinated exile-like persona. Such a collective effort would transform the land after two millennia of decay. Accordingly, the white, often elevated cubes of the New Architecture did not grow out of the soil, they were a product of its transformation.

Mendelsohn, on the contrary, believed this soil had always been fertile. Stressing 'the unity of Man and Nature' which underlined 'the organic culture of the East', he did not wish to transform the land, but rather to let his Western implants be nourished by it. Against the Zionist myth of conquering the desert he said:

> It has been said, that the re-entrance of the Jews into the Arab world does not mean anything else but civilization fighting the desert. That opinion originates from the political mentality of the rational Western world, which political development since the last war has made obsolete.[101]

The barrenness of the desert was the focus of Mendelsohn's admiration. If in Europe he was motivated by the sublime modernity of the Western metropolis, here it was the desert and 'heavenly Judaea' that overwhelmed his emotions. Such artist-soil-creation exclusivity was foreign, needless to say, to the Chug's ideology.

The activities of the Chug were geared not only to instill the New Architecture in the public consciousness, but to institutionalize it, thus legitimizing their enterprise. As a collective, they aspired to establish a set of norms by which to regulate architectural production through professional mechanisms such as architectural competitions, building regulations (via representation in municipal statuary committees) and publications.[102] This quest for institutions was intended to remedy the transient life in exile. Henceforth, the Modern, that radical challenge of any preconceived notion, the 'cutting edge' of European architectural production, became classic in Palestine.

Such illustrations of Modern Architecture agreed neither with Mendelsohn's expressionistic modernism, nor with his Buberian Zionism. Mendelsohn was left out of a similar formulation of the International Style in the US not simply because his architectural form did not comply with the rules Johnson and Hitchcock determined, but also because his architectural style resisted indoctrination. Disillusioned with European materialism, Mendelsohn envisioned a Palestine that would function as an incubator for the style of a new age. Activated by the creative deed of the Zionist enterprise, he saw the new emerging out of the existing. The architectural style for the new age, Mendelsohn insisted, should be a cultural hybrid of East and West, ratio and spirit. Recycling and appropriating an existing architectural vocabulary could not approximate his mission: 'The new Palestine', he professes in 1937, 'has still to be created'.[103]

Mendelsohn deviated from the mainstream Zionist project. He succeeded in reading the existing situation provocatively, but failed to recognize the vigour of the emerging nation state. Against the political insecurity of the nation-building years, the religious accent of Mendelsohn's Orientalism, his relation to 'the Arab', and his opposition to nation statism made his architectural position incompatible with mainstream Israeli architecture. In the 1980s this ideological deviation was blurred, allowing the inclusion of Mendelsohn's architecture as a major contribution to Israeli architecture. This attitude adhered to the quest for a unified, non-conflictual architectural heritage, a symbol for the secular Hebrew culture of Israel.

Mendelsohn departed for the US during the war and did not return despite his continuing interest in Israel. In the context of the 1930s rapidly growing socialist hegemony, Mendelsohn's convictions contested the Yishuv's narrowing architectural discourse by introducing a different voice, which was later weakened by his erroneous inclusion as a central proponent of the 'International' or 'Bauhaus Style'. Today, architectural production in Israel has to contend with increasing challenges to the unified Zionist narrative. In order for architecture to reflect the complexity of the society to which it gives physical form, it should be understood as a plural, critical and stylistically undetermined practice, as modernism still was in the 1930s.

Notes

1   This chapter was originally published in *Architectural History*, Vol. 39, September 1996, pp. 147–80. Among more recent publications on the subject, which are not recounted here, the most relevant is: Regina Stephen and Charlotte Benton (eds), *Erich Mendelsohn: Architect, 1887–1953* (New York, 1999). A further development of the historiographical argument I present here can be found in: Alona Nitzan-Shiftan, 'Whitened Houses', *Theory and Criticism*, 16 (2000), pp. 227–32. Illustrations were drawn from the Zionist Archive (Figure 1.4), Isaac Kalter collection (Figure 1.2), the Archive of the Weizmann Institution (Figures 1.6 and 1.7), and the Visual Archive of the Hebrew University (Figure 1.8).

2   It is thanks to Royston Landau, whose memory I warmly cherish, that this chapter was initially written and submitted to the Society of Architectural Historians of Great Britain. I thank them for awarding a prize to an earlier manuscript. This work was conducted at the History, Theory, and Criticism section of MIT Department of Architecture, where I benefited greatly from the insights of Sibel Bozdogan, Akos Moravanszky and Mark Jarzombek. Further thanks go to Gwendolyn Wright, Sarah Ksiazek, Zeev Rosenhek and Hadas Steiner.

3   The Yishuv, literally 'settlement', is the politically organized Jewish community in Palestine established by the Zionist waves of immigration since the end of the nineteenth century. A selected list of recent books about the architectural production of the Yishuv primarily during the Mandate period includes: Jeannine Fiedler (ed.), *Social Utopias of the Twenties: Bauhaus, Kibbutz and the Dream of the New Man* (Germany, 1995); Gilbert Herbert and Silvina Sosnovsky, *Bauhaus on the Carmel, and the Crossroads of Empire* (Jerusalem, 1993); Richard Ingersoll, *Munio Gitai Weintraub: Bauhaus Architect in Eretz Israel* (Milan, 1994); Irmel Kamp-Bandau et al., *Tel Aviv Modem Architecture 1930–1939* (Berlin, 1994); Nitzah Metzger-Szmuk, *Batim min ha'hol* (Tel Aviv, 1994).

4   For criticism of such literature see: Daniel Bertrand Monk, 'Autonomy Agreements: Zionism, Modernism and the Myth of a 'Bauhaus' Vernacular', *AA Files*, 28 (1994), pp. 94–8.

5   This quote is taken from an essay articulating the position underlying the 1994 events. See Michael Levin, 'In Praise of the White City', *Ha'ir*, 20 May 1994, pp. 31–3, p. 37.

6   This chapter does not attempt to cover the larger picture of architectural production in the Mandate period, to which the books cited above contribute greatly. Mendelsohn and the Chug are two significant examples with which I attempt to demonstrate ideological oppositions within the architectural community.

7   The distinction to which I refer between cultural and political Zionism relates primarily to the conceptual difference between Ahad Ha-'aim's aspiration to throw light unto the nations ('or la'goyim') and Herzl's will to be a nation like all nations ('am ke'chol Ha'amim') as will be explained shortly.

8   Henry-Russel Hitchcock and Philip Johnson, *The International Style: Architecture since 1922* (New York, 1932).

9   Michael Levine's landmark exhibition had two catalogues: Michael Levin, *White City: International Style Architecture in Israel, A Portrait of an Era* (Tel Aviv, 1984), and *White City: International Style Architecture in Israel, Judith Turner: Photographs* (Tel Aviv, 1984).

10   Leaders from Left and Right endorsed the style, but promoted it differently. For Shimon Peres, for example, who was the Foreign Affairs Minister in the Labour government at the time of the 1994 celebrations, the style was the heritage of the Labour Zionism that founded the state. His opening speech at the 1994 conference is partially quoted in Monk, 'Autonomy Agreements'. Rom Milo, Tel Aviv's right-wing Mayor, found in the International Style architectural heritage a foundation for tourism and a springboard for future economic development of a city with a strong tradition. See Milo's introduction to the celebrations of 'Bauhaus in Tel Aviv' in the accompanying brochure.

11   Recent scholarship confirms this bond. See Fiedler, *Social Utopias of the Twenties: Bauhaus, Kibbutz and the Dream of the New Man*, and Ingersoll, *Munio Gitai Weintraub: Bauhuas Architect in Eretz Israel*, above. Gilbert Herbert argued that the greatest influence on 1930s modernism in Palestine was that of Gropius, the Bauhaus, and the various Siedlungen projects. See: Gilbert Herbert, 'On the Fringes of the International Style', *Architecture SA* (Cape Town, September–October 1987), pp. 36–43.

12   Eugene (Yohanan) Ratner, 'Architecture in Palestine', *Palestine and Middle East Economic magazine*, 7–8 (1933), pp. 293–96. Reprinted in Sosnovsky, *Yohanan Ratner*, p. 25e.

13   A representative list of influential writings on modem Israeli architecture in the 1950s and 1960s includes: Aba Elhanani, 'Directions in Israeli Architecture', *Handasa ve'Adrichalut (Engineering and Architecture)*, Vol. 20 (September–October 1962), pp. 313–15; Avia Ha'Shimshoni, 'Adrichalut yisraelit', *Omanut Yisrael*, ed. Benjamin Tammuz (Tel Aviv, 1963); English edition: 'Architecture' in *Art in Israel*, ed. with Max Wykes Joyse (Philadelphia, 1967); collected writings of Yohanan Ratner reprinted in Silvina Sosnovsky (ed.), *Yohanan Ratner: The Man, the Architect, and his Work* (Haifa, 1992). An exception would be: Gershon Canaan, *Rebuilding the land of Israel* (NY, 1954). Canaan's book, advocating a more regional modernism, was written in the US, with an introduction by Erich Mendelsohn. Answering his imagined audience, the author explains that the book was not written in Israel because there people were too busy building the state.

14   Ha'Shimshoni, 'Adrichalut yisraelit', p. 227.

15   Sosnovsky, *Yohanan Ratner*, Hashimshony, 'Architecture'; Elhanani, 'Directions'.

16   Ha'Shimshoni, 'Adrichalut yisraelit', pp. 219, 224–8.

17   This wave of immigration is known as the third Aliyah (literally 'ascend'), and is considered to be more motivated by Zionist sentiments than the fifth Aliyah of the 1930s, which was a response to the crisis in Europe.

18   Biographical information on individual Chug members can be found in the Architectural Heritage Research Centre at the Technion, Haifa, and in the following books: Arieh Sharon, *Kibbutz + Bauhaus* (Stuttgart, 1976); Ran Shechori, *Zeev Rechter* (Jerusalem, 1987); Kamp-Bandau, *Tel Aviv*; Metzger-Szmuk, *Batim*.

19   Neufeld defines Organic building as a harmony between the needs of man on the one hand, and the built content and expression on the other, a harmony between social tendencies and form which ultimately results in a total architectonic organism. Neufeld acknowledges individual intuition but undermines the cult of 'names' who allegedly lead modem architecture. He rather advanced his belief in the *Kuntswollen* of his time. Neufeld, who interwove here the modern, organic, and beautiful, would rarely write for

the subsequent publication of *Habinyan* despite his continued active membership in the Chug.

20 When the Chug launched the first architectural publication in Palestine, *Habinyan Bamisrah Hakarov* (*Construction/Building in the Near East*) in December 1934, it did not only localize the journal geographically, but addressed its audience in an editorial translated into Hebrew, Arabic and English. Motivated by the lack of 'any effort towards analysis, direction or influence' on 'the building movement', it was staged as a forceful tool for the advancement of modern architecture and professional interest. The publication was pluralistically modern with an emphasis on social, economic, climatic and technical discourse and professional administrative struggles. However, the bias toward the Chug's work and its underlying ideology was challenged by architects and engineers from Haifa, who took over the publication. Most conspicuously, they attempted to diversify and regionalize the journal to include neighbouring countries. The subtitle *Itono shel Chug Adrichalim be'Eretz Israel* (the Paper of the Architects' Circle in the Land of Israel) was transferred to the more professional and less pluralist journal, *Habinyan*. A comparison between the coverage and the quality of the two leaves no doubt regarding the power and influence of the Chug. Habinyan focused primarily on housing and Zionist villages and lengthy articles demonstrated alternative layouts for the ultimate minimal apartment with which to accommodate the working family.

21 Gilbert Herbert and Ita Heinze-Greenberg, 'Anatomy of a Profession: Architects in Palestine During the British Mandate', *Architectura* (1992), pp. 149–62.

22 Julius Posener, 'One-Family Houses in Palestine', *Habinyan 2* (1937), p. 1.

23 Ibid.

24 Ibid.

25 Batia Donner, To *Live with the Dream* (Tel Aviv, 1989), English introduction.

26 Arieh Sharon, *Kibbutz + Bauhuas.*

27 Ibid., p. 29.

28 This position was offered to Sharon by M. Bentov, the minister of housing and construction, who was a member of *Hashomer Hatzair* (the young guard), the same organization to which Sharon's kibbutz belonged.

29 Zionist ideology not only repudiated the immediate past of exile as national uprootedness but also endorsed the remote past by means of a leap-into antiquity, when Jews last experienced 'authentic' national life on their promised land. A continuity between the biblical past and the premises of socialism, and eventually statehood would be necessary for the textual integrity of the Zionist narrative. For the uses of history and the formation of collective memory in the context of Israeli nationalism, see: Yael Zerubavel, *Recovered Roots: Collective Memory and the Making of Israeli National Tradition* (Chicago, 1995).

30 Anita Shapira, *Berl: the Biography of a Socialist Zionist, Berl Katznelson 1887–1944* (Cambridge, 1984), selections from the Hebrew edition (Tel Aviv, 1983).

31 Ze'ev Tzahor, 'Ben Gurion Mythopoetics', in *The Shaping of Israeli Identity: Myth, Memory and Trauma*, ed. Robert Wistrich and David Ohana (London, 1995), pp. 61–84. For a general discussion of the relation of Labour Zionism to the 'Arab question' see: Yoseph Gomy, *Zionism and the Arabs 1882–1948: A Study of Ideology* (Oxford, 1987); Anita Shapira, *Land and Power: The Zionist Resort to Force 1881–1948* (New York, 1992).

32 For a discussion of the relationship between Jewish nativeness in Palestine and the natural attributes of the land of Israel, see Yael Zerubavel, *Recovered Roots*, pp. 28–9.

33 Posener, 'One-family', pp. 1–2.
34 Posener, 'One-family', p. 2.
35 Posener wrote: *'Habinyan* equally refrain from romantic glorification of the wholeness of the fellah village as well as from criticism and denunciation. We will not say: we should build in such a stable and traditional manner, nor will we say it is forbidden to build in such an odd and bad way. The Arab village does not serve us as a model for imitation, nor is it a contradictory position to any alternative, which determines: this or that, old or new style.' See 'Villages in Palestine', *Habinyan*, 3 (1937), p. 1.
36 Ibid.
37 Hannes Meyer, 'Building', in *Bauhaus* (Cambridge 1968), p. 153.
38 Ibid.
39 Julius Posener, 'One-Family house in Palestine', *Habinyan*, 2 (1937), p. 1.
40 Whittick, *Erich Mendelsohn*, p. 133.
41 Erich Mendelsohn to Kurt Blumenfeld, July 1933.
42 Translated in Gilbert Herbert, 'Erich Mendelsohn and the Zionist Dream', *Erich Mendelsohn in Palestine, Catalog of an Exhibition* (Haifa, 1987).
43 Ita Hienze-Muhleib, *Erich Mendelsohn, Bauten und Projekte in Palastina (1934–1941)* (Munchen 1986). This is the most comprehensive study on Erich Mendelsohn's work in Palestine.
44 For Agnon's writings on Mendelsohn look at S.Y. Agnon, 'Ad hena', in *Ad hena*, pp. 83–5; 'Pitche dvarim' and 'Misha melech Moav', in *Pitche dvarim*, pp. 121–3, 150, 92–6. I would like to thank Avraham Vachman for drawing my attention to these writings.
45 In 1914 only 1.5 per cent of German Jews were Zionists, a situation that changed radically only after the Nazis seized power. See Hagit Lavsky, *Before Catastrophe: The Distinctive Path of German Zionism* (Jerusalem, 1990). Mendelsohn, however, is mentioned in 1919 in a list of potential immigrants to Palestine (I would like to thank Ita Heinze-Greenberg for this information).
46 An English translation is included in N. Glatzer (ed.), *On Judaism* (New York, 1976).
47 Erich Mendelsohn sent this book to Louise in 1915, writing that in it he identified 'the strict confession of my Jewishness. And indeed exactly at the mixture Buber attempts to realize'. Letter of 2 April, 1915, in *Briefe eines Architekten*, ed. Oskar Beyer (Munich, 1961), p. 35.
48 Buber's involvement in the Munich art circles is described by Peg Weiss, *Kandinsky in Munich: The Formative Jugenstil Years* (Princeton, 1979).
49 For Buber's early thought see: Paul Mendes-Flohr, *From Mysticism to Dialogue: Martin Buber's Transformations of German Social Thought* (Detroit, 1989).
50 Excerpts from Buber's early political writings are in: Martin Buber, *A Land for Two People: Martin Buber on Jews and Arabs*, ed. Paul Mendes-Flohr (New York, 1983).
51 Erich Mendelsohn 'The Problem of a New Architecture', lecture in the 'Arbeitsrat für Kunst', Berlin, 1919, reprinted in *Erich Mendelsohn: Complete Works of the Architect* (New York), p. 20.
52 Erich Mendelsohn, *The International Consensus on the New Architectural Concept, or Dynamics and Functions*, reprinted in 'Complete Works', p. 22.
53 Mendelsohn, *Palestine*, p. 15.
54 Paul Mendes-Flohr, 'Fin de Siecle Orientalism: the *Ostjuden,* and the Aesthetics of Jewish Self-Affirmation', *Divided Passions: Jewish Intellectuals and the Experience of Modernity* (Detroit, 1991), pp. 77–132.
55 Gershom Scholem, *On Jews and Judaism in Crisis* (New York, 1976), p. 138.

56   Martin Buber, 'The Spirit of the Orient and Judaism' in *On Judaism*, pp. 77–8.
57   Mendelsohn, *Palestine*, p. 19.
58   'A New Architecture in Palestine: an Interview with Mr. Erich Mendelsohn', *Palestine Review* (Jerusalem, 20 August 1937), p. 318.
59   Ibid.
60   Oskar Beyer (ed.), *Erich Mendelsohn: Letters of an Architect* (London, 1967), particularly pp. 40–43.
61   Ibid., letter of 9 July 1916, p. 44.
62   Mendelsohn, 'The International Consensus', p. 24.
63   Mendelsohn, *Palestine*, p. 6.
64   Ibid., p. 9.
65   Erich Mendelsohn, 'Il Bacino Mediterraneo a la Nuova Architettura', *Architettura* (December 1932), pp. 647–8.
66   To the social content of much of the modern architecture of his time Mendelsohn was largely indifferent. Seeing architectural production as a great spiritual deed, he did not express much interest in housing projects either in Europe or in Palestine.
67   Whittick, *Erich Mendelsohn*, p. 133.
68   Josef Hoffmann, 'Architektonisches aus der Östereichischen Riviera', *Der Architect I (1895)*, p. 37, reprinted in H.F. Mallgrave (ed.), *Otto Wagner: Reflections of the Raiment of Modernity* (Santa Monica, 1993), p. 221.
69   Edward F. Sekler, *Josef Hoffmann: the Architectural Work* (Princeton, 1985). Bernd Krimmel (ed.), *Joseph Maria Olbrich 1867–1908,* Exhibition Catalogue (Darmstadt, 1983).
70   Lian LeFaivre and Alexander Tzonis, a talk in Jerusalem, March 1996.
71   Erich Mendelsohn, 'Il Bacino Mediterraneo a la Nuova Architettura', *Architettura* (December 1932), pp. 647–8. Piacentini, the editor of the magazine, previously published Mendelsohn's work in his book. Publishing Mendelsohn's piece in 1932 was part of updating the magazine to encounter the pressure of the Rationalist group. To this effort Giuseppe Pagano, a young Jewish Italian architect who admired Mendelsohn's work, contributed. *Architettura* was the official architectural magazine of the Fascist party. However, in 1932 it was not connected in the German public opinion with the local emergence of the Nazis. For the Mediterraneità discourse in Italy see Brian McLaren, 'Carlo Enrico Rava, mediterraneità and the architecture of the Italian colonies in Africa in the 1930s', forthcoming in *Environmental Design.*
72   Wolf Von Echardt, *Erich Mendelsohn* (New York, 1960), p. 24.
73   Mendelsohn, *Palestine*, p. 7.
74   Ibid., p. 5.
75   Ibid., p. 6.
76   Mia Fuller, 'Mediterraneanism', *Environmental Design*, 9–10 (1990), pp. 8–9.
77   Mendelsohn, *Palestine*, p. 9.
78   Whittick, *Erich Mendelsohn*, p. 147. This statement was eliminated from the later edition which postdated the foundation of the Israeli state.
79   Mendelsohn, *Palestine*, pp. 13–14.
80   Ibid., p. 14.
81   Ibid., p. 11.
82   Ibid.
83   Ibid.
84   Whittick, *Erich Mendelsohn*, p. 133.

85   Ibid., p. 14.
86   *Palestine Review*, p. 318.
87   Mendelsohn, *Palestine*, p. 11.
88   Letter to Schocken, 27 July 1936, in *Erich Mendelsohn: Letters of an Architect*, ed. Oskar Beyer (New York, 1967), p. 145.
89   Letter to Louise, August 1936, ibid., p. 146.
90   Mendelsohn, *Palestine*, p. ii.
91   Ita Heinze-Greenberg, 'The Impossible Takes Longer': Facts and Notes About the Weizmann Residence in Rehovot', *Katedra*, 72 (June 1994).
92   Whittick, *Erich Mendelsohn*, p. 133.
93   Ibid.
94   Erich Mendelsohn, a letter in *Habinjan Bamisrah Hakarov*, 3 (February 1935), p. 4.
95   Posener, 'The Villages', p. 1.
96   Mendelsohn, *Palestine*, p. 7.
97   Carl Rubin, *Habinjan Bamisrah Hakarov*, 4, p. 1.
98   *Brith Shalom* (the Convent of Peace), the principal group to support a bi-national solution for Arab and Jewish co-existence, was initiated in 1925 by Arthur Rupin, the principal planner of Zionist settlement policy. Rupin brought Richard Kauffmann to be the architect in charge for this undertaking. Kauffmann, who studied with Mendelsohn in Munich under Theodor Fischer, played an important role in bringing Mendelsohn to Palestine in 1923 and 1934. Many members of *Brith Shalom* were part of the Hebrew University, a circle in which Mendelsohn interacted both socially and professionally as the architect of the Hebrew University.
99   Yoseph Gorny, *Zionism and the Arabs 1882–1948: A Study of Ideology* (Oxford, 1987).
100 Sibyl Moholy-Nagy, *Native Genius in Anonymous Architecture* (New York, 1957), p. 51.
101 Mendelsohn, *Palestine*, p. 12.
102 Much of the activity of the Chug and the Association of Engineers and Architects within which it operated focused on professional struggle. See for example *Habinjan Bamisrah Hakarov*, 4, p. 16, 5–6, p. 2, and particularly the editorial of issue 8, which declares the great achievement of incorporating the architects' representatives into the Tel Aviv Municipal Construction Committee. The same issue features Ratner's article: 'The Influence of Building Regulation of the Architecture in *Eretz Israel*', in which he explains the power of prohibition and creative building regulations to control and manipulate the architectural image.
103 *Palestine Review*.

Chapter 2

The Flight of the Camel:
The Levant Fair of 1934 and the Creation
of a Situated Modernism

Sigal Davidi Kunda and Robert Oxman

The Rhetoric of Design: Situated Modernism and Progressive Society

The Zionist leitmotif of the conquest of the land had for some generations placed priority on the symbolic and pragmatic significance of agricultural settlements. Slowly in the twentieth century the city and urbanism as a cultural strategy emerged from the situation of being of adjunct, and secondary, status. This phenomenon occurs with the crystallization of the physical and symbolic presence of Tel Aviv in the period of the 1920s and 1930s. The mythical status of the city, the White City, as the locus for progressive society – simultaneously its incubator and its manifestation – grows in the early 1930s. Two phenomena accompany and support this ideological transformation in the young Jewish settlement in Palestine (or Eretz Yisrael, as its Jewish inhabitants preferred to call it). The first is the broad background of socioeconomic and political upheavals in Europe; this contributes to the increase of immigration of a European professional and social elite. With them they bring the avante-garde affinity for the new modernist formal vocabulary as the symbolic expression of a dynamic social order.

A series of international fairs and exhibitions served to reinforce the rhetoric of 'new form for new social order'. Among these were 'Exposition des arts decoratifs' in Paris, 1925; Barcelona, 1929; Stockholm, 1930 and The International Style exhibition in New York, 1932. By 1932–33, the International Style in its diverse manifestations had become the preferred avante-garde expression of the cultural potential of modernization. The international fair had become the medium of choice for the manifestation of the semantics of being progressive. The promotion of modernism in a situated, or localized, form began to be the preferred strategy for the creation of an emerging impetus for national identity. All of these evolving cultural tendencies coalesced in the new city of Tel Aviv of the early 1930s. In 1934 these forces came together to create one of the most significant cultural events of the new city, and one which was to crystallize its commitment to the new architecture and urbanism. This event, the Fair of the East, or the Levant Fair of 1934 so convincingly established this idea of 'situated modernism' that Tel Aviv eventually developed as a unique international repository of Modernist architecture and urban fabric.

On 26 April 1934, the sixth Levant Fair opened in Tel Aviv. It was an international exhibition organized by the Company for Trade & Industry with the object of creating international economic and trade ties and, in particular, with the Middle East and Palestine. An area of about 100 dunams (25 acres) of sand, north of the then existing city, between the Yarkon river and Mediterranean Sea, was selected as the site for the fair. The site was built according to a carefully prepared master plan and was an important design and urban planning event, the first of its kind and scale in Jewish Palestine (Figure 2.1). Seventy-four pavilions were built and during the short period of its operation, approximately 600,000 people visited the fair. This was twice the total of the Jewish population in the country at that time, and a tremendous number, in particular, for little Tel Aviv with its population of only 100,000.

Figure 2.1   General view of the fairgrounds, 1934

Thirty countries participated in the exhibition, which ran for six weeks and was a great economic success.

Both objective and subjective factors influenced the design of the site, which was built specifically according to International Style principles. Private ownership of portions of land prevented building on the sea front. In addition, economic factors, a limited time for building, a shortage of professional builders as well as the limited supply of locally available building materials all of these factors affected the nature of both the design and the building. However, the main aim in planning the site was to present modern and innovative architecture in an urban context and, as such, to establish an appropriate identity of Jewish society in Palestine. In the following pages we will examine this question of the rhetoric of design in the Levant Fair. We will explore the ability of the Fair to achieve these objectives and the nature of its successes in promoting situated modernism as a new medium for the image of Zionist settlement as a dynamic society.

Exhibitions in Palestine

Trade exhibitions, both local and international, were inaugurated by the Jewish settlement, also known as the Yishuv, in Palestine starting from 1923 and they proved to be important factors in the development of industry. These exhibitions were effective promotional agents for increasing local production and marketing possibilities for export of local products, and for attracting money and new investors to the country. During the 1920s there was little faith in the possibility of significant industrial development in the country. There were those who viewed the call for private enterprise to develop trade and industry as a negation of 'Romantic Zionism' which had working the land and developing the agricultural settlement as its central ideology. Others simply did not believe in the potential for industry in a country which had neither raw materials, cheap electricity, developed means of transport, nor experienced experts in the various fields.

Until 1932 the exhibitions inaugurated by the Yishuv were initially to promote their products. They were all temporary exhibitions and eclectic in style. They served as part of the process of searching for the appropriate symbolic language of Jewish culture in Palestine. However, these early exhibitions served to establish the basic model, which was adopted and then dramatically adapted for the 1934 fair. For the first time it was decided to erect permanent buildings. The rationale was that this would be the locale for future fairs to be held in Tel Aviv every two years.[1] This decision increased the urban status and importance of the building plan, as it enhanced the significance of the architecture of the buildings of the fair.[2] The impact of the Fair of 1934 was in the production of a highly creative large-scale example of both planning and execution. As such it was unique in scale and conception in the local history of Jewish settlement in Palestine. Furthermore, it presented a new cultural conception which was remarkably broad and interdisciplinary. It promoted such diverse activities as graphic and landscape design in the service of this integrated vision of a new society.

International fairs have served as platforms for the penetration and dissemination of new architectural ideas, such as those of the Pavilion de l'Esprit Nouveau of Le Corbusier in the Paris Exhibition of 1925, or the German Pavilion of Mies van der Rohe in the Barcelona Exhibition of 1929. As in the case of all international fairs, so too the Eretz Yisrael fair offered the opportunity for promoting cultural agenda through a semantics of form, material, color and landscape. The impact of the 1934 fair resided in its integration of the different design disciplines, and in their holistic display as a complete and harmonious project. In the early 1930s the fair was the largest and most impressive public building project undertaken by the Yishuv. The number of buildings, their scale and high level of architectural and engineering quality, and the active dissemination of publicity in the country and in the world were all factors in achieving the objectives of the organizers.

The Search for Style

The building of the fair created the largest and most prestigious concentration of buildings executed in the International Style up to the mid-1930s. The collection of buildings is an outstanding example of the penetration of modernist architecture into the country. However, this condition was preceded by a complicated process of evolution towards the crystallization of an approach which saw modernism in general, and the International Style in particular, as representing the social dynamics of Jewish society in both rural and urban Palestine. The question of which style of architecture might be suitable for the buildings of the Jewish people in their rediscovered homeland had already been raised in the 1920s after the wave of immigrants arrived and Jewish building in Palestine had began. This process of the search for style as the medium of cultural representation raised many contradictions. Among these conflicts of value were the desire to integrate into the cultures and symbolic ambience of the Moslem Near East versus the need for developing an original Jewish style. A second focus of this expressive dilemma was the question of the nature of national identity, and whether it might be possible to adapt the modern style and approach to a national style particularly in the hot climate of the country. We have referred to this objective as the search for a *situated modernism*.

This evolutionary process resulted in buildings of the Yishuv in a variety of styles including an eclectic restatement inspired by Islamic building, the Classical, and a combination of modernism with local, eastern elements. By the early 1920s this had become a controversial subject and the need to establish guiding principles had already been raised and became a public issue. Many architects agreed that a uniformity of style should be found, and that this might be achieved by developing a new formal vocabulary suited to the nature of the country. However, in practice, they disagreed on its design and formal components. Until the final victory of modernism in the 1930s, no unanimous crystallized style had been achieved and this resulted generally in the eclectic design of buildings.

In this period of eclectic experimentation during the 1920s, the search for an architectural style to express the 'genius loci' of the Jewish homeland led many architects, one of the most outstanding among them, Alexander Baerwald, to reach a synthesis between their European roots and the architecture of the Middle East, by adopting and adapting eastern motifs and elements.[3] A different school of thought turned to neo-classical sources believing that they would be suitable for a Mediterranean country. Joseph Berlin was the proponent who, in his search for a local identity, negated the use of the eastern style in preference for the neo-classical.[4] The influence of the East was not ignored by modern architects; they also attempted to include eastern formal components in their buildings. Rather than eclectic design practice, they used formal motifs, a dome or an arch, or functional motifs such as the patio. No less a didactic modernist than Mendelsohn succumbed to the attraction of orientalism.

Towards the end of the 1920s, more modern architecture began to penetrate the local scene. The early modernists in the country rejected the past and attempted to crystallize a new vocabulary of forms suitable for a new country, including a flat roof cast in concrete, simple geometric volumes, and the elimination of expressions of historical origin such as decorated facades. Their goal was to build in the framework of advanced architectural culture, and the desire to be modern was taken to be an explicit identification with Western culture. The most important names were Richard Kauffmann, Ze'ev Rechter, Leopold Krakauer, Dov Kuzinski and Yohanan Ratner. In spite of their interest and activities, however, the impact of modern architecture in Tel Aviv was minimal during the 1920s. In the mixture of styles, with eclectic architecture still dominant, early modernism was merely another style.

The Ascendance of the International Style

In the early 1930s the architectural scene changed in Palestine as the International Style found fertile ground. The increase of interest in the International Style as a formal medium for building in Tel Aviv and other cities coincided with an increase in building activity. This process was enhanced by world architectural events of the time. In 1932 at the Museum of Modern Art in New York at the well-known exhibition, 'The International Style', curated by Henry Russell-Hitchcock and Philip Johnson, exhibited the work of the architects of the Modern Movement and defined the characteristics of the International Style. The activity of Le Corbusier in Paris of the 1920s was also an important source of inspiration for young Jewish architects and students who came to France from Europe and Palestine. Among the architects who worked in his office were Sam Barkai, Shlomo Bernstein and Binyamin Chelnov.

During the late 1920s and early 1930s, the arrival in Palestine of young architects who had studied and worked in Europe started the massive penetration of the International Style into the country. Until 1939 almost 86 per cent of the country's architects, whether they were immigrants or born in Palestine, whether they qualified before or after arriving in the country, actually received their professional qualifications overseas and not in Palestine.[5] Even later, until 1948, despite the increase in the number of graduates from the school of Architecture of the Haifa Technion, this factor remained as high as 81 per cent. Architects in Palestine numbered mainly immigrants from countries such as Russia, Poland and Germany where they had been exposed to Western values and a modernist professional education. Along with this, they had been witness to dramatic socioeconomic and political transformations.

Most of the prominent architects of the 1930s had either received their qualification before coming to the country or had gone to Europe to study during the 1920s. Most of them studied in Germany (seven were Bauhaus graduates), Vienna, Czechoslovakia, Paris, Belgium and Rome. Many architects also remained to work in Europe for short periods after completing their studies. Despite the diversity of countries where the Yishuv's architects studied, the major influence was that of Germany where about one quarter of them had been students. The profile of the architectural community of the 1930s was that of very young architects with few years of professional experience. This generation was also characterized by their pioneering zeal to build a new country. As they began to assume this dramatic responsibility for building a new culture, most of them had been in the country less than ten years and were still in the process of adapting. In 1929 half of the architects of the Yishuv were under the age of 30 and an additional 32 per cent under the age of 40. The image of the architect as a young pioneer and an arbiter of culture influenced their work as well as their social institutions.

The lack of a strong local building tradition, or a well-defined selection of precedents or models, created a condition in which the new architecture that emerged in the period of their European education could easily be assimilated. This was strongly the case with respect to the large body of residential construction for the growing population of Tel Aviv and other urban centers. Style symbolized the connection with Western culture and many architects continued to keep in touch with Europe by visiting and by reading books and professional literature. The desire to build according to advanced Western standards and the urge to be modern overcame the desire to adapt and to be integrated into the culture of the East.

The promotion of modern architecture was not a passive process of stylistic imitation. A great effort was made to adapt a modern approach to the special demands of place and with particular reference to the climate. In Europe the style was characterized by large glazed openings which were unsuitable for the climatic conditions of Palestine. Local architects experimented with shading devices and smaller openings. The development of the localized modern architectural language was manifested not only in smaller windows, but also in the design of the building section. Within the emphasis of horizontal proportions they exchanged the strip windows of Western Europe for long balconies with deep upper overhangs in order to provide shade. Much consideration was given to the building's orientation and direction of the main facades relative to prevailing winds. Architects invested great effort into locating the correct angles for positioning windows in order to allow for maximum ventilation (usually East-West). They thus created a *situated modern tradition* of climatic sensitivity and great plasticity in the strong sunlight of Palestine.

The Planning and Organization of the Fair

Introduction

'Mischar & Ta'asiah' (Trade & Industry) was the commercial company, which developed the idea and the concept of industrial fairs in Palestine and realized a series of them in the 1920s and 1930s. The company was founded in 1922 with the objective to encourage local production in industry and agriculture and attract private capital to invest in Jewish Palestine. From its founding the company worked at developing industry and establishing the local economy and all its actions had originality and were pioneering in spirit. Even though the Levant Fair was a private economic enterprise it received large public support. Since its success was of national interest the support came from all the Jewish institutions: the Zionist institutions, the Tel Aviv municipality, heads of industry and agriculture, and even the British Mandate government. The British High Commissioner, Sir Arthur Wauchope, became the fair's patron.[6] A public committee was formed consisting of fourteen members with representatives from all the local economic fields, including bankers, merchants and politicians.[7] Their job was to further the interests of the fair locally and worldwide and to give backing and support to the fair's management by promoting economic and trade relations. The involvement of these persons and the influence of their activities was of great importance to the building of the fair. The president of the committee was Meir Dizengoff, the mayor of Tel Aviv and his deputies were Yisrael Rokach, the deputy mayor, and Hoofien the manager of the Anglo-Palestine Bank.

The great success of the fair can be mainly attributed to the efficiency of this organization in which both private and public bodies were partners. In addition to erecting an economically successful fair, it was also their aim to build pavilions, which would exhibit the achievements of modern architecture in Palestine. These were the goals of the policy makers who themselves accepted the new modernist architecture as the fitting symbol of a dynamic and progressive society. Their devotion to the cause resulted in a magnificent exhibition, perfectly planned and one which displayed a crystallized image of the Jewish settlement as a dynamic entity.

To emphasize the fact that the fair represented only Jewish Palestine, the planning and building was executed by the Jewish population exploiting only local industrial products, 'Totzeret Ha'aretz' (local products). In addition, it was a display of strength of the Yishuv to the Arab population of the country who boycotted the fair and took no part in it.[8] All the buildings, among which were the pavilions of the international community, were planned by a body of Palestine's active Jewish architects. Even the British pavilion was planned by a Tel Aviv architect, Josef Neufeld, although a British architect might have been chosen from those available in the country.

The planning department of the fair or its 'Technical Bureau' was in charge of all planning including architectural and engineering works, landscaping, interior design and the organization of the displays of all of the pavilions. This department became influential in promoting the rhetorical program of the organizers, and guaranteeing the internal consistency of all artifacts and exhibits. In addition, they were also in charge of drawing the plans and of the management of the buildings on site. All in all, the bureau included 20 planners, designers and draftsmen (four of whom were women), eight architects, seven engineers and two electrical engineers.[9] The bureau was headed by the artist-architect Arieh El-Hanani (1898–1985) and the engineer Willie Weltsch (1887–1978). Besides the planners of the technical bureau, leading architects from the Yishuv were invited to plan the main building of the fair and to advise in the planning of the remainder of the buildings. The large scale building in the fair called for a reorganization on the part of the Tel Aviv municipality to found an advisory committee to the planning department of the fair. Its function was to instruct and confirm the preparations for the fair and to answer technical and organizational questions.[10]

An investigation into the personal and professional background of the planning team revealed that it was extremely heterogeneous. It included both new immigrants and veteran settlers from diverse backgrounds. Josef Neufeld, Arieh Sharon and Louis Redstone who were members of HaShomer Hatzair ('The Young Guardian')[11] had been in Palestine at the beginning of the 1920s and had left to study and returned at the beginning of the 1930s. Genia Averbuch was raised in Palestine, had studied architecture in Europe and returned at the beginning of the 1930s. Weltsch and Harry Lurie, experienced planners, immigrated several months before the work started while El-Hanani, Richard Pacovsky and Kauffmann had been active and well-known in the country for about a decade. The group as a whole was a team of young, highly talented architects most of whom had returned from Europe, and were already well known and appreciated in the country. They planned in the spirit of the new European avante-garde. Their outspoken belief that the new style was appropriate to represent Jewish society in the country was the reason for their selection as the fair's designers.

The Jewish Worker

One of the most impressive achievements was the speed with which this huge building project was completed. The construction required only eight months which is a very short time considering that seventy pavilions were built, of which more than a dozen were major buildings.[12] This achievement is to be credited to the Construction Office, or 'Misrad Kablani' which was The

Cooperative Contractors' Office of the Jewish Laborers in Tel Aviv. Aside from the great satisfaction derived from the success of building the project within the short time schedule, the building of the fair was a Zionist symbol and served as a public demonstration of the capabilities of the Jewish worker:

> I saw work in many countries and did not see the joy of working and the enthusiasm to learn new ways, or such emotional connection to the work as I saw in the building of the fair here. It was a pleasure to work with Jewish laborers and the management of the co-operative of the Misrad Kablani and to be part of the open and demonstrative denial of the terrible fallacy that Jewish work is not good or efficient. In Germany people could not imagine that Jewish labor could be so successful (Weltsch, 1934).[13]

The success of the Yishuv's laborers in building the fair, was a source of national and Zionist pride. One of the first urban statues in Palestine was erected at the fair and dedicated to the workers. El-Hanani designed the statue 'Hapoel Haivri' (The Jewish Worker) and erected it in the central square of the fair. This concrete statue, the largest and most important of the fair, is the figure of a Jewish worker carrying on his back a heavy beam. Both Cubist and Constructivist influences can be seen in the sculpture; however, in addition to its aesthetic qualities, it had a high level of symbolic significance for the pride of the young Yishuv in its new city.

The buildings of the fair were the only large complex of buildings of their kind in Palestine in the beginning of the 1930s. The halls were among the largest in the country with roofs cast in reinforced concrete.[14] The total number of the public buildings on this one site was enormous by almost any contemporary standard. In the framework of its international scale, the pride of the planners and builders resided in the fact that the building was based only on local capabilities, was planned by Jewish architects and engineers, and built by Jewish workers using mainly locally made building materials[15] and utilizing building systems developed in the country.

The Involvement of National Institutes

National institutes were heavily involved in trying to promote communication with other Middle Eastern countries and to encourage them to take part in the fair. The Yishuv wished to present Palestine as the best channel for the world to the Middle East and to convince that the Levant Fair could become an optimal alternative for strengthening trade relations with the Middle East.[16] As such, they sought to promote modern architecture as a potentially new language for a modernized East, and one representing progress and openness. This was especially obvious in light of the building of a competing Arab fair in Jerusalem; in this fair the prevailing tone was Orientalism and traditionalism in both appearance and content.[17]

The Jewish agency hoped to take advantage of the economic success of the fair to promote political issues and to alleviate regional disputes. The head of the agency considered that the participation of the Arab states would be an important political success in the attempt to initiate contact with them. The state department of the agency, headed by Moshe Sharet, joined on its own initiative in that foreign policy through its representatives in the neighboring Arab states: Lebanon, Syria, Jordan, Iraq, Persia and Egypt.[18] Dizengoff was also active in foreign policy and being the chairman of the fair, made special trips to Lebanon and Egypt in order to promote it.[19]

The political view of the agency on the role of economics in solving the political issues of the Middle East was voiced by Haim Arlosorov, head of the state department of the Jewish agency, in an interview he gave about the building of the fair:

> Eretz Yisrael will not be built by political action alone … Economic projects like those promoted by this company [The Company for Trade & Industry] are also of great political importance, as economics is the foundation for life and political activities … We must pay attention to the fact that all over the Middle East there are many problems and questions that appear to be political, but actually have no solution other than an economic one … The distribution of correct information by means of fairs with explanation through economic meetings and the attempt to find mutual interests – these are methods I appreciate and prefer to other types of contact and explanations (Arlosorov, 1933).[20]

Although the objective of the Levant Fair was profit, the political consequences of its success were of great importance and contributed to the establishment and strength of the position of the Jewish settlement in Palestine and in the Middle East. Though the fair had the patronage of the British High Commissioner and the participation of the Government of Palestine, is was necessary to have the participation of the neighboring countries to gain the recognition that Eretz Yisrael was a major economic force in the Middle East. This was difficult to achieve due to the boycott by the local Arab population and their call for a boycott by all Arab states.[21]

Architecture and Representation

Situating Modernism: Contextualizing the International

The architecture of the fair was consistently planned and designed in the International Style. Its formal character was totally modernistic with much thought invested in the adaptation of the design to the context and cultural character of place (Figure 2.2). The fair contributed to a local evolution of modern form and details; this effort formed the basis for the definition of the content of situated modernism and its promotion in Palestine. Special attention was paid to the design of the buildings in response to the climatic conditions, including appropriate sensitivity to the wind and the direction of the sun in the sea side site. Though the buildings were designed in accordance with International Style principles, these were adapted so that the design of the building mass, facades, windows and openings, entrances, and details made optimal use of shade and breeze. This was a visible extension of the modern style, but one of simpler masses, smaller openings, and shaded areas. In

Figure 2.2   Galina Restaurant, 1934. One of the most outstanding pavilions in the fairgrounds. Architects Averbuch, Gidoni, Ginzburg.

addition to the architectural character of a situated modernism, contextualism was manifested in planning the integration of the fair area into the urban texture of northern Tel Aviv.[22]

The desire to create a Jewish Palestinian architecture of modern representation is clearly shown in the words of the fair's chief engineer, Weltsch, who explained the theme behind the planning principles and the architectural style of the fair:

> The soul of a people subjected to the deepest degradation abroad while living through a period of unprecedented revival in Palestine must of necessity find its expression not in stillness, but in movement, in creative energy leading to the most startling contrasts, in spite of a craving for unity. Any Jewish architecture developing during such a period must be dynamic. It is our duty not to suppress its dynamic methods of expression, but to bind them together into a rhythmic whole.
>
> The Fair was to provide a picture of modern Jewish architecture and it was our aim to invite a number of architects to draw up plans for individual buildings and constructions, and then to work these plans into a corporate whole. Thus we managed to create something which is as rich in variety as is the composition of the Jewish people, and yet shows a unity of architectural line, a unity of purpose (Weltsch, 1934, p. 10).[23]

Weltsch's words testify to the meaning given to the symbolism and expression of the fairs' architecture. Since it was clear that the representation should be modernistic and in the spirit of the International Style, the choice of architects was from a well-defined sector. These included mainly young architects who had returned to Palestine from studying and working in Europe and who had the ability to imbue the Fair's buildings with the best of the modernistic and avant-garde tendencies that they had absorbed in Europe. Their choice testifies to the desire to represent the Jewish settlement in Palestine through a fair directly connected to current world architectural events and to Western culture.

Being mainly large open-space buildings, they allowed great freedom of creativity. Although intended exclusively for fairs, these were permanent buildings and consequently considered to be of urban importance. The adaptation of the international style to the climatic conditions received top priority. An advertising brochure (1933) issued by the fair's management prior to the building stated: 'the best of the builders will identify advanced architectural themes with the demands presented by the country's climate'.

The pavilions were all built from materials produced locally and building technology developed locally and adapted to local needs. All the buildings were plastered and painted white. Although there was a demand for emphasis on, or high-lighting of, certain of the pavilions, it was decided that all would be consistently painted white.[24] This application of a specific and consistent architectural approach in such a large-scale of public building, enhanced the spread of the idea of modern architecture among the population, and promoted it as the appropriate 'national building style'[25] for the Yishuv. The fair's buildings served as a model for the building of apartment housing in Tel Aviv in the following years.

The modernistic power of the Levant fair was a symbolic statement of the direction of the development of Jewish Palestine and it represented the entire Jewish population. There was representation of the working class of the Yishuv, its organizations and factories, the middle class merchants, industrialists, and capitalists as well as state representation achieved by active participation of national institutes. The Fair of 1934 presented a visual and cultural image and portrayed the Zionist concept that the Jewish settlement wished to show the world. Its rhetoric promoted the image that in Palestine a secular, Western, advanced and strong society, liberated from past traditions, was developing.

The Yishuv and the Production of an Image for Jewish Society

The Yishuv was widely represented in the fair and exhibited in six pavilions. Here they exhibited the achievements of the Jewish society in Palestine. Even in the fair's uniform, modernistic style, these pavilions represented a unique and highly creative level of design. As would be clear in the future, the chosen designers of these pavilions presented the brilliant future of early Israeli creative design. The master plan of the fairground divided the pavilions into three main zones: Eretz Yisrael, Britain, and Foreign Countries (Figure 2.3). The zones were divided in such a way that the Eretz Yisrael zone was the central and dominant area. The Yishuv's pavilions were divided into two zones, urban and agricultural. The main urban zone had four pavilions: the Palestine Industries Pavilion ('the Palace of Local Products'), the Jewish Agency and National Institutes Pavilion, the PICA Pavilion (Palestine Jewish Colonization Association) and the Tel Aviv Pavilion. The agricultural zone was less formal and included the Farmers Federation Pavilion, the Histadrut Pavilion (the General Federation of Jewish Labor in Palestine), and the Galina Restaurant. The planning of the most important Jewish pavilion, the Palestine Industries Pavilion, was placed in the hands of Richard Kauffmann who was a well-known architect and identified with the planning and design of settlement projects.[26]

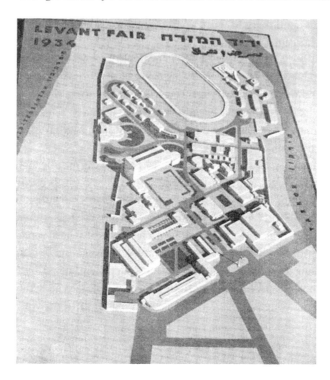

Figure 2.3   Cover of catalogue for the Levant Fair, 1934. Designer unknown.

This was the largest and most elaborate pavilion of the fair with an area of 3,000m^2 in which about 200 Jewish Palestinian companies exhibited. There were great hopes for this pavilion as part of the general effort to advertise and market local products in the world market. Kauffmann planned a simply designed building with a large rectangular closed mass with an accompanying tower. In the context of the layout of the fair, this large horizontal mass with its accompanying vertical tower had a strong symbolic meaning. The tower became a special image and served as a landmark in Tel Aviv and in the fairground. It was illuminated at night and functioned as an advertisement for local products, appearing in all the advertisements and posters so that it became one of the symbols of the fair.[27] This building's architecture was among the best of the International Style in Palestine and remains one of the most important buildings designed by Kauffmann.

It was only natural therefore, that the well-known Kauffmann who had planned the new Jewish settlement for the Zionist Institute, should also plan its representative pavilion, the Jewish Agency and National Institutes Pavilion.[28] All the significant Zionist institutions exhibited in that pavilion; the Jewish Agency, the Jewish National Fund (Keren Kayemet), Palestine Foundation Fund (Keren Hayesod) and the Jewish National Council (Vaad Leumi). The latter included the exhibition of the Hebrew University, Hadassah, Bezalel and the Technion. WIZO, the World Union of Zionist Women, exhibited in a small separate pavilion.

For the first time all of the Zionist institutions exhibited in the same pavilion and thus demonstrated their responsibility for all of the national activities in Palestine.[29] The Zionist management placed great importance on the symbolism of the position of its pavilion in the fairground. At their request the pavilion was situated close to the Palestine Industries Pavilion and the P.I.C.A Pavilion.[30] This position expressed the strong ideological connection between the Zionist institutions and the exhibition of the achievements of the Yishuv.

The Fair as *Gesamtkunstwerk* and the Representation of Cultured Society

Gesamtkunstwerk: The Total Work of Art

Beyond the commitment to advanced architecture and planning, the planners of the fair viewed this as an opportunity to create an exposition of the integration of all of the cultural disciplines in the creation of a total work of urban art, and one which might be representative of the integration and progressiveness of society. In all fields of art and environmental design such as landscape, graphics, painting, sculpture, music and sports, the ambitious cultural creation also served to glorify the Zionist activities in Palestine in the cultural presentations. But it was the totality and integration of the various manifestations as a total work of art, a *gesamtkunstwerk*, in the modernist sense, that was to foster a representation of the flowering of a new society.

This representation characterized the Yishuv's society in Palestine as a Western culture. The music played in the fair was mainly classical, opera and jazz with very little Jewish traditional music and folk dancing.[31] The painting exhibition exhibited the work of painters who had immigrated in the beginning of the 1920s and painted the landscape of Tel Aviv and its surroundings in a modernist spirit;[32] the graphics designers used abstract means, invented new typography, and created posters which expressed the new architecture in boldly dynamic graphic means.

Landscape Design

One of the young and emerging new arts put into the service of this cultural manifestation was landscape design. The ideology behind the master plan and the landscaping design changed in the 1934 fair under the influence of this move towards a new cultural representation. The Yishuv chose to present to the world a well-developed urban image as part of the cultural status of Jewish society in Palestine. In previous exhibitions erected in Tel Aviv small pavilions had been set within greenery and the area had a rural atmosphere. The new fairground layout was urban in character containing a wide avenue, a central square with streets and landscaped squares surrounded by buildings. The landscaping and gardening were planned accordingly, with the garden areas following the geometric order of the buildings and squares.

All of the fair's gardening was planned and executed by the gardener, Yaakov Nitzan (1900–2000). He was a pioneer in the field of landscape design in Palestine and had an important role in the design of the Yishuv's gardens and in the development of the local gardening culture. Nitzan, too, had been a member of a Zionist Youth movement along with which he had immigrated to Palestine at the beginning of the 1920s. He was chosen to plan the 1934 fair's landscape, since he had by then accumulated much experience and practical knowledge of local plants and local climatic conditions especially in Tel Aviv. Nitzan was active in adding local plants to public gardens; at the same time he also introduced to the country plants and trees from all over the world. He was the first to bring cacti and succulents and the first to plan rockeries.[33]

From a climatic point of view, the fair's site was very difficult. The soil was sandy and the open area between the sea and the Yarkon river was affected by winds from the sea and sand storms.[34] Nitzan's achievement under these difficult conditions was highly appreciated and won a gold medal from the Judicial Committee, which chose the excellent designs of the fair from the various arts.[35]

Interior Design and Display

The creativity and boldness of modern design was especially prominent in the interior design and display designs of the fair's pavilions. The inventiveness of modern design was emphasized in the pavilions created by the Jewish Settlement such as the Palestine Industries Pavilion, National Institutes Pavilion and especially the Workers Union Pavilion. The rhetoric of these modern displays expressed the strong desire to build a dynamic and modern image for the Jewish society developing in Palestine.

The display at the Histadrut Pavilion was outstanding in its innovative presentation. Moshe Raviv-Verobeichic (1904–95) was recruited to design it.[36] He was one of the most important avant-garde artists to come to Palestine.[37] He saw himself as a Zionist photographer and was active for the Zionist cause and promoted its ideology through his work. He started working intensively for the National Institute and exploited the photographic techniques and methods of display he had developed in Europe for presenting and propagating the Zionist idea. Photography became one of the great media of the promotion of social and cultural dynamics.

Among his other works, Raviv-Verobeichic created many collages assembled by photo-montage and these created a unique medium of display. He exploited this technique to describe the development of building in Eretz Yisrael in the agricultural settlements and the cities and to elevate the image of the settler as pioneer. Although the displays were based on the two-dimensional qualities of photography, the complexity of the photomontages created a three-dimensional illusion of space. The works were multilayered and this spatial quality accentuated the message of dynamism. In addition to his work for the Histadrut he created photomontages for the catalogue and advertisements of the fair. Here too, his personal, unique style contributed to, and served the national need to present a Western, modern, cosmopolitan image of modern Jewish society.

Graphic Design

The development of modern tendencies in graphic design in the Yishuv of the 1920s and 1930s evolved along with the development of modern architecture. This mutual evolution can be seen in the graphic representation of the posters prepared for the fairs from 1924–36. As a whole, the more modernistic the architecture, the more abstract, dynamic and asymmetrical was the graphic representation of the posters. Along with architecture and photography, graphic design had become one of the great modernist tools of communication in the social developments of the early twentieth century. The fair exploited this potential of graphic design as a communications medium and both the content and form created innovative approaches to serve the Zionist idea.

The influence of the formal style and abstraction of architecture on graphic design practice can also be observed in the design of fonts used for the posters advertising the fair. Designers appear to have adjusted the typographic approach to harmonize with the purity and simplicity of the building style of the time. This too can be seen in the posters designed for the exhibitions from 1924–36. Those designed for the first exhibitions used traditional fonts sympathetic to the eclectic and orientalist building style that prevailed at the time. An example was the traditional 'Sephardi' font whose biblical source had been converted for daily use. Its use represented tradition and in the 1930s it appeared unsuitable to represent a developing society in a new country.

The posters designed for the 1934 fair – the first modern fair in Palestine – used a new modern style of Hebrew fonts probably influenced by the modern forms of the new architectural style of the fair. Its square form and simplicity allowed for easy reading and accentuated its modernist sources. This was evident in the poster designed by the Shamir brothers for the fair of 1936 in which the connection and confrontation between the two building styles, Eastern (1924) and modern (1936) and the two font styles was accentuated (Figure 2.4).

Beginning with the 1929 exhibition, poster design changed dramatically and the multitude of information in posters was now transferred to a style of communication via abstract visual messages. The text was minimal and its graphic presentation and composition became an integral part of the artistic design of the poster. Furthermore, the design of most of the posters made use of architectural elements in order to equate the idea of progress with the forms of modernism. This approach to design for the posters and advertisements was initiated for the 1934 fair and reached a climax with the posters for the 1936 fair, in which the text itself was also designed in the form of architecture. The image of 'the Palace of Local Products' designed by Kauffmann appeared as a symbol of modernism in all the posters, many of which illustrated building designs. The formal dynamics of the interplay of the vertical tower with the horizontal main facade appears to have captivated the imagination of the graphic designers of the fair.

The graphics chosen for the title page of the fair's catalogue as well as the leaflets distributed by the fair's management also had a visual message. Architectural figures expressing an Eastern atmosphere alongside the image of modern buildings attempted to illustrate the modernity of the fair and to symbolize its contribution to the development of Palestine. 'The Flying Camel', the emblem of the fair, floated above the elements representing Arab Palestine – the dome and the arch, the palm tree, the cactus plant and the minaret of the mosque. Coexisting with this contextual background, modern buildings with their strong geometric simplicity, dynamic play of forms, vertical masses and horizontal windows represented the new world and its building in the Zionist settlement (Figure 2.4).

**Figure 2.4   Poster for 1936 Levant Fair (left). Designers, the Samir
brothers. Poster for the 1934 Levant Fair (right). Designer
unknown.**

'The Flying Camel', the emblem of the Levant Fair, became one of the best
known and most successful images in the 'war of symbols' of international
fairs. The Flying Camel was the idea and design of El-Hanani for the 1932
fair.[38] He presented it as a symbol of the East while the image of a flying camel
represented the changes taking place in the Zionist settlement: the slow and
traditional East becoming advanced and in the flight of new development. The
Flying Camel fired the imagination of the Jews in the country and in the world
at large. Many institutions wished to adopt it as their emblem; it appeared in
many and varied ways and its significance became much more than simply
that of a symbol for a fair with only local and ephemeral meaning.[39] El-Hanani
had cleverly designed a symbol expressing the romance of the Orient as well
as a vision of innovation and dynamism. The great success of the emblem,
and perhaps of the fair as well, was its ability to emblematically present the
evolution of tradition into modernism.

Memory and Transition

The Levant Fair of 1934 lasted barely 45 days. Its physical traces are today all but obliterated. However, it must be portrayed as an apogee of the cultural statement of both the achievements and hopes of the young Zionist society. The complex and collective work of art, which was planned, built and operated so successfully radiated a remarkable optimism regarding the future of this new society and its ability to coexist with tradition and the presence of the past. It is an almost historically unparalleled statement of the ability of an avante-garde modernism to symbolize and support the cultural aspirations of a highly contextualized society.

That model spread in the coming years to become a more general model of urban development in Tel Aviv and elsewhere in Israel. The white exhibition became a white city; it no doubt influenced the acceptance of modern architecture in Palestine, and later, in Israel. The memory and suggestive symbolic idealism of the fair also lived on in the body of remarkable designers who found their first opportunities in this transitional event. The search for a situated modernism continued and inspired several generations of young architects to create in Tel Aviv one of the largest urban ensembles of International Style architecture in the world.

Notes

1   M. Novomyesky, 'For the Opening of the Exhibition', *Ha'aretz*, 27 April 1934.
2   'Pictures from the Levant Fair 1934', *Building in the Near East*, pp. 12–13, December 1934.
3   See: Alexander Baerwald, 'The Art of the Homeland', *Mischar Wetaasia Magazine*, Special issue on Architecture, 15 February 1925, pp. 95–6.
4   See: Joseph Berlin, 'Architecture in Eretz Yisrael', *Mischar Wetaasia Magazine*, Special issue on Architecture, 15 February 1925, p. 96.
5   For more information, see: Herbert Gilbert and Ita Heinze-Greenberg, 'The Anatomy of a Profession: Architects in Palestine during the British Mandate', in Gilbert, Herbert. *The Search for Synthesis*, pp. 303–13, The Jubilee Edition, Technion, June 1997 [1992].
6   As patron of the fair, the British High Commissioner was actively involved: he laid the corner-stone, visited the site during construction and spoke at the opening and closing ceremonies of the fair. His speeches express his appreciation for the modern architecture of the Fair and his support for its development in Palestine; *Tel Aviv Municipality Magazine*, Vol. 11–12, August–September (1933), p. 410, TAMA; *Ha'aretz*, 27 April 1934.
7   See list of members: *Tel Aviv Municipality Magazine*, 10, July 1933, p. 357, TAMA.
8   Letter: Moshe Sharet to Fair's management, 27 April 1934, CZA S/25 7323.
9   See Catalogue of Levant Fair 1934.

10   'Building of the Exhibition', *Ha'aretz*, 29 June 1933; Samet, S. (1933) 'Building of the Exhibition', *Ha'aretz*, 29 November 1933.

11   HaShomer Hatzair (The Youth Guardian) is a secular-Zionist-socialist youth movement, founded in Europe in 1913. It is the oldest youth movement currently active in Israel. Chapters of the movement's history include landmarks in Zionist history, such as the immigration of thousands of its members to Eretz Israel during the 2nd immigration, and the foundation of the Kibbutz Ha'Artzi movement in 1927 and its 85 Kibbutzim.

12   'Specified Cost at Specified Time – Labor Cooperative's Triumph', *The Palestine Post* (Levant Fair Opening, Special Supplement), 26 April 1934, p. 4.

13   Willie Weltsch, 'Labor and Style in the Fairground', *Davar*, 26 April 1934.

14   Ibid.

15   Building materials unavailable in Palestine were usually imported from countries participating in the Fair, mainly from England, Belgium, Romania and Yugoslavia.

16   The economic aspect of the participation of the Arab countries was important not only for Palestine but for other participating countries as well. The Federation of British Industries emphasized the importance of securing the visits of Middle Eastern merchants to the Fair. Letter: Dizengoff to Johnson, Treasurer, Government of Palestine, 18 October 1933, TAMA 4–3177(A).

17   'Arab Fair a Fair Bazaar', *The Palestine Post* (Levant Fair Opening, Special Supplement), 26 April 1934, p. 8.

18   Letter: Moshe Sharet to Fair's management, 28 September 1933, CZA S/25 7323.

19   Letter: Yevzerov (Fair's management) to Dizengoff in Cairo, 14 November 1933, TAMA 4–3177(A).

20   Haim Arlosorov, 'The Economic and Political Value of the Levant Fair', CZA S25/7323. published in *Davar*, 10 March 1933.

21   Letter: Fair's management to Moshe Sharet, 8 January 1934, CZA S/25 7323.

22   The master plan had two existing main streets of Tel Aviv running into the entrance square, which continued into a wide avenue leading to the central square of the Fair.

23   Willie Weltsch, 'The Eight Month Wonder – The Architecture of the Fair', *The Palestine Post* (Levant Fair Opening Special Supplement), 26 April 1934, p. 10.

24   The colour scheme, see: Jacob Shiffman, 'The Levant Fair, Tel Aviv', *Town Planning Review*, XVI (3) (1934–35), p. 194.

25   'National Building Style', defined by Ratner in his article 'Towards the Original Style' from *The Annual Building Book* (1935), pp. 34-6, 75.

26   Kauffmann (1887–1958). Born in Frankfurt, Germany. Studied architecture in Darmstadt and Munich. In 1920 Kauffmann was invited by Dr Arthur Rupin, Head of the Settlement Department of the Jewish Agency, to come to Palestine and to take an active part in designing the Urban and Agricultural Zionist settlement. He was head of the Design Department of the Jewish Agency in Jerusalem. At the same time he worked intensively in the private sector and was one of the first modern architects in Palestine.

27   See poster illustration in Figure 2.3.

28   Two artists were chosen to design the exhibits: the painter Ya'acov Steinhardt (1887–1968) and the sculptor Ze'ev Ben Zvi (1904–1952). It was important for National Institutes to introduce artistic values in the presentation of their exhibits; CZA KH4B/5567. Ben Zvi made a huge bronze sculpture depicting a pioneer spreading seeds, which was artistically impressive and dominated the pavilion. However, many did not understand its symbolism which intended to show the important part played by the national capital in building and

developing the Zionist settlement. Letter: 'Report of the National Institutes Pavilion in the Tel Aviv Exhibition 1934', May 1934. CZA S25/7323.

29   Letter: Epstein (Jewish National Fund) to Kaplan (The Zionist Management), 17 November 1933, CZA S53/1231(B); Letter: 'Report of the National Institutes Pavilion in the Tel Aviv Exhibition 1934', May 1934, CZA S25/7323.

30   Letter: Yevzerov to Kaplan (the Zionist management), 15 October 1933, CZA S53/1231(B). Letter: Yevzerov to Olizur (Palestine Foundation Fund), 12 December 1933, CZA S53/1231(B).

31   Artistic program: 'Kunstlerisches und Vergnugungsprogramm fur die Levant Fair 1933', 1 October 1933, TAMA 3176–4 (D).

32   'The Exhibition of Eretz Yisrael's Artists in the Levant Fair', TAMA 3177–4 (B).

33   He designed and exhibited in the 1929 exhibition and in the 1932,1934, and 1936 fairs. In the Anglo-Palestine exhibition (London 1933), he was chosen to exhibit plants and flowers growing in Palestine and described himself as 'Landscape and Rock Gardener of Tel Aviv and a specialist in Palestine wild flowers'.

34   Letter: The Company for Trade & Industry to Nitzan, 25 December 1936, from Asa Nitzan.

35   Certificate Award, Tel Aviv 1934, from Asa Nitzan.

36   Raviv-Verobeichic (1904–95). Born in Vilna. Studied in the Bauhaus and was influenced by the avant-garde design of Kandinsky and Moholy-Nagy. In 1934 Berl Kazenelson, the head of the Histadrut, invited him to come to Palestine to design their exhibits for the Levant Fair. He lived in Tel Aviv and worked as a graphic designer and photographer, for the Zionist institutes and independently.

37   See Rona Sela, *Photography in Palestine in the 1930s–1940s* (Herzliya Museum of Art, Hakibbutz Hameuchad Publishing House Ltd., 2000).

38   As graphic design became more modernistic and dynamic, so too the flying camel underwent a process of abstraction. Its original realistic three-dimensional figure by 1934 became flat and dynamic, and by 1936 was a completely abstract icon.

39   'The Emblem and the Levant Fair', *Ha'aretz*, 12 July 1933; 'The Flying Camel for Children', *Ha'aretz*, 4 December 1933; 'The Flying Camel-Symbol of the Revival of Hebrew in the Diaspora', *Ha'aretz*, 10 December 1933; 'Levant Fair for What?', *Ha'aretz*, 4 March 1934; 'The Emblem The Flying Camel', Harusi, Imanuel, *Ha'aretz*, 27 April 1934; Arieh El-Hanani and M. Carpas, 'Story of an Industrialist', September 1934, CZA S8/1579.

Abbreviations

CZA – Central Zionist Archive.
TAMA – Tel Aviv Municipality Archive.

References

Barkai, S., 'La Foire du Levant de Tel Aviv', *Architecture d'Aujourd'hui*, 10 (1934–35), pp. 75–7.

Bauhaus Archiv and Droste, M., *Bauhaus 1919–1933* (Berlin: Benedikt Taschen, 1998).

Gilbert, H. and Heinze-Greenberg, I. (1992), 'The Anatomy of a Profession: Architects in Palestine during the British Mandate', in H. Gilbert, *The Search for Synthesis*, The Jubilee Edition, Technion, June (1997), pp. 303–13.

Gillian, N., *The Bauhaus Reassessed, Sources and Design Theory* (London: The Herbert Press, 1985).

Levant Fair 1934 Catalogue (1934), Levant Fair Publications.

Levin, M., *White City* (Tel Aviv Museum, 1984).

Mischar Weta'asia Magazine, Special issue on Architecture, 15 February 1925.

Redstone, G.L., *Louis G. Redstone – From Israeli Pioneer to American Architect* (Ames: Iowa State University Press, 1989).

Riley, T., *The International Style: Exhibition 15 and The Museum of Modern Art* (New York: Rizzoli, 1992).

Sharon, A., *Kibbutz + Bauhaus* (Massada: Givatayim, 1976).

Shiffman, Y. (ed.), *The Annual Building Book 1935* (Tel Aviv: The Company for Trade & Industry Publications, 1935).

Shiffman, Y., 'The Levant Fair, Tel Aviv', *Town Planning Review*, XVI (3) (1934–35), pp. 191–4.

Chapter 3

Mold

Zvi Efrat

Everything will be systematically worked out in advance. In the elaboration of this plan, which I am capable only of suggesting, our keenest minds will participate. Every achievement in the fields of social science and technology of our own age and of the even more advanced age, which will dawn over the protracted execution of the plan must be utilized for the cause. Every happy invention that is already available or will become available must be used. Thus the land can be occupied and the State founded in a manner as yet unknown to history, with unprecedented chances of success (Theodor Herzl, *Der Judenstaat* (*The State of the Jews*) 1896).[1]

The great revolution is not over yet, and its essential functions have barely begun. In the near future we must lay foundations that will stand for decades and possibly centuries to come. We must shape the character of the State of Israel and prepare it to fulfill its historical mission (David Ben-Gurion, *The War Diary*, January 1949).[2]

As it is – crowded, heaped up, frantic at the center; diffused, dissociated, monotonous at the margins – the fabricated space of Israel is a 'manner as yet unknown to history' whereby 'land can be occupied and the State founded'. Contrary to common belief and to visual impression, it was not born of haphazard improvisation, emergency solutions or speculative entrepreneurship and certainly not of spontaneous diachronic development – but rather of the unprecedented objective to put into practice one of the most comprehensive, controlled and efficient architectural experiments in the modern era.

Indeed, this is Israel's singularity among nation states: it was 'systematically worked out in advance', formulated a priori by means of arithmetic, planimetric and demographic formulas, drafted in pencil, ink and watercolor by planning professionals from various disciplines, who were called upon to actualize Ben-Gurion's prescript 'to transform the country, the nation, our entire modes of life'; or to put it in a more technical language – to engineer and redesign (no less) the country's geographic, ecological and agronomic mold; its patterns of urbanization, socialization and employment; the overall framework of production and service; the character of its public life as well as its patterns of domestication in the new Israeli state. An all-encompassing planning ambition such as this is not self-evident even in the context of its own period – one that revered the 'planning sciences' and perceived mega-architects, infra-engineers, macro-economists and sociology-masters as the omnipotent agents of progress itself.

Precedents for the Israeli Project can be found in Stalin's Five-Year Plan for the Soviet Union, in the American New Deal infrastructural projects and public works of the 1930s, in the German National-Socialist regional plans for the occupied territories of Poland and in the post-World War II British schemes of New Towns, but its sweeping vision, exceeding all instrumental models and circumstantial explanations, is rooted in the utopian imagination and the topian (to use Martin Buber's neologism) praxis of the Zionist Movement. These roots call for some elaboration, so as to re-enact the bedding upon which the Master Plan of Israel was drafted immediately upon the 'outbreak' of the state.

The very notion of 'HaMiph'al HaZioni' (the 'Zionist Enterprise' or the 'Zionist Project') enfolds the highly institutionalized, explicitly synthetic and actively constructive nature of the process of land appropriation and nation-building by the Jews in the twentieth century. Any attempt to 'normalize' Zionism by overemphasizing aspects such as spontaneous immigration, organic settlement or market forces, misses the point: the artificial essence of Zionism, the grounding of its rhetoric on the notion of 'negation', 'inversion', 'synthesis' or 'combination'; its self-definition as a constrained, corrective, redemptive (Ben-Gurion even used the notion of 'messianic') intervention in historical time and geographic space. In this context, controlled centralized planning is the ultimate trope that binds Words and Things: it is the Zionist spirit itself, emanating from layers of fictional prose, ideological manifestoes or programmatic protocols and printed on the landscape over and over again with every new spatial move or architectural object.

The most conspicuous use of centralized planning – not merely for territorial organization and control of settlement dispersal, but rather as an apparatus molding a new ethos – is manifested in the consistent efforts to shift the political, cultural and economic weight from the city to the countryside and from the center to the periphery. As a rule it may be said that in its first 50 years, the Zionist movement devised and developed a range of pioneering models of agricultural settlement supported by sophisticated logistics of manufacture, organization and marketing, but never imagined, planned, or actually built a city. In fact, the modern metropolitan city was consistently portrayed in both literary utopias and direct propaganda as an anathema to the Zionist concept of land redemption; a parasitic growth threatening to undermine the fundamental values of the re-emerging Hebrew civilization.

Thus, revolutionary objectives, totally controlled planning and a well-coordinated course of action had characterized the Zionist Enterprise from its very outset. However, the country's crucial 'conversion' was obtained with the founding of a sovereign state. The end of the British Mandate in 1947 and the ensuing administrative vacancy; the War of 1948 and the ruinous grounds it created; the exchange of refugee populations during the war and immediately following it; the confiscation and nationalization of over 90 per cent of the country's land; the emergency legislation (most of which is still valid today) and the austerity decrees; the virtually absolute monopoly of Mapai (Israel Labor Party) over all state and Histadrut (Federation of Labor Unions in Israel) apparatuses; the moral and material support provided by the world's superpowers for the new state (specifically, the budget allotted for development, which was separate and equal in size to the comprehensive national budget) – all these came together to provide an opportunity and ostensible legitimization for a project of construction (and obliteration) more daring than any of its literary precedents.

Only a few weeks after the Declaration of Independence – during the War of 1948 and as a means of pushing forth its goals and achievements on ground – Arieh Sharon, a Bauhaus graduate and one of the prominent architects of Israel's Labor Movement, was commissioned to establish the governmental Planning Department.[3] Within about a year, this department presented an overall master plan for Israel (known as the Sharon Plan) and provided the political leadership of the time a powerful tool for molding a new landscape and dictating the shape of things to come.

In the opening session of the Government Districts and Zones Planning Committee (6 December 1948), Sharon presented not only his own view as an architect who had literally chanced upon a country to plan, but also – mainly – the spatial perception of his patron, Ben-Gurion, and in fact the very essence of Zionist rhetoric:

> I am glad to have been granted the opportunity to present to you some of our work and discuss with you the problems and difficulties that we encounter. For many years those engaged in the field of planning have felt the lack of central and national planning – a factor that hinders our work and restricts our capacity to build the country …
>
> We are told that various, wealthy countries have existed and thrived for many years without central planning institutions. To this argument one should

reply by stressing that affluent nations can perhaps afford experimentation and lack of planning. At the same time, one must also note the incurable chronic illnesses in the big cities of these countries.

We know that we are incapable of withstanding such maladies. If we were to spawn such entities as the big cities with their grave conditions, we wouldn't have the scope to rectify the situation at our disposal. The health and welfare of the people dwelling in these entities would deteriorate, until entrepreneurs would have no power left for creation and construction. We can see that after one, or two generations at the most, of the country's construction, the results of unguided construction are highly burdensome. We live in apartments overlooking our neighbors, or at best – hot and narrow streets. Our homes do not have a view of rolling green lands that may pacify us.The new ownership of the land thus makes it possible to put in order and re-arrange the space and to ensure the people's physical and mental well-being through central planning. The 'Old World' (which, for the purpose of this argument, comprises not only Europe, but also the migration countries of the 'New World') is already degenerate, sick, spawning urban monsters. Here there exists an opportunity for a fresh start, on a *tabula rasa*, as it were. Here, as opposed to there, there is not enough space, there is no leeway for uncontrolled developments, there is no room for degeneration.

The pressing national task assigned to Sharon and his team of planners was providing temporary housing solutions for the masses of new Jewish immigrants and settling the country's borderlands, in order to stabilize the 1948 cease-fire lines, prevent territorial concessions and inhibit the return of Palestinian war refugees. The planners accomplished this by drafting a statewide network of civil frontiers composed of transit camps and outpost agrarian settlements, as well as by resettling deserted Arab villages with new Jewish immigrants (mainly those coming from Asia and North Africa). Concurrently, a long-term mission was outlined: preparing a plan for 'the country's intense and comprehensive development, which would reach all its corners'.

Just one year after its processing began, the first draft of the Sharon Plan was presented. Its objectives were targeted toward a local population of 2,650,000 inhabitants (a target obtained in 1966), which would be dispersed throughout the country, thus adjusting the 'anomaly', or the 'colonialist pattern,' as the planners dubbed the development of the Jewish community in the country during the British Mandate. (Upon the establishment of the State of Israel, two-thirds of the Jewish population was concentrated in the three large cities: Tel Aviv, Jerusalem and Haifa. 82 per cent lived along the coastal plane. The aim of the Sharon Plan was that only 45 per cent of the urban population would dwell in the big cities, while 55 per cent would settle in the new medium-sized and small towns.)

The plan was divided into five categories: agriculture, industry, transportation, forestry and parks, new towns. Agricultural settlement was perceived as the major factor for developing and ensuring economic independence. Accordingly, the target set comprised of 600,000 farmers who would live in 120,000 farms, in units of 20–25 dunam (5–6.25 acres) per family, and who would provide 75 per cent of the population's food. An irrigation network would conduct water from Northern Israel and the Shfela lowland to the Northern Negev desert and the Judean Mountains, thus enabling the Negev's de-desertification and inhabitation project.

It should be pointed out that in general, a Heimat ideology and countryside logistics characterized most modern reform movements in most modern nation-states whether the conceptual progress moves radially from the city to the suburb to the country, or builds up vectorally from the hinterland to the metropolitan nodes. Zionism is no different, only that its provincial ideology and conception of the New Village (or *kibbutz*) as the basic building block for the country's construction, was literally, consistently and forcefullly translated on ground.

The planning of industrial areas in the Sharon Plan was also reminiscent of social and ecological reform movements prevalent in the Industrial countries from the mid-eighteenth century onwards, which maintained that appropriate geographical dispersal of factories would disinfect the city and improve the workers' living conditions. In Israel, a country without any history of industry and its discontents, only the reformist after-effects were enacted, as could be gathered from texts accompanying the Sharon Plan: 'The medium-sized, more efficiently organized town would facilitate the lives of industry workers, furnishing them with comfortable and cheap public housing, convenient transportation to their work places and to parklands for recreation.'[4]

A nationwide transportation network was delineated, including a new port, two airports, railroads (mainly for freight trains) and a rhizomatic road network to provide access to the dispersed villages and small towns. (As opposed to a sensible main railroad or highway system that would directly and efficiently connect between population centers and the big cities.)

The Plan's vision of landscaping entailed a comprehensive afforestation campaign (which in fact, encompassed all territory not viable for agriculture or construction), as well as a designation of areas for national parks and nature reserves according to their anticipated archaeological value, unique natural environment, or reverberation as former battlefields or commemorative sites for the Holocaust. (The national parks system was perceived as a didactic instrument, figuratively illustrating Jewish history while cultivating a new Israeli culture of recreation.)

Professionally speaking, the Sharon Master Plan is by no means an innovative document. Rather than original ideas, it presents an assemblage of models, theories and experiments; some of which had been developed locally during the British Mandate era, mainly by members of the 'Settlement Reform Forum', others were imported from Europe as ready-mades and abruptly naturalized. In fact, the Plan is unique only in its scope; in its ambition to put forth at once a single layout, a single vision, a single stately concept at a scale of 1:20,000. Such ambition could have remained anecdotal – too rational, instrumental and progressive; or alternatively, too dreamy, redemptive and reactionary – had it not been implemented verbatim, almost entirely, often by short-cutting standard planning procedures; always through the systematic reproduction of zoning doctrines, building typologies and construction methods.

While it had no statutory status (or perhaps, precisely because it did not have to face legislation procedures),[5] the Sharon Plan instantly, within less than a decade, transformed from a principles document to a mega-project embracing dozens of cities and towns and hundreds of rural settlements *ex machina*; extensive woodlands, national parks and nature resorts *ex fabrica*; networks of roads, electricity, water, ports and factories *ex nihilo*. Soon after the official publication of the plan in 1950, Sharon and his team of planners realized that they had provided the government with an all too readable building manual. In spite of their enduring commitment to overall centralized planning and their loyalty to the party, they made efforts to slow down the literal application of their plan on ground. However, the genetic code they had devised on paper in their laboratory had already been cloned and disseminated throughout the country, proving its durability *in vivo*, even under the most adverse conditions.[6]

Deterritorialization (effective upon the local Palestinian population but by and large futile vis-à-vis veteran Jewish population) and decentralization (according to the strategic precept of using civilian settlements as military outposts *de jure* or *de facto*, a precept developed already in the early pre-state days and valid to this day) – formed a consecrated cause, which dictated all moves and procedures of the national plan, even if at times they were inconsistent with professional discretion, even if they entirely failed the test of economic logic, even if they turned the 'melting pot' rhetoric against itself and created, in effect, severe geographic and social segregation between veteran residents and new immigrants. Mass immigration was both the problem and the solution. *The problem*: since the immigrants' predispositions regarding the choice of their place of residence were known in advance and considered by the authorities a threat to the Zionist settlement policy.

Thorough studies of the colonization patterns in 'New World' countries carried out by the planning committee, indicated that in Israel too, without decisive state intervention, the first generation of immigrants would undoubtedly choose to crowd the coastal cities, thereby deepening the Mandatory 'anomaly' even further and accentuating the emptiness of the rural regions. *The solution*: since, were it not for the statistical body of new immigrants, the historical opportunity to re-invent the Israeli space would not have emerged. Indeed, in spite of the considerable propagandist endeavors invested by the government in the attempt to induce a population migration to the periphery, it was clear to leaders and planners alike that the Sharon Plan would not be voluntarily implementable. Eliezer Brutzkus, one of the senior conceivers of the Plan, described its achievements *post factum* vis-à-vis the relevant model – namely, the construction enterprise of the new workers' towns in the Stalinist Soviet Union:

> Truth be told, these results were obtained here too, against the free will of the settled subjects – namely the immigrants – through a method whose underlying principle was 'straight off the boat to development regions'. We must not forget the basic fact that the creation of the new towns, and the settlement of the peripheral regions, was done primarily through directing the new immigrants, and only marginally by attracting the 'veteran' population.[7]

The Soviet project, efficient as it was in constructing and inhabiting, top-down, hundreds of new provincial towns and edge cities, was not the only model from which the Israeli planners drew inspiration.[8] The rehabilitation projects of Western Europe after World War II, especially the new satellite towns around London founded by the British Labor Government, were also thoroughly analysed by Sharon and his team (the planner Sir Patrick Abercromby, the 'father of the New Towns', was even invited to Israel and met with Ben-Gurion). In the Israeli lab – and this is the enigma of its proletarian charm – a new paradigm was created, based on the unlikely marriage of a suburban garden-city in a Western welfare state and a peripheral industrial town in the outskirts of the Bolshevik Empire. Such juxtaposition embodies the two formative paradoxes ingrained in the Sharon Master Plan: the attempt to concoct, synchronize and monitor a rational mechanism for 'organic', 'regionalist' and quasi-historical settlement; and the unequivocal ideology of anti-urban urbanization (as many towns and as little urbanity as possible).

The basic tool suggested by the plan for achieving this original construct was the redivision of the country into 24 districts, designed mathematically to contain an equal number of residents. The districts were determined according to geographic characteristics and were planned as arrays of agricultural settlements clustered around central villages and served by a regional town. Size, dimension, and amount (of population, of area, of employment) were perceived as reliable criteria for obtaining the desired interactions between center and periphery, city and country, industry and agriculture.

Over 400 agrarian settlements were founded during the state's first decade according to the Master Plan's guidelines (more or less the number of indigenous Palestinian villages evacuated, deserted and ruined during and after the War), but its epitome was the creation of the District Town – infamously known as the Development Town – whose optimal size was the subject of lengthy academic discussions among the planners. Ultimately, the preferred model was of an intimate town, housing between 20,000 and 50,000 residents, assumed to be exempt of the disorientation, alienation, social injustice, speculative realty and other urban malaises associated with the cosmopolitan city. (Was it also a return, *mutatis mutandis*, of the repressed shtetl, the Jewish diasporic provincial town, still haunting the planners?)

In order to prevent at all costs the development of unruly colonization and socialization patterns typical of New World countries, the Sharon Plan chose to emulate the European historic layout, whereby the majority of the population dwells in small and medium-sized towns integrated into the agricultural hinterland, and only the minority lives in the big cities. The origins of this hierarchic web lie in pre-industrial agrarian culture, and reflects centuries of moderate organic growth. The planners of Israel tried to squeeze this process into a single heroic decade, backing their ambitions with intricate pseudo-scientific theories that analysed the link between settlement patterns and endurance during times of crisis.

Especially authoritative for the Israeli planners was the 'Theory of Central Places' formulated by German geographer Walter Christaller through his 1939 doctoral research. Christaller constructed a mathematical model accounting for the population dispersal in Europe, meticulously analysing the distances between centers. According to his theory, the Great Depression that affected both Europe and the United States during 1929–33, attests that the big cities and distinct agricultural areas suffered economic destruction and unemployment, while the medium-sized and small towns located in rural areas, that sustained a mixed economy of agriculture, industry and craft, maintained relative social and economic stability due to mobility of employment.

This theoretical foundation of Israel's Master Plan was, in all probability, transferred out of pure academic and professional conviction – yet in retrospect, it must be said that perhaps today, time has come to discuss a particularly uncanny allusion re-enacted by the planners of Israel; a diabolic jest of history, as it were. For it was the Jews, the Jews of Poland, who were in fact the primary victims of the application of Christaller's spatial theory. Those among them who survived the population relocation project, the encloser in urban ghettos and the transfer to labor and death camps, would become, only a decade later, among the subjects of this very theory's permutation in their newly founded homeland.

In his essay 'The Nazi Garden City', Gerhard Fehl maintains that the major source of inspiration for the Nazi planners was a diagram of 'The Social City' – a humanist utopia of well-balanced regional development, sketched by Ebenezer Howard, the progenitor of the British Garden City Movement.[9] Fehl explains the ostensible discrepancy between the democratic model and its totalitarian simulation by maintaining that *Raumordnung* was, in fact, a sterilization of the Garden City's social-reformist contents and the blown-up application of the principle of regional hierarchy on the scale of large territories. The justification for such misuse of localist concepts was the supposed shortage of living-space (*Lebensraum*), created by the fact that Germany – unlike the colonial empires that dominated a 'world of empty spaces' – was a 'space-less country'. (Such horrific imagery of voids and spaceless-ness is invoked in slogans of certain Zionist circles who propagated the notion of 'land without people' for a 'people without a land'.)

Beyond the direct theoretical ancestory of the Christaller Plan for Poland and the Sharon Plan for Israel, certain parallels can be drawn between the Nazi and the Zionist spatial ideology and territorial practices. In general terms, both movements share a common aspiration to monitor currents of inner migration, to dictate patterns of settlements and to curb spontaneous urban growth, in order to accelerate national bonding and authenticate as it were land ownership. For these ends, both denounced the city as the source of spiritual degeneration; propagated the village as the 'building block' of the new 'spatial order'; appropriated the landscape by means of extensive afforestation; and claimed their acts of social engineering and civil occupation of frontiers, to be based on 'the Anglo-Saxon model of decentralization'.

Such blasphomous analogy, cross-breeding quintessential Left and Right nationalisms, is worthy of print only if it conveys the futility and the horror of a totally planned and controlled territory. The reader, however, is not entitled to infer that the Sharon Plan in particular, and the Zionist spatial ideology in general, conceal a hidden agenda of 'racial order,' 'ethnic cleansing,' mass deportation' or genocide, whatsoever.

* * *

The seemingly paradoxical Zionist attitude of activating a regressive revolution, or a pioneering Old World, may be discerned not only in the dispersal of towns and settlements on the map, but also in the attempt to base the architecture of the towns themselves on a conceptual crossbreeding between mechanistic planning methods striving to render the traditional city more efficient in terms of mass housing and motor vehicle traffic on one hand, and more picturesque conceptions, on the other, willed to tone down the city by deconstructing it into small, autonomous communities, protected from street life, zoned off from industrial sectors and wrapped by green pastoral surroundings. The planners believed that through critical study of urban history they had managed to develop an innovative method for ideal city planning, as could be gathered from the writings of Eliezer Brutzkus:

> The structure of the New Towns is determined by their division into Neighboring Units. This method differs from conservative methods of urban planning still prevalent in the old cities in Europe as well as in Israel. These cities are built as a monotonous continuum of houses, streets and residential neighborhoods, dragging on endlessly and making the lives of their residents gray and dull. [The author goes on to explain the conceptual superiority of cities such as Ofakim, Kiryat Shmona or Ashdod, over Paris, Berlin, or Vienna. Z.E.][10]

The Neighboring Units mentioned above (or the 'eggs' as they were called in the professional milieu) are the structural organizing principle of the new towns. In theory, they were intimate urban sections with biomorphic contours that rejected orthogonal grids and endowed the 'instant' towns elasticity and vibrancy. In reality, the separation into autonomous units created a jumbled grouping of disembodied organs containing a limited variety of housing types and self-contained in terms of commerce, education and leisure services. The town is a cluster of neighboring-units, assembled around a civic center with municipal institutions. The developing town aggregates modular units, thus preserving its neighborly character. The size of each unit was determined in relation to the estimated capacity of schools and kindergartens, the optimal dimensions of the commercial center, and the desirable length of paths in the neighborhood.

The units were planned in such a way that would separate motor traffic from pathways within the neighborhood, enabling pedestrian access to all daily service at a distance of up to 250m without having to cross a street. The smooth, plexing lines of the units; the abundance of open space within and between them; the placement of education and recreation facilities at the heart of the units amidst lawns or woods; the distanciation of industrial areas from living quarters and their separation by green belts; the design of repetitive social housing on undivided land, rather than normative parceling and speculative construction – all these forge the most deceptive illusion of all: the new Israeli town was meant to be a blown-up kibbutz based on homogenous community, collective and egalitarian, without private capital or unanticipated market forces.

However, unlike the kibbutz, or even the pre-state Cooperative Worker's Housing in the well-established towns, which were created as exclusive and hegemonic structures by and for the members of a social avant-garde movement, the new town came into being superficially and coercively – a professional and bureaucratic doctrine forced upon a population of unsuspecting newcomers used as passive subjects of a national experiment. With the foundation of the first New Towns, it became apparent that the progressive zoning principles and the generous 'ecological' aptitude simply do not work. The detached, sparsely-populated ready-made town weighed down disproportionately on the national budget due to the huge amounts of infrastructure they demanded.

The supply of capital and entrepreneurship (from both public and private sources), required for the creation of jobs in those out-of-the-way locations, lagged behind the pace at which the immigrants were sent to the New Towns (in Kiryat Shmona, for instance, the first factory was built a full decade after the town was created). The veteran urban population remained in the cities and ignored the national challenge. The veteran agrarian population of the communal or semi-communal villages already had a well-organized marketing network of its own (including such monopoly cooperatives as Tnuva and Hamashbir, still existing as a lucrative brand today), having no use in the services provided by the New Towns and completely discounting the planners' regionalist vision. The vast expanses that had been water-colored green on paper and in cardboard models, were totally incongruent with the climate, the water resources, and the maintenance facilities in the country, and in reality became dead zones, severing the urban fabric for decades. The autonomous, inward-looking units and the separation of motorways and pedestrian paths, obstructed the development of street-life. The 'alienation, degeneration and low quality of life' in the big city, so consistently denounced by official state propaganda, was replaced in no time with homogeneity, remoteness, and deprivation. Criticism soon took over the prophetic positivism of the professionals. During a retrospective discussion on the planning of the New Towns which took place in 1964, architect Yitzhak Yashar concluded:

> There was an industrial world, there were enormous cities, but they sought dispersal there. It was in the center of London – where you cannot go in and you cannot go out, where traffic and noise are tremendous, where there is neither sunlight nor greenery – that the concept was born. The very same idea – not only in its qualitative but also in the quantitative sense, in its formal sense – was shifted to Beer Sheva. And in Beer Sheva, where one searches longingly for traffic and commotion, for a bit of social gathering – there, in the middle of the desert, we solved the problems of London … but obviously, what is good for five or eight or ten million people – is catastrophic when you have a mere ten thousand in the desert.[11]

Fifty years after its official publication, the Sharon Master Plan still holds its own. The vision of colonization and modernization laid out by the plan has, for the most part, been implemented. The country has developed at an unprecedented rate of growth. The New Towns – a well-intentioned hybrid of imported urban theories and physiocratic local ideology – still exist more or less as they were originely planned: barren Garden Cities, lethargic Work Towns, bypassed regional centers, homogenous melting pots, underdeveloped urban odds and ends still struggling to preserve their special Class A tax-reduction status, granted by the various governments to 'areas of national priority'. Just as the citizens of these New Towns played a historic role in realizing the logistic 'Reversal' of the 1950s, they have fueled the so-called Political 'Turnabout' of the seventies (the left Labor Party losing for the first time its hegemony) and the 'cultural revolution' of the nineties. With each metamorphosis, their distance from the heart of the country only increased.

But the Israeli cultural and territorial vortex is still sweltering, leaving a hope for yesterday's New Towns. Once the big sell-out and redevelopment enterprise of state-owned agricultural lands is over, and once the political settlements beyond the Green Line are road-mapped back to the legal boundaries of the state, the peripheral Garden Cities of the 1950s may become the most desired land reserves: the last option for suburban 'quality of life' so revered by a society that never really coped with its own socialist-agrarian rhetorics and never really sublimated the values of liberal urban life.

Notes

1   Theodor Herzl, *The Jewish State*, trans. Harry Zohn (New York: Herzl Press, 1970). HTML version: http://www.wzo.org.il/en/resources/view.asp?id=287).

2   David Ben-Gurion, January 1949, *The War Diary: The War of Independence, 1948–1949*, ed. Gershon Rivlin and Dr. Elhanan Oren (Tel Aviv: Ministry of Defense – Publishing House, 1982), vol. 3, p. 937 [Hebrew].

3   The planning department was first set up within the Labor and Construction Ministry, but Ben-Gurion transferred it to the Prime Minister's Office in order to enable him to work directly with the planners. The department was later moved to the Ministry for Interior, and eventually became the foundation on which the Ministry for Housing and Construction was established. Currently a new planning department is being established within the Prime Minister's Office, similar to the original Ben-Gurion model.

4   Arieh Sharon, *Physical Planning in Israel* (Jerusalem, The Government Printer, 1951) [Hebrew].

5   Only in 1965 was the 'Planning and Building Law' enforced.

6   An interesting example of the critical second thoughts expressed by the institutional architects can be found in the writings of A. Newman, one of the planners in the department: '... we lost our innocent belief in the automatic regulation of economic and social processes long ago. It seems that the belief in a single planning theory that can provide happiness for everyone is also naïve... we now know that there is always a component of restriction in planning, which tends to oppress the individual' (*Engineering and Architecture Journal*, January 1953, p. 17).

7   Eliezer Brutzkus, 'Transformations in the Network of Urban Centers in Israel', *Engineering and Architecture Journal*, 3 (1964), p. 43 [Hebrew].

8   Arieh Sharon himself, after finishing his studies at the Bauhaus in 1929 and following two years of work in architecture offices in Berlin, was invited by his former teachers, Hans Meyer and Mart Stam, to the Soviet Union to participate in the planning of the industrial city Magnitogorsk. Sharon chose instead to return to the land of Israel, and began his career as an Israeli architect by winning a competition for the planning of workers' housing in Tel Aviv.

9   Gerhard Fehl, 'The Nazi Garden City', in Stephan Ward (ed.), *The Garden City: Past, Present and Future* (E&FN Spon, 1992), pp. 88–103.

10   Eliezer Brutzkus, op. cit., n. 4.

11   'A Discussion of New Towns', *Engineering and Architecture*, 3 (1964), p. 15 [Hebrew].

Chapter 4

Horizontal Ideology, Vertical Vision: Oscar Niemeyer and Israel's Height Dilemma

Zvi Elhyani

Israel proper, without occupied territories, can be larger and stronger … on the condition that the entire conception of building spring upwards – a small, strong, and wealthy state with skyscrapers. … Not an undersized state aspiring to expand. … I see Israel on the whole turning from a village into a city, shrinking in width and springing upward. Investing energy in itself instead of in an area held by force (David Avidan, *Moznayim*, 1986).

Israel must be built upwards and its cities planned vertically – something that will be lauded in the future and will conserve the land … Israel is developing at such a rate that low-to-the-ground construction is unthinkable, for in no time a territory as small as this will be disproportionately covered with low buildings, denying it its natural beauty and vistas, and leaving it without open spaces essential to its development. Low-to-the ground construction knows not what it begets (Oscar Niemeyer, *Ha'aretz*, June 1964).

Introduction[1]

For ideological and strategic reasons, the Zionist conception of space was premised on sprawl and dispersal and the avoidance of dominant urban centers and monumental objects. Ideologically, Zionism affiliated itself with the international garden city movement, fostering agrarian, anti-urban, and anti-bourgeois utopianism at the center of which stood the productive, land-laboring 'New Jew'. Concentration and crowding were perceived as exilic, anti-pioneering trends that could result in a loss of the land. Categorically, the Zionist movement preferred horizontal, sparse, low-to-the-ground construction.

Even after 1948, when the state was established, Israel avoided vertical construction in its spatial evolution. This chapter discusses Israeli architectural and urban planning on a massive scale from its advent in the 1950s through its subsequent concession to real estate speculation and big business interests in the 1960s, following a brief period of active public debate over vertical construction.

This discussion aims to reveal the relationship between the centralistic state urbanism of the 1950s and the market-led urbanism of the 1960s, as well as to offer an understanding of Israeli built space, which has always privileged dispersal over concentration.

At the heart of this chapter stands the story of renowned Brazilian architect Oscar Niemeyer whose architectural activities in Israel included grandiose plans for exceptionally large buildings in Tel Aviv, Haifa, and the Negev desert. Those mega, vertical and high-densed proposals offered a dramatically different sense of place than the small scaled, horizontal and low densed Genius Loci the Zionist Movement wished to produce. Discussing Niemeyer's 'Israeli' period allows for a reexamination of his critique (implicit in his proposed projects) of Israel's built space.

Niemeyer, an avowed communist who exiled himself from Brazil after a military coup in March 1964, arrived in Israel that same month as the guest of entrepreneur and businessman Yekutiel Federmann, who was looking to up the value of two recently acquired plots, one on Mount Carmel in Haifa (later Panorama Center) and the other in the Nordia neighborhood of Tel Aviv (later Dizengof Center). Niemeyer stayed in Israel for six months, during which time he was involved in planning a dozen private, commercial, municipal, and governmental projects throughout the country.

Niemeyer recorded his impressions of the socialist Zionist enterprise and Israel's natural scenery in his diary, as well as his vehement criticism of Israeli planning conceptions, which he regarded as low and sprawling. Virtually all his proposed plans for Israeli buildings addressed what he saw as Israel's adolescent architectural identity crisis: hesitation between the modern and the vernacular, the urban and the agrarian, the horizontal and the vertical.

Zionist Space

The negation of the city stood at the heart of Zionist ideology and its settlement enterprise. This ongoing conflict between the village and the city resulted in the neglect of the city as such over the course of several generations.[2] Since concentration and crowding were perceived as diasporic, anti-pioneering trends, Zionism, as a national revival movement, sought to distance itself from the historic image of the dense and decrepit European city, seeing agrarian life as the basis for the creation a modern society in a new-old land. The Zionist vision of settlement found its ideal in the garden city model, a rural-urban crossbreed conceived by Ebenezer Howard in late nineteenth century England.[3] Howard's garden city developed in conjunction with other modern social ideas (including Zionism and national socialism), which saw the city as a threat to the homogeneous nuclear family and, in the face of urban degeneration, advocated for the welfare of the working class in the form of extra-urban green lungs.

The garden city model, a sympathetic mechanism for denying the dark, stagnant, inhuman face of western European industrial cities, was an attempt to organically fuse a rural-urban lifestyle in an ideal city.[4] This model, adopted in the 1930s and 1940s by Jewish urban planners in Palestine, also informed the planning of Israeli 'development towns following the establishment of the state. According to the ruling ideology, if a field won't liberate the Jew 'returning to his land' from the alleys of the ghetto, at least he will have a small garden adjacent to his home.

The realization of this conception was made possible with the first national master plan for the State of Israel prepared by the governmental planning division in the early 1950s. Just weeks after the establishment of the state, then Prime Minister David Ben-Gurion summoned architect Aryeh Sharon – a graduate of the Bauhaus and a senior architect affiliated with Labor Zionism – to establish a planning division in the framework of his office. This invitation spawned the national master plan which in turn shaped the country for the coming 50 years by establishing guidelines for the physical planning of Israel from the statewide to the local level.

The plan gave physical expression to the government's policy of dispersing the population, settling the frontiers, and turning the 1949 armistice lines into permanent borders. It also delineated a scheme for establishing hundreds of new agricultural settlements and 29 new 'frontier' and 'development' towns. Despite its lack of statutory status, within a decade the Sharon Plan became a meta-project encompassing scores of cities, hundreds of settlements, forests, national parks, nature reserves, roads, electric grids, water and sewage systems, ports, and industrial complexes. In a sense, the Sharon Plan was the culmination of 50 years of Zionist spatial planning.

The plan served a clear policy of fixing borders and seizing land. It transformed frontier regions not in accordance with political agreements, nor through mere military occupation, but through civilian settlement and the creation of unilateral facts-on-the-ground.[5] Though this was also true of the Jewish organizations responsible for settling Palestine before 1948, the establishment of the state and the demographic changes it brought about, enabled not only the drafting of the master plan, but its implementation in practice.

The Sharon Plan translated the government's policy of population dispersal, be it for the political objective of seizing land in the Galilee (north) and the Negev (south), or the social aim of preventing metropolises from developing along the coast and in the center of the country. Either way, the policy makers of the day were averse to high rise construction, concentration, and crowding – the three principles upon which most of Niemeyer's plans for Israel were to be based.

Niemeyer and the Israeli Horizon

Niemeyer arrived in a country molded by the Sharon Plan. For someone who came from a nation practically the size of a continent, Niemeyer could not help but notice the dissonance between Israel's actual size and the low, abundant, and diffuse conception of space upon which it was premised. Before he even reached the Haifa port, the plain urban scene revealed beyond the harbor dismayed him: '… Mount Carmel is very exposed, without buildings, lacking the necessary urban centers. … When the ship approached the shore, I began to imagine the Carmel adorned with impressive urban structures. Not with low buildings that negate its horizons and cover it in a mess where one can no longer make out its vistas, its past, and its struggles, but with tall edifices and turrets whose height and magnificence will give it the mark of distinction. Such buildings will leave many open spaces'[6] (Figure 4.1).

Niemeyer's large and tall proposals aroused heated public debate in Israel (even though such discussion had been taking place since the late the 1950s) because they not only diverted from the local look they contradicted it altogether.

As a stranger, Niemeyer wasn't aware of the ideological, economic, social, and political underpinnings of the Sharon Plan. Like Niemeyer, Aryeh Sharon was aware that Israel is 'impoverished in space', as he wrote in the introduction to his plan, 'and can not afford to repeat the mistakes of larger, richer nations that can rectify them in due time'.But unlike Niemeyer, Sharon believed in spatial planning that spans 'from one horizon to the other'[7] and in the 'intensive and comprehensive development of the entire country that will penetrate every corner, by way of efficient, precise planning'.[8]

**Figure 4.1   View of Mount Carmel and Haifa Lower Town, 1964.
On hilltop: Dan Carmel Hotel. Photograph: Fritz
Cohen, courtesy Government Press Office, Jerusalem,
Photography Department.**

The longer Niemeyer stayed in Israel, the more he articulated his discomfort
with its horizontalness. In June 1964, when presenting his plans for the
Panorama Center in Haifa, he justified its 25 storeys with a general ideological
treatise and a particular vision for construction in Israel. The Corbusian duality
of progress and nature, spirituality and architectural materialism, characterized
almost all of Niemeyer's important Israeli works:

> I think that Israel should be built upwards, and its cities planned vertically
> – something that will be lauded in the future and will preserve the land for
> agricultural, industrial, and recreational purposes, bringing man closer to
> nature, something urban planners must engender. ... Israel is developing
> at such a rate that low-to-the-ground construction is unthinkable, for in no
> time a territory as small as this will be disproportionately covered with low
> buildings, denying it its natural beauty and vistas, and leaving it without open
> spaces essential to its development. Low-to-the-ground construction knows
> not what it begets.[9]

Niemeyer's Shadow

Whether a proclivity toward laziness, a desire to get a second opinion, or a real need for assistance – the pattern of inviting in expert opinion was not new to pre- and post-state Israeli architecture. Experts in development, agriculture, and infrastructure, as well as architects and urban planners, were frequently invited to advise Zionist settlement organizations in the planning of national projects before 1948.[10]

Niemeyer's presence in Israeli architecture preceded his arrival. Screening and shading stand at the center of Niemeyer's early work, and constituted his primary influence, and the influence of Brazilian architecture in general, upon Israeli architectural circles in the decade between 1954–64. Sun breakers appeared in Le Corbusier's research back in the 1920s, but their refinement on a large scale only happened with his work alongside the planners of the Ministry of Education and Health in Rio de Janeiro in the late 1930s.[11] Le Corbusier was surprised to discover the extent to which the Brazilian modernists had developed the sun breakers, industrializing them so that they could be pre-cast in concrete. Casting opened up numerous possibilities for their design: vertical, horizontal, mobile, adjustable, or fixed; sometimes in the form of simple weaves and sometimes as complex geometric ornaments.[12]

Niemeyer developed a mechanism for several advanced sun breakers back in the 1930s and 1940s, which were then copied and disseminated in Brazil and beyond, though it was the celebrated 'brise-soleil' that adorned the northern facade of the Ministry of Education and Health in Rio (Figure 4.2). The facade came to symbolize modern Brazil, and was displayed, among other places, on the cover of a special addition of the French journal *L'architecture d'aujourd'hui* dedicated to Brazilian architecture in September 1947. In 1958, the famous façade was emulated on a smaller scale on the southern wall of the administrative building on the Hebrew University campus in Givat Ram in Jerusalem. This homage to the original (and to the French journal that made it famous) was featured on the cover of the Israeli periodical *Engineering and Architecture* in April 1958.

The advanced means of shading developed by Niemeyer and his contemporaries in Brazil were very influential on Israeli architecture, especially after 1948. In 1953, Dov Carmi described the climatic relevance of Brazilian sun breakers to Israel: 'the sunray breakers – Brazilian shutters or 'Brizim' as construction workers call them – have recently appeared in Israel and have become increasingly popular, and, in my opinion, justifiably so, since with

Figure 4.2 **Detail of Brise-Soleil of the Administration Building, Hebrew University, Givat Ram Campus (D. Carmi, Z. Meltzer, R. Carmi, Jerusalem, Israel, 1954–56) on cover of *Handasah V'adrichalut* (IAEA Journal), April 1956**

their help we can adapt our buildings to our climate'.[13] Carmi contemplated the genealogy of the sun breakers and concluded that:

> … [I]n any event, the sun breakers originate in the Near East, and the proof can be found in the Old City of Jerusalem where the shutters and pergolas are built in a warp and weft of thin strips of wood. … In Brazil the breakers have been given both a utilitarian and a decorative purpose, and today they are made of different materials such as wood and pre-cast concrete, while in the future they will undoubtedly be made of plastics. … Now the use of sunray breakers has returned from Brazil to our country.[14]

In June 1955, the exhibit 'Brazilian Architecture' opened at the old Tel Aviv Museum on Rothschild Boulevard. At the opening, then Tel Aviv Mayor Haim Lebanon called upon Israeli architects to travel to Brazil 'to study the contemporary modes of construction there and dare to adapt new forms, instead of sticking to the routine'.

Aryeh Sharon, then chairman of the Israeli Association of Engineers and Architects (IAEA), reported in an official ceremony on his recent visit to Brazil, emphasizing that 'what ties Israel to Brazil architecturally is primarily the similar climate of the two countries, and the importance of climatic protection, a problem that must be resolved in architectural forms'.

Climate aside, the main attribute common to Israel and Brazil was the fact that both countries, each in its own way, served as laboratories for the enterprise of modern construction. Both busied themselves throughout the 1950s with building new cities from scratch, be it in the cerrado (savanna) or the desert, on a scale unprecedented in modern historey. In 1956, Israeli architect and architectural critic Aba Elhanani traveled to Brazil as a correspondent for the Hebrew daily *Ha'aretz*. Elhanani, a promoter of Israeli architecture in countless international periodicals, visited Brazil's major cities, including the massive construction site in Brasilia, where he interviewed Niemeyer. Following his trip, Elhanani contended in an article published in the Israeli periodical Engineering and Architecture, that Brazil, like Israel, is a laboratory for new architecture. Elhanani saw Brazilian influences on Israeli architecture, including those of Niemeyer, in a positive light.

Le Corbusier's status as the principle influence on urban architecture and planning in Israel both before and after 1948, is uncontested.[15] Brazil's influence on Israeli architecture, even if understood as an indirect Corbusian influence, like that of Japanese concrete architecture or English new brutalism, is greater than one might imagine. Brazilian modernism in general, and shading practices in particular, can be found in abundance in 1950s and 1960s Israeli architecture: concrete laces and asbestos lattices appear in public housing projects, apartment buildings, single family homes, banks, schools, universities, theatres, shopping centers, concert halls, government buildings, youth centers, public works projects, hotels, hostels, kibbutzim, and moshavim.

Brazilian sun breakers, in their pre-cast concrete, metal, plastic, mobile, and immobile incarnations were the architectural detail of the day. Oscar Niemeyer cast his shadow on Israel before he ever set foot there.

The Large Building

Toward the end of the 1950s, as mass immigration leveled off, the economy stabilized and even began to grow, and a prosperous class started to form partly thanks to German reparation monies, Israel decided to change its image from a 'developing' nation to a 'developed' nation. Israel's aspiration to 'fall in line' with cosmopolitan trends and fashionable appearances, found expression in the emergence of public debate over the option of building upwards and soliciting the advise of reputable international experts, including architects, in order to do so.

Falling in line with international goings-on, and blind faith in the daring genius of foreign architects such as Pei, Bakema, Candillis, Niemeyer and others, were understood as the precondition for local development. However, beyond the belief in the magic of foreigners, there was also a practical side to importing 'big' architects. For shrewd real estate entrepreneurs, it represented a good business opportunity. The latter took advantage of the hype surrounding the star-from-abroad, the expert-of-the-day, in order to bypass regulations, sidestep local politics, and disregard the public's fear of audacious structures.

For Israeli society, which was just beginning to manifest a capitalist culture of consumption beneath its officially socialist market mechanisms, Niemeyer was a very compelling character. His architectural language evoked the iconography of corporate entrepreneurship and consumerism, while his rhetoric was deeply rooted in the terms of reference of utopian social-realism. The budding Israeli engagement with architectural size and real estate ventures can be traced to the economic changes of the late 1950s. In Israel's economic history, 1954–65 is considered the growth years.[16] After which, 1965–67 are considered the recession years – a period that saw a veritable freeze in the construction industry and explains, at least partially, why Niemeyer's proposals were never implemented.

1954–65 were 'the good years' of centralized, controlled economics that saw a steady rise in the gross national product. From the mid-1950s onward, the official policy was to develop heavy industry, though the agriculture and construction industries continued to flourish as well, thanks to an increase in local demand as a result of natural growth and a rise in the standard of living. These were the years of almost total overlap in political and economic activity. In other words, the development of the Israeli marketplace was the exclusive domain of the government.

By the 1960s, this hold had begun to erode and a capitalist economy was emerging. The large building, a marker of this change, was also an assertion of new architectural and urban aspirations: it questioned conventions of parcelization, challenged the traditional conception of the Israeli street, and called attention to the fact that land in Israel is a limited resource.

Parallel to the government's construction of massive public housing projects on empty and relatively large tracts of land, privately owned commercial and residential high-rises began to spring up in a number of Israeli cities. It is no coincidence that most of these buildings were erected in Tel Aviv. Already a bustling metropolis, Tel Aviv was and is Israel's cultural and financial capital – a relatively progressive, fashionable city that sets architectural trends throughout the country. Unlike the public housing projects, these first tall buildings raised questions, albeit not on the scale that Niemeyer's proposals would, about the preservation of land and the possibility of speculating on its worth by doubling and tripling its size from the ground up.

The development of a new type of urban residential building also expressed speculators' desire to 'liberate' the land from garden-city style parcelization.

Unlike the uninterrupted European block, the garden city was based on separate structures 3–4 storeys high, with uniform lines and surrounding yards (a yard is a garden, otherwise it's repetitive). The new apartment buildings were planned on a number of plots that had been combined into a single lot, which was then re-parceled. This process, enabled planners to deviate from the accepted building rights and create wide expanses for use as gardens and parking lots. Many feared that this revolution would erode the traditionally anti-urban character of Tel Aviv, if not the entire country. In the years 1959–64, many such buildings erected in Tel Aviv included commercial spaces on the ground level, such as supermarkets and department stores.

Before one could adjust to the changing horizon, two concrete columns were imprinted on Tel Aviv's skyline – the circulation cores of the Shalom Meyer Tower[17] whose 35 storeys rising to a height of 140m came to dominate the cityscape for the coming three decades. The first real skyscraper in Israel is a quintessential (albeit extreme) example of this new building style (Figure 4.3).

The newly liberated parcels in the heart of the city aroused the interests of private entrepreneurs and government officials alike, and both acted to reinvent the rules for the sale, purchase, and appreciation of urban land. The construction of the Shalom Mayer Tower in 1960 laid the groundwork for the development of speculative urbanism in Israel.[18]

Figure 4.3 **Y. Perlstein, G. Ziv Arch., S. Ben-Avraham Eng.: Shalom Mayer Tower, Herzl Square, Tel Aviv, 1959–66. Developers: Wolfson-Clore-Mayer. Aerial photo-montage. From the promotional brochure 'The Shalom Mayer Tower', issued by the developers.**

Urban Holes

Among the dozen or so Israeli projects Niemeyer was involved in, two were for elaborate residential, retail, office, and hotel complexes in Tel Aviv – one in the Nordia neighborhood and the other in Kikar Hamedina ('The State's Square') – both centrally located sites significant to the city's development.[19]

The phenomenon of empty lots, or 'urban holes' is typical to the Israeli city and plays a central part in its quasi-improvised development. Both Nordia and Kikar Hamedina, relics of Tel Aviv's historic design as a garden city, came to function as the city's municipal, economic, and architectural drafting (and redrafting) boards throughout the 1950s and 1960s. Kikar Hamedina remains so today.

Niemeyer's point of departure in planning both sites was fantastic thanks to Pei's precedent setting proposals. Like Pei, Niemeyer didn't limit himself to a city's approved building schemes. Niemeyer – the man and the name – was invited to raise the building rights to provoke the public's imagination, to free urban holes from the bureaucratic chains of planning commissions, and to elude legal and environmental struggles altogether.

The Nordia neighborhood, established in 1919 as a Jewish settlement on the outskirts of Jaffa, was planned as a garden suburb for 'our brethren, children of the diaspora, sons of the inner city ... who grew up in the dark and narrow urban back streets ... who too shall enjoy the live scent of the earth'.[20] This initial anti-urban approach, typical of Zionist planning, was antithetical to the ultra urban proposals advanced for the neighborhood only a few decades later, beginning in the 1950s, and culminating with the construction of 'Dizengof Center' in the 1970s.

In the early 1960s, on the eve of Niemeyer's arrival in Israel, scores of shacks still dotted Nordia. Endless negotiations over the neighborhood's status and frequent changes in ownership precluded new construction on the site. In those years, Nordia, which came to be synonymous with municipal failure, made for a curious reality: meters from the smart boutiques of Dizengof street one found a shtetl-like shantytown once destined to become the garden city of the future.

In December 1962, hotel mogul Yekutiel Federmann obtained the deed to the land. Federmann sought to appreciate its value by raising the construction rights as much as possible. Inviting in a famous architect from outside was the first step to expediting a convoluted planning and approval procedure and suppressing the struggle of residents destined for eviction. Oscar Niemeyer was the chosen man for the job.

Niemeyer's name alone raised the value of the land ten-fold, and the building rights to unprecedented levels. The city of Tel Aviv had already decided to develop the Nordia neighborhood skyward in 1958, when the 'Nordia Residences' company first entered the picture. But Pei's precedent introduced the possibility of reaching colossal heights. An expert such as Niemeyer was not invited for his design talents alone. He was expected to help Federmann rhetorically as well. After studying the situation, Niemeyer was asked to formulate the manifesto that his host wanted articulated. In his words to the press about 'low-to-the-ground construction [that] knows not what it begets', he may have had Israel's best interests in mind, but also those of its most ambitious entrepreneurs. This statement laid the foundation for the burgeoning towers of Tel Aviv and Haifa, commissioned by Federmann as projects whose stature could only be outdone by acclaim.

In May 1964, Niemeyer met with members of the Israeli Architects Association in Tel Aviv. After reviewing his work in Brazil, and Brasilia most recently (1956–60), he unveiled his plan for Nordia's 33 dunams. The design included three cylindrical towers of 25–40 storeys (65–100m) for residential and mercantile use. All three skyscrapers were to rise above two concrete surfaces that would house more than 1,600 commercial spaces. The following day, Niemeyer's crystal vision for a new Tel Aviv was featured prominently on the front page of the Hebrew daily *Ha'aretz*, which reported that, 'no large scale project of Niemeyer's has been executed outside Brazil, making it possible that the Nordia plan will put Israel on the architectural world map'.[21]

Niemeyer and his plans for Israel provoked an interest that exceeded relevant professional circles. The disproportionate journalistic attention paid to an architectural matter attests to just how radical and foreign, given the local scale, the Nordia plan was compared to the Shalom Mayer Tower, which had began to rise above the historic districts of Tel Aviv. Niemeyer rejected the municipal building plan, which prohibited high-rise construction, and refused to plan buildings of eight, ten, or 12 storeys that would waste land, in his words: 'Tel Aviv's low horizon lacks landmark structures. ... Our plan leaves open spaces, breathing spaces for a suffocating city.'[22]

Also in Kikar Hamedina Niemeyer sought an urban arrangement alien to Tel Aviv's low, spacious habits. Both plans, premised on the values of crowding, density, and height, contained clear markers of the modernist, functionalist principles of zoning. Both plans reflected the mega-structural ambition of late modernism to create a multi-layered urban site, or, 'a city within a city'. Kikar Hamedina developed as part of the General Plan for East Tel Aviv of 1938, one of the first major amendments to the Geddes Plan of 1925. The square (in fact a circus) grew in importance as Tel Aviv gradually spread northeast in the 1940s. The circle was patched together from mostly privately owned plots[23] (Figures 4.4 and 4.5).

Like in Nordia, Niemeyer suggested building the complex to great heights, within the contours of the existing lot. In contrast with other proposals for

Figure 4.4   O. Niemeyer (with A. El-Hanani and I. Lotan), Kikar Hamedinah Project, Tel Aviv, 1964. Developer: Tel Aviv Municipality. Early version with three towers. Photograph: Y. Lior, courtesy Y. Lior and *Ha'aretz* daily photographs archive.

the site-drafted beginning in the early 1950s (and based on the local scale), Niemeyer's plan constituted a significant jump in the building rights.

Niemeyer's design for the circle, together with Israeli architects Aba Elhanani and Israel Lotan, was based on an open urban space of some 70 dunams. At the center of the circle, three 40-storey square towers were to accommodate hundreds of offices and hotel rooms. On the ground level, open public spaces, both cultivated and paved, were planned alongside extensive retail spaces. The residential section, a ring of 900 apartments in five elongated buildings of 10-storeys each in the shape of a crescent were meant to 'correspond naturally with the apartment buildings of the area'. Finally, some 6,000 parking places were planned on four subterranean levels. This underground parking lot was to have a dual function: in the spirit of Israeli insecurity, Niemeyer's Israeli partners made a point of emphasizing how the parking lot can also serve as a bomb shelter for some 100,000 residents.

Figure 4.5   O. Niemeyer, Nordia Plan, Tel Aviv, early version, 1964, model photographed against the background of Tel Aviv beach. Photographer unknown, from the promotional brochure 'NP' (Nordia Plan), issued by the developer Y. Federman.

Niemeyer's proposals for Tel Aviv's urban holes aroused heated public debate, perhaps one of the most bitter and prolonged struggles Israeli planning has known. Toward the end of the 1970s, Nordia became Dizengof Center – Israel's first shopping mall, and the cradle of post-modern Israeli consumerism. Niemeyer's towers were a new kind of architectural object, not just for Israel – but for Niemeyer himself, who continued to refine them throughout the 1960s. Once back in Brazil, his scheme for the Nordia Towers evolved into the plan for the National Hotel in Rio de Janeiro. In Israel, Niemeyer's super-modern vision only resurfaced in the late 1990s in the form of the Azrieli Towers, a new business complex on the banks of the Ayalon freeway.

Kikar Hamedina is still empty. Its many reincarnations on paper bespeak the limitations of Israeli plans which fail to look beyond the specifics of a given real estate venture, become dated before they are ever begun, and disregard the good of the city and its inhabitants. Niemeyer's design for the circle, legally approved in 1969, still haunts the site – whose fate continues to be publicly debated to this day. In hindsight, it seems that it was Niemeyer's height ambitions that ultimately delayed the plan indefinitely. Both its altitude and its density have made it a target for scathing public and professional criticism. Niemeyer's proposals for filling Tel Aviv's urban holes were straightforward, uncompromising architectural positions perceived by Israelis as disruptive and threatening.

Brasilia on the Carmel

The University of Haifa campus, and the silhouette of the Eshkol Tower rising 103 meters above Mount Carmel, is the Israeli architectural creation most closely associated with Oscar Niemeyer, even if ultimately planned by others. The concentrated conception that characterized most of his designs was also true of the campus – his only plan for a public structure in Israel. The 'campus in a single building' was understood as a new typological interpretation of the traditional campus, and a critique of specific Israeli universities. Niemeyer's objection to the classic campus and its modern version in the form of pavilions, went hand in hand with his overall critique of the garden city and semi-rural scattered construction in Israel.

Abba Houshi, then mayor of Haifa, wanted to include Niemeyer in his vision for a university on 900 dunams of natural growth forest atop Mount Carmel. The area, zoned as a national park, was designated a monument to the founder of Zionism, Theodore Herzel, by the Jewish National Fund in the 1930s. Herzel saw Haifa as a locus for utopian-Zionist activity, envisioning it in his book, *AltNeuland* (1902), as the city of the future. This was the ground upon which Niemeyer was asked to design the city's newest institution of higher learning.

At the heart of the plan stood Niemeyer's desire to condense all academic activity into a single structure. This 'compact' thinking expressed his displeasure with the sprawling model of the 'student city' which, in his opinion, was to blame for the isolation and alienation of campus life. The monolithic campus expressed Niemeyer's vision for pluralistic, interdisciplinary academics.

Niemeyer's plan consisted of four main components: a central horizontal building whose roof will constitute the campus' central surface (completed in 1971); a vertical administrative building some 20 storeys high overlooking the horizontal building (ultimately 31 storeys built in the 1970s); two midsize horizontal wings for future buildings (never constructed); and an auditorium in the shape of an upside-down pyramid (never constructed). Finally, Niemeyer's plan included a number of scattered structures, meant for dormitories, clubs, restaurants, athletics, and performances on a slope north of the site. A winding road leading from these structures back to the main building, created an Acropolis-like effect. The central building's platform and the complex of smaller structures erected upon it, seemed to want to suggest a Parthenon, but in fact evoked an ivory tower.

The Haifa campus dealt with another aspect of concentrated construction, this time environmental-ecological. 'Upon studying the possible solutions, … we tried to avoid scattered construction … wherein the significance of the work disappears into a series of small buildings for the most part with no connection between them. … The amazing beauty of the site necessitates a compact, simple, but monumental solution.'[24]

In the non-urban context of the Carmel forest Niemeyer proposed a plan most like that of the impressionist-monumental formalism of Brasilia: geometric objects set on a neutral surface. In his campus plan, Niemeyer's trademark schematic sculpted composition took extreme form due to the physical characteristics of the site. Niemeyer's buildings didn't seek to 'integrate' into their environment, or 'respond' to the topography. Rather, they created a new artificial topography. As was his habit, Niemeyer decontextualized the site at the outset of planning, distinguishing it from the surrounding environment by delineating a cubic platform. He created a new, clear working surface – a tabula rasa – atop the dominant field conditions of the Carmel.

The dense, concentrated conception of the campus and the irregular stature of its main building ignited one of the most heated debates on the Israeli built environment Israel has ever known. Even if Niemeyer's name helped to expedite the process of getting such a plan approved, a protracted battle against it nonetheless ensued and the plan took years to implement. Among the objections were the questionable need for a second university campus; the campus' disconnection from the urban center; the magnitude of the plan; and its implications for the view.

Despite these oppositions, Niemeyer's concrete surfaces began to penetrate the public's visual consciousness. The invitation to the cornerstone laying ceremony for the new campus was designed as a cardboard strip with a rough sketch of one of Niemeyer's early plans. At the event, held in October 1965 on the eve of Haifa's municipal elections, then prime minister Levi Eshkol asserted that 'population dispersal necessitates the dispersal of institutions of higher learning'. Thus, even Niemeyer's compact design, which like all his Israeli proposals criticized Israel's diffusion practices, was enlisted in the service of the Zionist ethos of population dispersal.

Implementing the grandiose plan contradicted the austerity programs of the day, in force during the great recession of 1966–67. Arguments against the plan's flashiness dominated public discussion. By the end of a long approval process, Niemeyer's original design had been severely transformed by the Israeli architects who carried out the final planning of the campus' structures. Just as he removed his name from all his other Israeli projects, Niemeyer disowned the Haifa University campus, which was built in his spirit, if not to his liking.

The initial buildings erected on campus contradicted Niemeyer's basic conception of a monolithic plan. The 'stairs building', conceived by architect Shlomo Gilad in the early 1970s, represents the most blatant architectural break with Niemeyer's design. Gilad's system of graduated buildings violated the architectural language and anti-contextual topographical logic of the campus: no longer a neutral artificial plane, but a building that seeks to 'correct' Niemeyer's 'mistaken' reading of the area by adding a graduated, disjoined mass concordant with the site's elevation marks. The stairs building – including its name – is symptomatic of the changing topographical conception of international architecture since the 1960s in general, and post-1967 Israeli architecture in particular. This new trend in the treatment of the contact between structure and ground, was a rejection of the neutral relationship to topography so prevalent in the 1950s and 1960s – including Niemeyer's explicitly modernist proposal for the campus.

The stairs building opened the Niemeyer plan to further interpretations and amendments. The university's current expansion plan bears no trace of Niemeyer's original conception of a campus in a single building. The current master plan (by Rechter Architects, 1993) represent the final rejection of the concentrated principles at the core of the campus' original design, determining its future development on antithetical conceptions of academic, functional, and programmatic dispersal. Niemeyer's clear separation between the natural and the artificial have been replaced by a strategy of blurred borders and contrived blending with the environment.

The towers on either end of the Carmel peak – Haifa University's Eshkol building and the Panorama Center – are the only constructed remains of Niemeyer's 'Israeli period', even if he did not build them himself. The vertical edifices he envisioned upon approaching the city from the sea remain solitary and lonely on Haifa's conflicted horizon (Figure 4.6).

**Figure 4.6   O. Niemeyer, H.G. Muller, S. Rawett, G. Dimanche, Haifa
University, 1964. Developer: Haifa municipality. Photo
montage of the plan model on the site of the Carmel
National Park.**

Vertical Kibbutz: The Negev City

The prominent architect's presence in Israel tempted even the chief ideologues
and strategists of Zionism to think in terms of size, at least momentarily. David
Ben-Gurion, for example, upon meeting Niemeyer in the Negev – the ultimate
site of national renewal – suddenly lost sight of his doctrines of dispersal and
homesteading and promised the architect, 'Here you will be able to build as
high as the sky'.

In the late 1950s and early 1960s, parallel to the construction of
experimental housing projects, the Ministry of Housing decided to establish a
number of 'advanced' new towns.[25] The success of these towns depended on
the development of the construction industry, the training of skilled builders,
the expansion of architectural knowledge, and social research. Israeli planning
authorities were aware of the harsh criticism that their early 'development
towns' had engendered, especially in the Negev, and starting in the late 1960s,
tried to apply new approaches to public housing projects.

Upon touring the Negev with his hosts, Niemeyer asked to see the
already established desert towns of Yeruham, Kiryat Gat, Eilat, and the latest

neighborhoods of Beer Sheva. As the designer of Brasilia in the 'tabula rasa' of Brazil's backcountry, he saw the Negev as one of Israel's greatest planning challenges, and was compelled by the notion of 'cities built in the heart of the desert'. Though he saw Israel as 'a country with an optimistic culture, unafraid of the progressive solutions that wealthier industrialized countries are afraid of', he was disappointed by the conservatism and short sightedness of Israeli planners. Therefore, he asked to qualify his proposal for a Negev city in advance as a utopian-conceptual non-site-specific plan. In his memoirs, Niemeyer relays that when asked to present his design for the city, representatives of the Ministry of Housing and Construction ordered him to plan structures of 4–8 storeys at the most, but he refused. 'The project we have planned', he wrote, 'will not be a variation on Israel's built environment, but a different solution altogether … fantastic, but nonetheless logical and reasonable'.[26]

Niemeyer's Negev city – a town of some 40 skyscrapers for tens of thousands of residents – embodies his critique of the planning conceptions that transplanted the garden city to the desert, a sparseness born of another cultural climate altogether. Yossef Almogi, then Minister of Housing and Development and the commissioner of the plan, told the weekly paper Davar that Niemeyer was hired to advise the Ministry and 'concentrate his energies on the unpopulated areas of the country … where we have yet to ruin the landscape. One of his assignments is to build a new city in the Negev. He himself asked to do so, after we showed him the desert'.

The ideal city he planned, based on 30–40 storey skyscrapers, was described by him in the Brazilian periodical *Modulo* as 'a new kind of metropolitan kibbutz that has grown, expanded, and modernized without losing any of its human qualities – enthusiasm, solidarity, and idealism'[27] (Figure 4.7).

The circular shape of Niemeyer's city resonated with local memory of the ideal collective agricultural settlement Nahalal, planned by architect Richard Kauffmann in the 1920s.

Niemeyer's romantic-futuristic dialectic can also be found in the text accompanying the Negev city plan. A city, in his words, that aspires to bring man back to the human scale of cities past, where simple pedestrian movement was possible. 'The Negev city grew from the nostalgia for the small traditional cities we were all charmed by, those cities destroyed by the big metropolitan centers.'[28]

At the core of his design for the Negev city stood the dominant principle of a 'walking-distance' city. Municipal/public (e.g., city hall, hospital), commercial (e.g., hotels, banks, stores), and entertainment/recreation (e.g.,

Figure 4.7 O. Niemeyer, H.G. Muller, S. Rawett, G. Dimanche, Negev Plan, a proposal for a new city in the Negev Desert (southern Israel), 1964. Developer: Ministry of Housing and Development. Photograph of a model buried in the sands of Tel Aviv sea shores. Photographer unknown.

theatres, cinemas, restaurants) facilities were all planned in the city center. At the center of this center, there was to be a shaded public park. The pedestrian paths of this center were also designed as shaded groves, offering protection from fierce desert winds.

Automotive traffic and foot traffic were clearly differentiated in the city's scheme. A ring-like boulevard that drew traffic outward, where bay-like parking lots awaited residents, demarcated the town's limits. Another parking lot for some 10,000 vehicles was planned near the city center, as was a subterranean garage and bus depot to allow for unhindered pedestrian movement on the walkways above. Additional commercial spaces were also planned underground. The roof of this sunken structure was to be a broad esplanade marking the entrance to the city and the point from which its foot and vehicular traffic commence.

According to Niemeyer, low construction in the form of multiple horizontal blocks, would hinder the separation of automobiles from pedestrians, creating countless insurmountable problems. A low building – and Niemeyer considered six, 10, and 15 storeys to be low – would have precluded the possibility of designing a city wherein the maximum distance between a residence and a business is no more than half a kilometer, where cars are unnecessary, and most of the grounds are designated for public use.

Building vertically was one of the significant improvements Niemeyer sought to make after designing Brasilia. Despite Brasilia's famous image: a dominant tower and two concave and convex half spheres laid in purist composition upon a wide horizontal plane – it was a relatively low city and the buildings in its residential district, or 'Superquadras', were planned as 10-storey structures at the most. Niemeyer had no problem finding convincing arguments for building vertically in the Negev. The difference between a 40-storey building and a 10-storey building could, according to him, be found in the solutions dictated by a climate of sandstorms. Other advantages touted by Niemeyer included the prevention of air pollution and breathtaking desert views.

Niemeyer rejected claims that horizontal construction was more economic in the face of vertical construction that would require the installation and maintenance of elevators, among other things. Niemeyer saw such arguments as petty, irrelevant pre-conceived notions. He suggested seeing elevators in a 30 or 50 storey building as an integral part of the infrastructure – a vertical mechanical road if you will.

The apartments Niemeyer designed for these buildings aimed to meet what he identified as the local desire for single-family homes. The two-level apartments included attributes of real houses, in his words. The apartments were to be accessed through an internal corridor open on either end to the view. Each apartment would be entered into through a private 'hanging' garden, facing all its windows and doors. There were to be no openings on the building's facades due to sandstorms. Niemeyer saw the Negev city as a desirable housing solution for the modern city based on functional separation and equilibrium between man and nature.

Though the closed structure of the city didn't enable expansion, Niemeyer contended that it could be duplicated into a series of vertical 'Negev cities' along a central route at 10km intervals, with open, green spaces between them. Thus, he posited it would be possible to promote advanced urban life in the less-developed hinterland. In response to the claim that his city was planned for an area where there is no shortage of land and horizontal construction is feasible, Niemeyer asserted that his plan is relevant for urban schemes across Israel.

Niemeyer was aware of the radical, controversial nature of his plan. In any number of occasions he expressed his awareness that his conception, unproven as of yet, has failed to override the preconceptions of Israeli urban planners. Already in the 1960s he foresaw the accelerated suburban sprawl that would plague Israel in the twenty-first century: 'I have always objected to architectural monotony, and especially tried to caution against the dangers of non-concentration, unplanned sprawl, and horizontal crowding that enable concrete and cement to encroach on human life.'[29]

Niemeyer found Israel's empty spaces a neutral, optimal work surface for perfecting his compact urban plans, new to him as well. He arrived in Israel four years after Brasilia – an artificial city erected in half a decade according to a master plan designed by Lucio Costa – was designated Brazil's new capitol. Niemeyer, the city's chief architect, did not participate in its inaugural festivities. All around the city planned a symbol of humanity and progress, there are shantytowns that sprung up to accommodate the tens of thousands of workers brought in to build it. Many of these people are still living in the same squalid conditions today.

Niemeyer's Negev city was a critical manifesto against the hegemonic urban planning of Israel. But it was also his chance to rethink the controversial planning of Brasilia – his failed life's work. The Negev city's compactness, its correspondence with the medieval city on the one hand and the kibbutz on the other, its vertical roads, and its hanging gardens, provided Niemeyer with an opportunity to atone for Brasilia's horizontal concrete planes, vast distances, and harsh functional separations that made for inhuman living conditions. If Brasilia was perceived as the city-of-tomorrow, Niemeyer wanted the Negev city to be the socialist city-of-tomorrow.

In retrospect, Niemeyer's criticism of Israeli urbanism didn't penetrate public and professional discourse. Garden apartments on the 40th floor, in other words transplanting the amenities of suburban living to metropolitan settings, preoccupy avant guarde circles in other crowded countries like the Netherlands, but not Israeli planners. The ongoing success of community settlements in Israel, characterized by suburban low-to-the-ground houses surrounded with yards, attests to the triumph of the garden city conception in Israel.

The Negev city sought to connect the traditional village-city with progressive, modernist principles. Even if unrealizable in its day for cultural, social, and economic reasons, and perhaps today as well, the issues it raised are cardinal to statewide planning in Israel.

Conclusions

Niemeyer's proposals were never implemented in Israel as planned, and in practice not one of his structures was erected as such. Israeli architects later built some, but with far reaching changes. The recession of 1965–67, and complicated licensing procedures, kept Niemeyer's extravagant ideas on paper. Nor did the economic boom following 1967 bring about the realization of his plans. The occupation of the West Bank and Gaza and the new atmosphere of a borderless space, rendered Niemeyer's proposals even less relevant in the eyes of policy makers. His vertical socialism contradicted the 'Greater Israel' push for horizontal expansion, and the swift departure from the ethos of a workers' society.

Niemeyer's ideas were perceived in Israel of the 1960s to be radical and fanciful and his plans not only triggered professional debate but a real fear of heights. They were foreign to the provincial, peripheral fabric of the Israeli city and threatened the official rhetoric of the welfare state and the petit-bourgeois lifestyle of its citizens. Modernist Israel retreated in the face of Niemeyer's monumental, iconographic, mannerist, flashy modernism, which ironically forestalled, rather than advanced, local discussion of crowding and construction.

Niemeyer's ideas, at one time dismissed, re-emerged in 1990s Israel due to accelerated population growth (following the immigration of some one million people from the former Soviet Union between 1989–2001), and increased preoccupation with Israel's capacity to bear its own weight in the future. Influenced by recent global discourse on architectural size, crowding, concentration, uncontrolled urbanization, and the imperative of conserving open spaces, Israelis have also began expressing concern about these issues in documents like the National Outline Plan and the Master Plan for Israel in the twenty-first century ('Israel 2020').

Niemeyer's apocalyptic description of the flat and scattered echoes throughout the plan drafted by a commission headed by architect Adam Mazor in 1997 ('Israel 2020'). This new Masterplan analyses the forecasted accelerated development and increased crowding, unbalanced usurpation of land and diminishing land reserves in the center of the country, degraded quality of urban life characteristic of the Israeli city, and irreversible damage caused to the unique scenery and environment of entire regions. The plan proposes, among other things, concentrated and crowded development, abstention from the establishment of new settlements, and further crowding the existing urban centers – three concerns raised by Niemeyer in his sojourn in Israel in the mid-1960s.

Since the early 1990s, horizontal is no longer a vision but a default option in Israel. Alongside the Zionist settlement tradition of spreading out and establishing new cities, suburbs, and collective communities, one sees a rise in pragmatic, un-ideological 'commercial vernacularism' that literally aims for new heights out of pure economic considerations. In its finer moments (for example the Azrieli Towers in Tel Aviv), it creates colossal silhouettes surprisingly reminiscent of the futuristic crystals Niemeyer tried to sell developing Israel in the mid-1960s.

Frenetic vertical construction throughout Israel at the close of the 20th century appears to be informed by three parallel processes: a reckoning with Israel's limited spatial reality; political agreements that heralded a return to the pre-1967 borders (with minor adjustments), and a sharp increase in the population. Today, Israel's borders are once again at a question, as is her social, cultural, and spatial nature. In the coming years, Israel will be forced to settle these existential questions, which in turn, will redefine her constructed character once again.

Notes

1   The ideas presented here have been excerpted from the thesis, 'Oscar Niemeyer and the Beginnings of Speculative Urbanism in Israel After 1960' submitted to the Department of Architecture and Urban Planning at The Technion (The Israeli Institute of Technology), Haifa, 2002.

2   Erich Cohen, 'The City in Zionist Ideology', in A. Shachar, D. Weintraub and E. Cohen (eds), *Towns of Israel: A Reader* (Jerusalem: Akademon Press, 1973).

3   Ebenezer Howard, *Garden Cities of Tomorrow* (London: Faber and Faber, 1960 [1898]).

4   Erika Spiegel, *New Towns in Israel* (Stuttgart-Bern: Karl Kraemer Publishers, 1966).

5   Adriana Kemp, 'Borders, Space and National Identity in Israel', *Theory and Criticism*, 16, Spring (Jerusalem: The Van Leer Institute, 2000).

6   Niemeyer, quoted in *Ha'aretz Daily*, 3 June 1964

7   ArieSharon, 'Physical Planning in Isreal' (Jerusalem: Government of Israel Publication, 1951).

8   Ibid.

9   See note 6, ibid.

10   Among the foreign experts called in were reknown modernist architect Erich Mendelsohn and Sir Patrick Geddes in the 1920s; in the 1940s expert of housing and climatic planning Architect Alexander Klein from Berlin and American moderist Louis Kahn who was invited on the eve of the establishment of the state of Isreal to propose solutions for cheap, temporary housing in preparation for the massive immigration. After 1948 British urban

planner Patrick Abercrombie was invited by Israeli government to advise the planning division on the design of new cities. Most activities of international architects did not bear architectural fruit, and the real impact they had on Israeli architecture seems to have begun once their proposals were mixed. Mendelsohn's early visits to Palestine brought about far-reaching changes in the architecture of the Jewish community and its vocabulary. In a sense, Mendelsohn's case was a precursor to Niemeyer's. Though Niemeyer arrived in Israel with an international reputation, his endeavors outside Brazil, including those in Israel, certainly furthered his experience and notoriety.

11   Michael Levin, 'White City: International Style in Israel' (Tel Aviv: Tel Aviv Museum of Art, 1984).

12   William Curtis Jr, *Le Corbusier: Ideas and Forms* (London: Phaidon, 1986), pp. 116–17.

13   Dov Carmi, 'The Brise-Sollei', *IAEA Journal*, Tel Aviv, January (1953), pp. 14–15.

14   Ibid.

15   The Corbusian brise-soleil was largely absent from the international-style architecture of 1930s and 1940s Palestine. Certainly when compared with the explicitly Corbusian influences of stilts, flat roofs, and narrow horizontal windows (Levin, 1984). International-style architectural shading practices were achieved primarily through the shrinking of openings, the diminution of shaded balconies from the built mass, and the construction of small awnings over openings to keep the sun from entering.

16   Nachum Gross, 'Israeli Economy', in Tzameret, Zvi and Yablonka, Hana (eds), *The First Decade* (Jerusalem: Yad Ben Zvi Publishers, 1998).

17   In order to build the first Israeli skyscraper that would also house the first Israeli department store, the Gymnasia Herzeliya, a prestigious high school built in 1909, had to be demolished as it occupied that largest of the lots combined to accommodate the massive structure. While the demolition of the Gymnasia did not provoke public outcry until years later, it nonetheless marks one of the first attempts to establish a new Israeli architectural identity at the cost of obliterating its past.

18   Ultimately planned by Israeli architects Yitzhak Perelstein and Gideon Ziv, the Shalom tower is nonetheless associated with the name of Chinese-American architect I.M. Pei. The young Pei arrived in Israel in 1960 upon the invitation of a group of entrepreneurs headed by the Meyer family while the tower was still on the drafting table. Pei was asked to plan a luxury apartment complex in Jerusalem, and several other commercial and residential projects in Tel Aviv. Like Niemeyer, not one of Pei's proposals was ever implemented as such.

19   Planning the Nordia site for Yekutiel Federmann's company 'Nordia Residences' was the cause for inviting Niemeyer to Israel in the first place. The Tel Aviv municipality commissioned the Kikar Hamedina plan once Niemeyer was already in the country. The city was hoping to take advantage of the big architect's presence to salvage the last empty square in Tel Aviv from the architectural dead-end it had hit in the early 1950s.

20   Quoted in Gideon Biger (1987), 'Nordia – A Vanishing Neighborhood', *Ariel Journal*, 48–49, Jerusalem.

21   *Ha'aretz*, 11 May 1964

22   Oscar Niemeyer, 'Nordia Development', *MODULO Journal*, 39, March–April (1965), pp. 16–26.

23   The large number of lots, 91 altogether, are held by hundreds of heirs – one of the biggest obstacles to implementing a building plan at the site to this day.

24   See note 22, ibid., pp. 27–35.

25 The development towns were established gradually over the 1950s, and the authorities that planned them drew conclusions as they went. The first wave of towns was inspired by the European garden city. In the second wave, the open spaces between structures were reduced (as these became neglected and barren), and crowding was increased. In the third wave, the towns were designed on an urban scale and included social, climatic, and ecological considerations as well.
26 See note 22, ibid., pp. 1–12.
27 Ibid.
28 Ibid.
29 Ibid.

PART II
FRONTIERS

Chapter 5

Trapped Sense of Peripheral Place in Frontier Space

Erez Tzfadia[1]

Introduction

The issues surrounding the study of 'place', 'space' and 'sense of place' have in the last few years become principal items on geographical and architectural agendas. The questions of how 'space' and 'place' differ from each other, how these concepts relate to each other and how the interaction between them produces a sense of place, have contributed to the construction of intellectual frameworks within the geography in relation to the study of localities. Furthermore, concern about how the built environment, the architecture of localities, and changes in the physical setting of localities shape a sense of place has become a major focus of architecture. These debates have been conceptualized within a growing body of theoretical literature. Drawing upon this literature, it seems that the construction of 'place' is an endless process, which is subject to power relations within the 'place' and between the 'place' and the 'space' (where 'place' is a small portion of the 'space'). In-place and place-space relations construct and shape the particular sense of place of a community with respect to a specific locality, where the community resides. Since the construction of place and of a sense of place is an endless process, different aspects of the sense of place of a particular group of people to a particular locality are readjusted and reshaped constantly according to the balance of power in that locality.

Yet, efforts to subject the question of place and sense of place to sustained empirical analysis have been rarely made in peripheral localities or among peripheral groups. Moreover, while the theoretical literature assumes that a sense of place is constructed through interaction rather than through closure, almost no empirical attempts have been made to study how national projects of space production shape sense of place. Therefore, the main goal of this chapter is to follow the construction of sense of peripheral place and to look at how this is formed and readjusted in response to the production of (national) space.

I argue that in settler societies the process of producing a national space via territorial expansion and ethnicization processes is a major force that shapes peripheral places that were established during the production of the national space. Hence, sense of place in frontier localities is constantly readjusted, primarily in response to policies and practices, which are dictated by the space. Empirically, the chapter focuses on the construction and the readjustment of sense of place among Mizrahi[2] immigrants who were settled in development towns on the Israeli frontiers/peripheries in the years 1952–2000.[3] First, several notes on social construction of sense of place will be presented, mainly in relation to nation-building projects.

Constructing a Sense of Place – Theoretical Perspectives

Place and space are treated as essential concepts for addressing the question of creating a sense of place. These two concepts can hardly be explained separately. Two schools of thoughts have tried to stress the linkages between space and place, by asking how place is distinguished from space, how the two concepts stand in relation to each other and what the elements are that make places different from one another. The first school, presented by Albert Einstein (1970), aims to distinguish place from space and regards both space and place as objective physical concepts. Accordingly, place is a small portion of space. In other words, space is a container of places, which are all material objects identified by names. Thus, the idea of space-as-container gives rise to the notion of space as a system of places that can be transformed simply into mere locations that relate to each other, while each space is independent from other spaces.

The second school of thought regards place as a subjective feature of our experience, as a meaningful space, as a human, and therefore subjective, constructed. Space is more abstract than place and hence space tends towards objectivism. In other words, place is space with human attitude, and when space is thoroughly familiar to us it has become place (Tuan, 1977). To advance this idea, Massey argued that place is socially constructed and should be understood as a unique point of connection in a wider series of streams within the space. She interpreted places as 'bounded, enclosed spaces defined through counter position against the other who is outside' (Massey, 1994, p. 168). That is formulating concepts of space and place in terms of persistent development and changes in social relations, which are reflected in the place (Massey, 1994).

When place has meaning it triggers a 'sense of place'. This notion refers to a subjective element – human attachment to a specific place, feelings that people carry around about a place. Sense of place emerges from an interaction of social, cultural and natural settings that groups of people experience in a particular place. When different people or groups of people have feelings for the same place there is more than one 'sense' to any particular place. This represents different understandings of what the place is who dominates and represents it and who has the power to define its past and future (Massey, 1993; 1995). Hence, the multiplicity of senses of place is a source of competition between groups that inhabit and have feelings towards the place. Sense of place is regarded as a reflection of the way social relations are shaped in a particular place and these social relations always contain a dimension of power and competition within the place (Hansen, 1999). The power and competition that relate to sense of place highlight the psychological importance of attachment to place as a resource for organizing the community, mainly when members of a local group share a common ethnic and class affiliation.

Given that the sense of place is regarded as a reflection of the way social relations are shaped in a particular place, the sense of place of a particular group can neither remain fixed nor can it end. Just as place construction is a process that involves continuous adjustment to changing social relations and transformations in its built environment, sense of place is a process that symbolizes transformation in human attachment to place. In a world of globalization, which represents openness of places to common interactions, places rarely remain static. Places are transformed by physical development, by the actions of its residents, by the (global and national) market forces and by the arrival of new residents. However, place does not disappear as a value, so much as change (Wasserman et al., 1998). For its veteran residents attachment to place might become deep-rooted in a well-known place, symbolizing certainty within the uncertain space.

The idea that space and place are socially constructed, together with the idea that space is an abstract place attracted Taylor's attention in relation to states and nations. Taylor argued that if space is more abstract than place, spaces should be related to rationality and state bureaucracy. Hence, states are producers of space when their boundaries are drawn up and recognized. Places, by contrast, are locations created and divided by state elites in order to reproduce material life. For this particular reason places might also become sites for grass roots oppositions, built upon place-space tension. But this is not a simple process: when states' boundaries become familiar and embedded in society they have their own effect on the production of material life, hence the state becomes a place by itself (Taylor, 1999). In other words, state and national territories have become meaningful spaces in the modern period for the members of the nation-state, and can compete for human feelings with other places within the bounded territories. Following Murphy's (2001) definition of the attachment to state territory, I will adopt the notion 'sense of territory' to describe this attachment.

Yet, any geographical enquiry into place, sense of place and transformations in these two notions requires sustained empirical analysis. Several analyses have been conducted, but have rarely been made in peripheral localities or among peripheral groups, especially in regard to space-place tension. Since the chapter aims to focus on peripheral towns in settler society, several different facets of this issue will be presented now.

The Production of Space and Places in Settler Societies

The term 'settler societies' refers to societies that have generally been established by Europeans ('founders') who settled other continents and dominated indigenous peoples by seizing and ethnicizing their space, economy and politics (Stasilius and Yuval-Davis, 1995). The ethnicizing process begins when the European founders forcibly expel the indigenous peoples from their localities and replace them with members of the dominant group (McGarry, 1998). This is what happened in Israel in 1948, when large numbers of Palestinians were driven out of their towns and villages – out of their places. Taylor (1999) relates this process to the issue of space and place:

> States have been more successful in eliminating places in their imperial and frontier expansions. Worlds of non-European places were converted into new worlds of European spaces. As well as being 'peoples without history', the occupants of these areas into which Europeans moved were also converted into peoples without geography. They and their places were invisible to the state surveyors and cartographers carving up the 'empty space'. (Taylor, 1999, p. 15)

With 'empty space' there is no choice but to produce new place(s). The production of new space(s), which is the transformation of space into place(s), joins the objectives of settler societies to advance the project of nation- and state-building. Generally, to achieve these goals, settlement projects in frontier regions are promoted, and are known as territorial ethnicization. Such processes appropriate Agnew's (1987) argument that there can be modern places created after the destruction of traditional (indigenous) places.[4]

Among the settlers, immigrants who share common racial, ethnic, or cultural traits with the founders, are incorporated in the ethnicizing project, by settling or establishing new settlements. Sometimes the new settlements are established on the frontiers on the ruins of the indigenous people's places. This means that the immigrants are distanced from the centers of capital and political power, but at the same time they are contributing to the national project of settlement, which gives them a sense of belonging as well as certain material benefits (Tzfadia, 2000). The ambiguity of immigrant marginalization-inclusion transforms them into an ethno-class (or a cluster of ethno-classes) situated between the 'founders' and 'indigenous people'. In due course other immigrant groups join the project and create new axes of ethno-class tensions and struggles. Through this spatial-economic process the immigrants become 'trapped', as it were, between the founding group and the excluded 'natives'. Their identity thus develops in several parallel 'layers' – a quest for full integration with the 'founders' in the national arena, alongside an emphasis on 'difference' at a local level (Tzfadia and Yiftachel, 2001). The 'difference' is expressed via developing a sense of place to the new settled site, which organizes the immigrant group to oppose their marginality in the national space.

However, these contradictory tendencies highlight the importance of place and the immigrant group's need to intensify its attachment to the place it was settled in, to the point of claiming exclusive control over the place. But sense of place among a trapped group cannot be an exclusive attachment, because exclusiveness contradicts the trapped position. The entrapment refers to the ambiguity of the sense of place, as the immigrants share their feeling about the settled place and for the state territory. In other words, the immigrants possess both a sense of place and a sense of territory. This entrapment, I will claim, serves state elites in the reproduction of material life that preserves the immigrants' low status by employing practices that maintain the peripheral position of the settled place. In the empirical section I will attempt to explain how entrapment affects the sense of place among the immigrants.

Transforming Places into New Places in Israel/Palestine

The Israeli/Palestinian case served McGarry's research on 'demographic engineering' claiming that the expulsion of Palestinians during the 1948 war characterizes the formation of many new states. In this war and after it, the Palestinian places were demolished (Benvenisti, 2000; Morris, 1987). The human landscape that had existed was scattered to the winds, the life, history and places that had been built in the space were terminated, and the natural bond between the individual and his land was severed. The Palestinian landscape was appropriated by a different human landscape, partly foreign to the space, as settlement and development expanded and created a new, modern place(s).

By 1956 the Jewish community in Israel had tripled compared to 1948 because of the wave of immigration, only a minority of which came from Europe. Most of the immigrants came from Arab countries in Asia and North Africa. The imbalance in the spatial dispersal of the Jews,[5] along with the almost unlimited control of the European (Jewish) founders over the immigrants, set the stage for a population-dispersal policy-referred to as a policy for a balanced distribution throughout the country of population, economic activity, services and welfare (Efrat, 1998). This policy was implemented for the most part by means of the 'Sharon plan', named after the founder and head of the planning department of the Prime Minister's Office at that time (Sharon, 1951). The plan depicted the deployment of settlements as a pyramid, in which five main kinds of settlements would emerge on a hierarchical basis. Major forms of settlement that were not part of the settlement network before statehood were urban communities and medium-size communities of 6,000–60,000 people (Troen, 1994). The latter would come to be called 'development towns'. Along with the plan's recommendations, which were implemented almost completely, settlement on the Israeli frontier, mainly the Galilee and the Negev, was reinforced with hundreds of small, remote, agricultural Moshavim (cooperative and semi-cooperative rural colonies). These, like some of the development towns as well, were built on the ruins of 350 abandoned Palestinian villages and towns, and were mainly, though not only, populated by Jewish immigrants and refugees from Arab countries, who came to be known as 'Mizrahim'.[6]

The establishment of new towns and Moshavim in the 1950s and 1960s fostered social and geographic separation and isolation in Israeli society. Most of these communities suffered from economic backwardness and deprivation. Their remoteness and the training of their residents for blue-collar professions or agriculture led to discrimination against these people in the labor market (Swirski, 1989). Thus, by means of relegating the Mizrahim to settlements in the periphery, where opportunities are limited, and exclusion from the mythical ethos of settlement, 'population dispersal' – the Zionist term for the territorial ethnicizing process, became a key mechanism for maintaining ethno-class stratification in the Israeli-Jewish settler society. In the long term, this ethnic-class stratification became one of the main criteria for the distribution of wealth in the space and the society (Yiftachel, 2000). Uneven distribution of wealth, argues Massey (1994), is one key factor in producing place and differences between places, mainly when the uneven distribution parallels differences in ethnic and class terms.

Yet, among all the spatial Judaization projects undertaken during more than a century of Zionism, no project has managed to transfer so many Jews to regions where Palestinians compose the majority (i.e. Galilee and Negev). By the mid-1960s, soon after the establishment of all the development towns, 190,000 Jews lived in them. Of these, 83 per cent were of Mizrahi origin. Four decades later, in the mid-1990s, the number of Jews living in the towns had risen to 460,000 (Central Bureau of Statistic, selected years). Predictably, this project, which was 'a project of elimination and building that was more audacious than all literary precedents' (Efrat, 2000, p. 205), is regarded as the flagship project in the production of Israeli-Jewish space.

Israel's policy since the 1950s, i.e. both the creation of isolated towns and the directing of their Mizrahi population and development, has had a significant influence on the evolution of a sense of place in the towns. The sense of place, combined with cultural preferences, religious customs, ethnic affiliation, socioeconomic position, social contacts and political leanings led to the emergence of a Mizrahi ethno-class in the towns, 'trapped' on the margins of Israeli-Jewish society. The entrapment can be understood in the light of two processes that the development towns underwent. First, the exclusion of the Mizrahi Jews from the centers of economy and wealth, of political power, media and culture, and the perpetuation of their low status within the Jewish society. Second, the inclusion of the Mizrahim into the Israeli-Jewish national collective, mainly by transforming the immigrants into settlers in frontier regions. This transformation of immigrants into settlers should be considered as the highest level of inclusion in a society that produces a 'new space' by settlement projects. Yet, the ambivalent attitude toward the development towns and its Mizrahi population is the main reason behind the entrapment. What is the meaning of the entrapment, and how was the entrapment engraved in the Mizrahims' sense of place? The next section is devoted to answering these questions.

Possessing the Towns – Strong Feelings for Place as a Response to Exclusion

Recent research on the emerging sense of place among the Mizrahim in the development towns is based on a face-to-face survey that was conducted in six development towns on a representative sample of 264 adult Mizrahim. Of these 264 interviewees, half were first generation immigrants, and half were second generation, in other words, born in Israel. The opinion survey traced the influence of place, economic development, culture, social network, political orientations and inter-generational dynamics on the evolution of a collective sense of place in the towns. The six sampled towns are: Ofaqim, Dimona and Kiryat Gat in the Negev, Beit She'an, Shlomi and Ma'alot-Tarshikha in the Galilee.[7]

Several factors were found to be relevant to the issue of constructing a sense of place among the Mizrahim in the towns. The first concerns the reason for their arrival in the town in the post-1948 period. At that time the Judaization process (the production of space) was central to the nascent state's policies and this, together with the xenophobic attitude among European Jews to Arab culture, was the main reason behind the forced dispersal of the Mizrahim to the development towns. Several other steps were taken by the Israeli government to ensure the Mizrahim remained in the development towns. This included granting public housing to the Mizrahi immigrants in the development towns, without the possibility of choosing another location for dwelling (Lewin-Epstein et al., 1997); creating an education system aimed at training students for blue-collar jobs that became commonplace in the towns (Swirski, 1989); maintaining a low-class labor market, based mostly on manual jobs, a low salary, poor chances of promotion and a high level of unemployment.

In the face-to-face survey interviewees were asked: what was the main reason behind your arrival in the town? After eliminating those who answered: 'I was born here', 52 per cent of the interviewees claimed that they were taken to the town from the airport or the harbor, without been consulted. Less than 3 per cent claimed that they wanted to play a part in the settlement project. It would be reasonable to assume that compelling the Mizrahi immigrants to live in development towns would engender a sense of anger and antagonism toward the place, and more specifically to the Judaization project that was the main reason behind their settlement. However, most of the interviewees (57 per cent) claimed that the establishment of the development towns in the 1950s, which could be considered a major step in the production of Israeli-Jewish space, was unavoidable. Let's take this attitude as a first sign of the entrapped position of the Mizrahim. The inclusion of the Mizrahim represented the extension of Jewish-Israeli domination over the territory. Yet, they were dispersed to peripheral development towns, which perpetuated their low status within the Israeli-Jewish nation.

Furthermore, the Mizrahim in the development towns are aware of the discrimination against their towns: 72 per cent of the interviewees claimed that the development towns are deprived, especially in comparison with other type of settlements in the frontier, such as settlements identified with European Jews. Yet the discrimination is not a subjective feeling of the Mizrahim in the towns. In his 50th annual report in 2000, the State Comptroller found that there is constant deprivation of the Negev's development towns in financial terms, particularly in comparison with the settlements in the West Bank (State Comptroller, 2000). Ironically, this knowledge that the towns are discriminated against reinforces the Mizrahim's attachment. The sense of place among the Mizrahim is derived from a 'common destiny' of people who belong to the same ethnic group, are located in the same position in the social structure and were forced to live in peripheral towns. In addition, the sense of place is derived from the social interaction within the place.

The varies dimensions of the Mizrahim's sense of place were measured by adopting Osgood's (1967) semantic differential measurement.[8] Twelve pairs of contrasting adjectives that describe different possible characteristics of the town were presented to each interviewee. The interviewees were then asked to mark each pair with a score ranging from 1 to 7. The contrasting adjectives and the mean scores for each pair of adjectives are presented in Table 5.1.

The results of the semantic differential measurement indicate the Mizrahim's strong sense of affinity to the development towns, by claiming that the towns are friendly places, safe, developed and have a decent population. But the Mizrahim are also aware of the problems of the towns that emerge in the interaction between the place and the national space: they are aware that the towns are not favored in Israel, although they are bonded to the Israeli-Jewish space. And, the policy of depriving the towns of capital and resources is reflected in the low score given to the adjectives 'rich place – poor place'. The strong attachment of the Mizrahim to the towns complements another set of activities, which has a bearing on the sense of place among the Mizrahim. These activities include constructing local cultural symbols, establishing local newspapers and magazines, and cultivating a local leadership that has managed to govern the town without external intervention, as used to be the case until the 1970s (Ben-Zadok, 1993; Ben-Ari and Bilu, 1987). It should be emphasized that 60 per cent of the interviewees believe that the development towns are excellent places to live in.

**Table 5.1   'Describe your sense of the town in which you reside':
mean scores for the semantic differential measurement**

'Your town is ...'

Contrasting adjectives	Mean (Range 1–7)
Friendly/hostile	5.84
Safe/insecure	5.38
Attractive/ugly	4.94
Decent population/dreadful population	4.88
Get better/getting worse	4.86
Developed/undeveloped	4.81
Bonded to the state/severed from the state	4.62
Honorable/disgraceful	4.57
Preferred/neglected	4.01
Favored in Israel/disfavored	3.86
High quality of life/low quality of life	3.70
Rich place/poor place	3.21

N = 264.

The strong attachment of the Mizrahim to the development towns is combined with an awareness that the towns have poor facilities for mobilizing their residents, that the image of the towns is poor in the Israeli space and that the development towns are at the margin of the Judaization project. This awareness is supported by several statistical indications such as: the average income per household in the towns is 25 per cent lower than the average income per household in Israel as a whole; unemployment rates are generally higher in the development towns; and educational achievements are normally low (Swirski and Konor-Attias, 2001). As a result, a dynamic of leaving the towns has emerged among the Mizrahim, mainly among the young generation born in the towns: 67 per cent of all the interviewees wish to leave the towns, mainly to the central district of Israel. Among the second generation the rate is 72 per cent.

Returning to the principal contention of this chapter, it is clear that the sense of place among the Mizrahim in the development towns has been constructed in response to the state's settlement policy. The sense of place also reflects the entrapped position of the Mizrahim. Yet, in spite of the deprivation and the marginality of the towns, they provide a friendly and secure atmosphere for their Mizrahi residents, who have managed to develop a sense of community there, and emphasize the connection of the Mizrahim and the towns to the project of producing Israeli-Jewish space. The marginality of the Mizrahim and the towns in the Israeli-Jewish space leaves no room but to anchor Mizrahi social life in the towns. Yet, at the same time, this marginality prompts the Mizrahim to leave and seek other places.

In the following section I would like to follow the transformation of the sense of place among the Mizrahim in response to the policy of dispersing and settling Russian immigrants in the towns during the 1990s. This policy, implemented by the government of Israel, transformed the towns from single ethnic places to bi-ethnic places.

Adjusting the Sense of Place – Russians in the Towns

During the 1990s Israel was given an opportunity to stimulate the production of Israeli-Jewish space when almost a million immigrants from the Former Soviet Union (FSU) were encouraged to immigrate to the country. Soon after the beginning of the wave of immigration the FSU immigrants became 'fresh fuel' for the Judaization project. Nearly 25 per cent of the immigrants were settled in the development towns because the government decided to subsidize housing for immigrants who chose to live there, and to grant housing in the towns to needy immigrants (Tzfadia, 2000).[9] As a result, several development towns doubled their population within three years and others recorded significant growth. Most of the educated and skilled immigrants settled in the central district (Hasson, 1992) while many unskilled workers and senior citizens found their way to the development towns. This process preserved the old spatial organization, in which power, wealth and knowledge were concentrated in the central district with the development towns remaining dependent upon it (Lipshitz, 1998). The fact that each development town absorbed thousands of immigrants caused radical changes in their ethnic composition. The Mizrahim living in the development towns were suddenly exposed to a different ethnic group, at the onset of the absorption process (Tzfadia, 2000).

The exposure to a different ethnic group in a familiar place was not the only event that influenced the change in sense of place among the Mizrahim. The construction of thousands of apartments in the towns, mostly in new neighborhoods, and the settlement of immigrants in the new neighborhoods created intra-town ethnic segregation, where the immigrants mostly inhabited the new neighborhoods and the Mizrahim remained at the old ones – mostly constructed in the 1950s and 1960s. Yet, this intra-town segregation did not make an impression on the public spheres of the towns because in a small town there is no room to segregate the daily outdoor activities of the immigrants. Instead, the immigrants became visible in the towns, having ('Russian') shops, schools and educational activities, a cultural life and politics that threatened the domination of the Mizrahim in the town. In other words, in the development towns 'place' became a resource to compete for (Tzfadia and Yiftachel, 2001). The question that arises from this description is: how did the policy of dispersing and settling the FSU immigrants in the development towns, as part of the production of space, influence the sense of place among the Mizrahim in the towns?

A second face-to-face survey investigated the influence of the absorption and settlement policy of the immigrants during the 1990s on patterns of ethnic relations between Mizrahim and Russians in development towns. For the survey 610 adult Mizrahim who live in these six towns were interviewed. One of the main sections in the questionnaire refers to sense of place. Once again Osgood's (1967) Semantic Differential measurement was adopted, but this time the interviewees were asked to score each pair of contrasting adjectives twice: one score relating to the period *prior* to the arrival of the FSU immigrants, and the second score to the period *after* the arrival of the immigrants. A paired-samples T test was used to compare the mean scores (before and after). The mean scores and the differences are presented in Table 5.2.

The arrival of the FSU immigrants eroded the attachment of the Mizrahim to the development towns, mainly in relation to social/human perspectives: the towns now were less united, less friendly, and became less familiar places. Yet, the scores do not suggest a despairing reaction to the radical changes in human and ethnic composition of the towns. Rather, the minor decrease in scores implies that the strong attachment of the Mizrahim to towns has been maintained, with a little less attachment to the human aspects of the towns. The pairs of contrasting adjectives associated with physical aspects (developed/ undeveloped; pretty/ugly) were awarded higher scores after the arrival of the FSU immigrants than before their arrival. This tendency could be explained in terms of the construction of new neighborhoods in the development towns in order to absorb the immigrants, which affected the overall appearance of the towns. The construction of these new neighborhoods and the addition of residents obligated the planners to replace most of the old infrastructures in towns, to renew the roads, sidewalks and gardens, to rebuilt schools and kindergartens – even in the old neighborhoods.

The absence of either a despair reaction or an extreme decline in the sense of place among Mizrahim in the development towns contradicts the literature

Table 5.2 **'Describe your sense of the town that you reside in, before and after the arrival of the "Russian" (FSU) immigrants':
mean scores for the semantic differential measurement**

'Your town is ...'

Contrasting adjectives	Pre FSU immigration	Post FSU immigration	Difference	Sig. (2-tailed)*
United/divided	5.28	3.92	−1.4	5.95E–22
Friendly/hostile	6.41	5.96	−0.5	1.29E–15
Decent population/ dreadful population	5.87	5.15	−0.7	5.92E–22
Warmhearted place/ cold–hearted place	6.33	5.89	−0.4	1.17E–12
Familiar place/ unfamiliar place	5.67	5.32	−0.3	1.94E–05
A good place to live at/ a bad place to live at	5.75	5.33	−0.4	5.81E–12
Rich place/poor place	3.42	3.21	−0.2	0.005
Developed/undeveloped	3.80	4.21	0.4	8.3E–06
Pretty/ugly	4.50	5.08	0.6	2.3E–13

N = 610

* Resulted from paired-samples T tests.

on spatial relations between veterans and immigrants who settle in the same locality. Usually, an influx of immigrants results in the departure of the veteran residents (Sarre, Phillips and Skelington, 1989) or in a loss of hope that the locality will regain its former glory (Marris, 1986). The prospect of leaving the towns occupied a section in the questionnaire whose results showed that only 7.4 per cent of the interviewees have considered leaving the towns since the FSU immigrants arrived. Another 16 per cent of the interviewees had heard of Mizrahim who had left the towns since the FSU immigrants arrived.

However, these findings show that the affinity of the Mizrahim for the development towns is significantly high. Indeed, in the second half of the 1990s Mizrahim adopted several methods to protect the towns from spatial 'Russification'. They undeniably managed to prevent a 'Russification' of the local authorities during the 1998 municipal elections. In several development towns the deluge of old, chronically ill Russian immigrants was slowed down as part of a Mizrahi protection of 'place' (Tzfadia and Yiftachel, 2001).

However, the Mizrahim had no success in restoring the towns to the places they were before the arrival of the Russian. In other words, the remarkable changes that took place in towns during the 1990s, and the trapped position of the Mizrahim within Israeli settler society obligated the Mizrahim to readjust their sense of place. This happened according to the changes that took place in towns as a result of practices and policies aimed at Judaizing the territory of the state. The entrapment does not enable the Mizrahim to leave the towns because they lack the required resources to finance the move. There was also no point despairing at the changes in the towns because they could not move to any other place. Moreover, they could not resist the settlement of the FSU immigrants because settlement is a major practice in the production of Israeli-Jewish space, in which the Mizrahim played an important role.

Conclusions

With regard to changing places and spaces in settler society the theoretical literature identified two major transformations: non-European places were converted into new worlds of European spaces, and traditional places were replaced by modern places. In this chapter I argued that the process of constructing place, and hence sense of place, is a continuous process. I also claimed that in a settler society, in relation to peripheral places, the construction of places and senses of places generally results in practices of (national) space production. The case of the development towns in Israel supported these claims and emphasizes that the entrapment of the settlers in the frontier, entrapment that has its origin in the inclusion-exclusion practices in settler societies, compels the immigrants-settlers to readjust their sense of place according to practices of producing a space.

In relation to the Mizrahim – several practices were found to be relevant to the construction of a sense of place.

i)   An ongoing policy of discrimination against the development towns, but in conjunction with gradual inclusion into the nascent Zionist nation.
ii) The continuation of the Jewish-Arab conflict and the Judaization of Israeli/Palestinian space.
iii) The deepening socioeconomic stratification. These practices created an attachment and a loyalty among the Mizrahim to the development towns. Thus, the sense of place of the Mizrahim in the development towns crystallized in the 'twilight area' between inside and outside, between

included and excluded. The result is a strong attachment to the place.

The continued production of Israeli-Jewish space in the 1990s, i.e. the absorption and settlement FSU immigrants, affected the Mizrahim's attitude to their towns. While the towns were created as peripheral and impoverished places in the attempt to Judaize the land they have now become a significant – and threatened – ethnic and political resource. In other words, place emerges as a major source of communal identity and political power and is constantly being reshaped by social processes. At the same time, sense of place is constantly being reshaped according to external and internal processes that persistently aim to advance the production of space and place.

Notes

1   The author would like to thanks the Israeli Ministry of Science, who support this research, and to thanks Oren Yiftachel, who conferred a permission to use data from a joint research project.
2   *Mizrahi* Jews – *Mizrahim* in plural ('Easterners', also known as Oriental, Sephardi, or Arab Jews) – are Jews with origins in the Arab countries of Africa and Asia. Most Mizrahim immigrated to Israel in the early 1950s and were characterized by low socioeconomic status, and as having joined an already-established state and culture.
3   The terms 'frontier' and 'periphery' refer to borderlands. The 'frontier' is conceived as a zone formed through settlement expansion as part of nation- and state-building projects. The 'periphery' is viewed as a marginal area in social, economic, and political terms, produced by processes of uneven development and unequal exchange. The conceptualization of borderlands as 'frontier' or as 'periphery' serves different images of settlements and settlers, and represents relations between different groups within the Israeli society. See Hasson, 1996.
4   Taylor (1999) claims that Agnew misses the point that places do not disappear with modernity, rather they change their nature, or continue to be alive in peoples' minds.
5   In 1947 about 70 per cent of the Jewish population was concentrated in the three large cities: Tel Aviv, Haifa, and Jerusalem, among them 43 per cent in Greater Tel Aviv. Israeli planners at the time regarded this polarization as a severe geopolitical and security problem, and the effort to overcome it continues to this day. Needless to say, the planners perceived the Arabs as at best immaterial, and usually as a hindrance to the Zionist settlement project (see Kipnis, 1974).
6   That is not to say all Mizrahim were directed to settlement in the periphery; the majority of them settled in the central region, mostly in poor neighbourhoods. However, compared to earlier and later immigration waves than that of the 1950s, the proportion of Mizrahim who were sent to settle in the periphery was relatively large.
7   The survey is part of research on policy and identity in development towns. It was conducted by Yiftachel and Tzfadia in the *Negev Center for Regional Development* in the years 1997–2000. Permission from the co-author was given.

8   The Semantic Differential (SD) measures people's reactions to stimulus words and concepts in terms of ratings on bipolar scales defined with contrasting adjectives at each end. Usually, the adjective marked 1 is labeled as 'extremely bad'. The adjective marked 7 labeled as 'extremely good'. Usually, results are averaged to provide a single factor score for each couple of adjectives. In the case of the development towns the couples of adjectives relates to characters of the development town. The average scores are ranged between 1 to 7, where 1 implies on a low attachment to the place, and 7 on a high attachment (see Osgood, 1967, for more details on the measurement).

9   The criteria for receiving public housing, as published by the Ministry of Immigrant Absorption (1999) includes elderly couples, single parent families a family which contains a chronically ill member or member with at least a 75 per cent disability.

References

Agnew, A.J., *Place and Politics* (Boston: Allen and Unwin, 1987).

Ben-Ari, E. and Bilu, Y., 'Saints' Sanctuaries in Israel Development Towns: On a Mechanism of Urban Transformation', *Urban Anthropology*, 16 (2) (1987), pp. 243–71.

Ben-Zadok, E., 'Oriental Jews in the Development Towns: Ethnicity, Economic Development, Budgets and Politics', in E. Ben-Zadok (ed.), *Local Communities and the Israeli Polity* (New York: State University of New York Press, 1993), pp. 91–123.

Central Bureau of Statistics, selected years, *Statistical Yearbook*, State of Israel, Ministry of the Interior, Jerusalem.

Efrat, T., 'The Plan', *Theory and Criticism*, 16, pp. 203–11 (2000) (Hebrew).

Einstein, A., 'Foreword', in M. Jammer (ed.), *Concepts of Space: the History of Theories of Space in Physics*, 2nd edn (Cambridge: Harvard University Press, 1970), pp. xiii.

Hansen, K., 'Emerging Ethnification in Marginal Areas of Sweden', *Sociologia Ruralis*, 39 (3) (1999), pp. 294–310.

Hasson, S., 'Frontier and Periphery as Symbolic Landscapes', *Ecumene*, 3 (2) (1996), pp. 146–66.

Hasson, S., 'How are the Immigrants? Where and How are They Living? Where Will They Live in the Future?', *Israel Studies*, 5 (1992), pp. 19–24.

Kipnis, B. , *The System of Interrelations between a New City and its Region as a Basis for Development* (Jerusalem: The Hebrew University of Jerusalem, 1974) (Hebrew).

Lewin-Epstein, N. Elmelech, Y. and Semyonov, M. (1997). 'Ethnic Inequality in Home-Ownership and the Value of Housing: The Case of Immigrants to Israel', *Social Forces*, 75 (4), pp. 1439–62.

Lipshitz, G., *Country on the Move: Migration to and within Israel* (Dordrecht: Kluwer Academic Publishers, 1998).

Malpas, J., 'Finding Place: Spatiality, Locality and Subjectivity', in A. Light and J.M. Smith (eds), *Philosophies of Place* (Lanham, MD: Rowman and Littlefield, 1998), pp. 21–44.

Marris, P., *Loss and Change* (London: Routledge and Kegan Paul, 1986).

Massey, D., 'Questions of Locality', *Geography*, 78 (339) (1993), pp. 142–9.

Massey, D., *Space, Place and Gender* (Oxford: Blackwell, 1994).

Massey, D., 'Places and their Pasts', *History Workshop Journal*, 39 (1995), pp. 182–92.

McGarry, J., 'Demographic Engineering: The State-Directed Movement of Ethnic Groups as a Technique of Conflict Regulation', *Ethnic and Racial Studies*, 21 (4) (1998), pp. 613–38.

Ministry of Immigrant Absorption, *Aliyah Pocket Guide*, 7th edn (Ministry of Immigrant Absorption, Jerusalem, 1999).

Morris, B., *The Birth of the Palestinian Refugee Problem, 1947–1949* (Cambridge University Press, Cambridge, 1987).

Murphy, A., 'The Territorial Underpinnings of National Identity', paper presented in the International Symposium on *Symbolic and Practical Dimensions of the Territorial Discourse in the Middle East: A Comparative Analysis*, European University Institute, Florence, December 16–19, 2001.

Osgood, C.E., *The Measurement of Meaning* (University of Illinois Press, Urbana, 1967).

Sarre, P. Phillips, D. and Skelington, R., *Ethnic Minority Housing: Explanation and Politics* (Avebury: Aldershot, 1989).

Sharon, A., *Physical Planning in Israel* (Government Printing Office, Jerusalem, 1951) (Hebrew).

Stasiulis, D. and Yuval-Davis, N., 'Introduction: Beyond Dichotomies – Gender, Race, Ethnicity and Class in Settler Societies', in D. Stasiulis and N. Yuval-Davis (eds.). *Unsettling Settler Societies* (London: Sage, 1995), pp. 1–38.

State Comptroller, *Annual Report 50b* (State of Israel, Jerusalem, 2000).

Swirski, S. and Konor-Attias, E., *Israel: A Social Report 2001* (Adva, Tel-Aviv, 2001).

Taylor, P.J., 'Places, Spaces and Macy's: Place-space Tensions in the Political Geography of Modernities', *Progress in Human Geography*, 23 (1) (1999), pp. 7–26.

Troen, I., 'The Transformation of Zionist Planning Policy: From Rural Settlement to an Urban Network', *Planning Perspective*, 3 (1994), pp. 3–23.

Tuan, Y.F., *Space and Place: the Perspective of Experience* (University of Minneosota Press, Minneapolis, 1977).

Tzfadia, E., 'Immigrant Dispersal in Settler Societies: Mizrahim and Russians in Israel under the Press of Hegemony', *Geography Research Forum*, 20 (2000), pp. 52–69.

Tzfadia, E. and Yiftachel, O., 'Political Mobilization in the Israeli Development Towns', *Politika*, 7 (2001), pp. 79–96 (Hebrew).

Wasserman, D., Womersley, M. and Gottlieb, S., 'Can a Sense of Place Be Preserved?', in A. Light and J.M. Smith (eds), *Philosophies of Place* (Lanham, MD: Rowman and Littlefield, 1998), pp. 191–213.

Yiftachel, O., 'Social Control, Urban Planning and Ethno-class Relations: Mizrahi Jews in Israel's Development Towns', *International Journal of Urban and Regional Research*, 24 (2) (2000), pp. 418–38.

Yiftachel, O. and Tzfadia, E., *Policy and Identity in Development Towns: The Case of North-African Immigrants, 1952–1998* (Beer-Sheva: Negev Center for Regional Development, 1999) (Hebrew).

Chapter 6

The Political Construct of the 'Everyday': The Role of Housing in Making Place and Identity

Rachel Kallus

Introduction

Gilo, the Jewish neighborhood built as part of the Israelization process of Jerusalem following the 1967 war, has been under fire since the beginning of the second wave of the Palestinian Intifada.[1] Initial attempts to defend public spaces in the neighborhood by constructing protective walls around public buildings and along the streets facing the Palestinian town of Beit Jalla, were followed by the defense of private dwellings located on the edge of the neighborhood. As part of this process, all of the windows in each apartment facing Beit Jalla were shielded with bulletproof glass to protect the residents from sniper shots. This fortification, its actual construction and the discourse around it, portrays the concurring roles of the home, revealing the process by which the residential environment, the base of everyday civilian life, becomes the guardian of national territory, by actually serving as a borderline. This renders the political construct of the everyday and is the subject of this chapter.

The Gilo neighborhood, built initially as a potential fortress (Figure 6.1), has found itself turned into a real one. With walls confining the neighborhood (Figure 6.2) and bulletproof windows framing the scenic views (Figure 6.3), ignoring the contradictions inherent in Gilo has become more difficult. In a recent interview, a Gilo resident told a reporter: 'relatives and friends call and ask after our well being, as if we were living in the occupied territories – but, for God's sake, I am living in Jerusalem'.[2] Confirming her situation, she further claimed that she 'bought an apartment in a Jerusalem neighborhood and not in a remote settlement in the territories'.[3] This exchange, articulating the ambiguity of Gilo's existence and the uncertainty of its residents' everyday lived experience, further indicates the duality of the home as a personal space

Figure 6.1   A view of Gilo housing (Gilo housing cluster number 11, architect Salo Harshman)

Figure 6.2   Decorated defensive wall on Anafa Street, Gilo, 2001

Figure 6.3   View of Beit Jalla from a living room window of an apartment on Anafa Street, Gilo, 2002

and as a public territory. Gilo's case exposes the state's intense involvement in the everyday, as well as reiterating national traditions of collective identity and territorial definition. It further suggests a need to reconsider the notion of the everyday and its practice as a potential vehicle of resistance.

In present day debates, the everyday figures simultaneously as an analytical category, as well as a conceptual instrument. However, in the complex reality of state-constructed everyday, the role of the housing, in establishing territory and identity, reveals the everyday as a problematic reality. Through the discussion of housing and its role in the production of the everyday, the chapter develops the notion of the residential environment as a political arena, exposing the space of the everyday as a battlefield where both national and personal struggles take place. To fully understand this phenomenon, the chapter briefly considers the notion of the everyday, shortly discusses the nature of housing, and then moves to examine the formal strategies – the planning policies and the architectural practices – used in state-constructed everyday. In the conclusion some of the contradictions inherent in Gilo and its lived experience are discussed, in order to begin considering the socio-cultural implications of state-constructed everyday.

Architecture and Everyday Practice

Although the fascination with the everyday can be traced to the pre-modern time,[4] it is currently acknowledged as a critical practice resisting modernity, typically referring to the political analysis of consumer society and to the nation-state as challenged by its subjects. Lefebvre's comprehensive theory about the relationship between everyday and modernity has pointed out the potential of the everyday 'spontaneous conscience' to stand against the oppression of daily existence.[5] De Certeau, ignoring the monotonies and tyrannies of daily life, has stressed the individual capacity to manipulate situations and create realms of autonomous action as 'network of anti discipline'.[6]

The consideration of the everyday as a critical construct has gained architectural attention, mainly as a reaction within a domain that has increasingly become allied with universal forces of globalization.[7] It has represented the attempt to put forward architectural resistance to commodification and consumption. Confronting, in Lefebvre's words, 'the bureaucracy of controlled consumption', everyday life has been suggested as a lived experience of political struggle against forces of late capitalist economy and their complicity with governmental authority. As pointed out by Harris 'the resistance lies in the focus on the quotidian, the repetitive, and the relentlessly ordinary'. Accordingly, the definition of the everyday has been 'that which remains after one has eliminated all specialized activities', deriving everyday anonymity from its 'insignificant quality'.[8]

However, it seems as if current architectural interest in the everyday has degenerated mainly into aesthetics, celebrating picturesque qualities and setting its analysis in a rather romantic framework. The progressive avant-garde consideration of the everyday has not been able to surmount the postmodern scarch for authenticity, thus promoting the everyday as a commodity, often using it as a means to frame the exotic 'other'. Attempts to celebrate 'ordinary' 'banal' and 'less photogenic' environments, in opposition to the postmodern emphasis on monumentality and professional machismo,[9] have confronted grandiose architecture with domestic situations.[10] But, even the claim that the everyday interest in facts could confront modernist utopia,[11] has seemed to further encourage professional programs of power, promoting architecture practice again as a special and unique orchestration of bodies in space.

Recent accounts of the everyday tend to be oblivious to geo-political circumstances, often unaware of the limitations and/or potentials of everyday practices in highly politicized contexts. They have been unable to explain the political construct of the everyday as part of post-partition[12] internal social, economic and political forces taking place in non-western formations of the nation-state, where constant struggles are not only against old and new forms of western dominance, but also against new patterns of local control. The architecture of the everyday consciousness to politicized cultural production,[13] although far from benign in the west, has further consequences in the non-west. Though having an immediate impact on professional awareness of political consequences of professional practice, recent critiques have tackled the connection of politics and architecture in a rather limited way, neglecting to point out the intricacy of power relations, and thus overlooking the dangers of any 'regime of knowledge', including that of architectural critique.[14] It seems as if the non-west requires a different consideration of its architecture of the everyday. This generates new possibilities for encountering hegemonic knowledge, mobilizing a more sensitive awareness to local nuances of power relations, and observing internal complexities and multi layered hierarchies of colonization. Examination of the role of housing and the use of the home in nation-state building, serves to reveal the construction of the everyday, not only as a modernist project, but also as an ongoing national ideological and political process, assisted by the complicity of professional practice. This demands a reassessment of the meaning of everyday and its potential as a space of resistance in order to further politicize its meaning, by confronting it, not only with western architectural discourse and its adaptation in the non-west, but also with nation-state forces in the non-west.

Housing: Private Home or National Territory?

The twentieth-century phenomenon of housing is to be seen as a public institution that grew in tandem with the modern nation-state. Being a physical asset providing the residential services necessary for households in the modern industrial society, housing is considered private goods. But, as the society as a whole tends to be affected by its externalities, housing is often operated as public goods. Thus, currently in most countries, housing is among the few remaining public policies, based on government acknowledgement of certain responsibilities, mainly assuring citizens of agreed-upon minimum living standards. Public housing – the provision of planned residential setting by the state – is one of the identifying features of the modern welfare state. It is usually considered one of the 'public goods' bundles provided by the state to those 'eligible' who, it is assumed, could not have acquired a place to live in the private market. It usually involves one or more of the state's three main activities: production, subsidy and regulation. But, public intervention in housing is also based on government action concerning services, such as street pavements, the supply of water, sewage and garbage disposal, standardization of material and building conventions, zoning ordinances, health, fire and safety codes, regulations of labor, finance, taxation, and law.[15] This means that an in-depth analysis of the phenomenon of housing cannot merely suffice with evaluating the efficiency of service provision, but must expose the latent and manifest agendas of governments when intervening so massively and effectively in housing, determining the material, political and social realities of the everyday. Marcuse has already pointed out 'the myth of the benevolent State', showing the role that governments play in interacting with private interests and shaping public policy.[16] Others have also questioned the genuine motives, which guide governments in making decisions as to how, and where people should live,[17] questioning the very notion of housing as a public policy.[18]

The role of housing as a first-rate national instrument has been evident in nation-state building projects, playing an important role in strengthening the relationship between a community, sometimes utterly imagined[19] and a territory.[20] Thus, in the production of the political entity called the nation-state, national identity (in the form of citizenship), although meant to erode local hierarchies, is very much a locally bound construct.[21] Being a sociopolitical construct,[22] the modern nation-state has relied extensively on architecture and in particular on modern architecture imagining itself and presenting its identity through a variety of formal institutions.[23] However, the development of national traditions is derived not only by means of monumental architecture, but also through the creation of relevant architecture for the everyday.[24] Balibar has argued that the nation-state derives its collective meaning through creation of 'fictive ethnicity' based on family and school, in which the home and the residential environment at large are most effective.[25] Planned neighborhoods are thus viewed as ideal settings to mold the mass into a national community. These are the everyday settings in which struggles over representation, access and entitlement among differently empowered groups take place.[26]

The use of housing in the spatially sociopolitical project of nation-state building is not unique to Israel. As in the case of other settler societies, such as nineteenth century Australia, Canada and the USA, housing has been instrumental in defining and constructing national territories, while giving land benefits to dominant settler groups.[27] Examples from present day Sri Lanka and Estonia similarly show how current housing policies are used by governments to shape territorial boundaries and for constructing citizenship identity of hegemonic ethnic groups around the space of living. In the case of Singapore and Hong Kong, public housing policies are seen in the context of planned economic development, while maintaining the interests of specific hegemonies.[28] These cases have been instrumental in understanding the role of housing policy in the concurrently forming local and global economy and its part in forming new societies. Recent examples from Belfast and South Africa also demonstrate the social instrumentality of housing policies, which, in these cases, serve to mediate conflicts and alleviate ethnic and racial tensions.[29] All these examples demonstrate how a basically spatial concept – housing – plays a role in a socioeconomic arena. They show the ways in which public intervention in housing yields socioeconomic and geo-political benefits for the nation-state and thus become an important public policy measure. These examples further demonstrate how planning, building and administration of housing become nation-state mechanisms to construct the everyday. To fully grasp this phenomenon, the following discussion attempts to examine the ways by which a state-constructed everyday is manufactured, controlled and managed, and the intricate ways with which the nation-state engages in the production and the consumption of the everyday.

The unique appropriateness of housing is that it can serve to design concurrently both space and society. It thus gives a concrete form to national goals and at the same time, shape the image and identity of people. Here housing serves as an efficient and powerful tool since it allows for formal construction of a sense of place through the control of the informal practices of the everyday. Thereby it is turning the residential system not only into the provider of shelter and a roof over one's head, but also into a context in which a complete redefinition of relationships between the individual and the state take place.

State-constructed Everyday: The Role of Housing in Israel

The guiding principles of housing policy in Israel, set forth during the State's early independence years, have been geared towards two national-interest goals – immigrant absorption and population dispersal. Both interests are underlined in Ben-Gurion's declaration in the unveiling of the first government, stating the need for safety and security, by a 'swift and balanced settling of the country', and the need for absorption, by 'housing all immigrants'. A third private-interest goal, the provision of satisfactory housing to every household, is again underlined in Ben-Gurion's declaration, referring to the need to 'eliminate the chronic problem of over-density and resultant sickness within slum dwellings'.[30]

It has often been argued[31] that the first two national-goals were dominant in the formative years of Israel's housing policy, when, during nation-state-building stages, the government and various other state institutions[32] played a central role in housing supply, being deeply involved in the actual settlement, development, physical planning and construction. According to this argument the third private-interest goal has come to the forefront only when the focus has turned from quantity to quality and the state has started to retreat gradually from its deep involvement in the housing market. However, as the fortification of Gilo demonstrates, although the state's involvement in the housing market has changed, it is still in control to the present time. This involvement enables it to fulfill national-interest goals through its housing policy, by various activities of government and government-controlled apparatus. As previously discussed, this public involvement in housing is manifested in the many measures taken by the government, not only in the provision of the actual dwelling, but also in the provision of related services, as well as in the controlling of development through regulation and standardization of material, spatial, financial and legal matters. This government involvement has informally controlled the development of residential environments even before the establishment of the state and is still in use today. Here Gilo is not a unique example. The initiation and construction of about 50 new Jewish settlements in the Galilee in the 1970s and 1980s is the result of government instigation, based, as in the 1950s on land confiscation and construction, assisted by the Jewish Agency.[33] In strategic areas, such as the West Bank, the government has consistently been in charge of initiation, physical planning and actual construction.[34] The Ministry of Housing and Construction is still one of the major land development agencies and is budgeted accordingly by the state.[35] The government also keeps the mortgage market under full control, preventing the development of a secondary mortgage market, thereby maintaining its power of direction and management.[36] Likewise, all planning institutions are centrally administrated and controlled, subordinating all planning decisions to the government.[37]

This intense and continuous involvement of the government in the housing market might be better explained by referring to the non-official goals underlining the state's activities and the role of public housing in achieving these goals. In its drive to shape its national space the state of Israel, from its early days, has been aiming to achieve the territorialization of space, i.e. the establishment of sovereignty over territory, and the spatialization of this territory – the establishment of control, power, authority and domination over that space. Spatialization of territory has often been achieved by constructing and giving form to the physical landscape by virtue of the actual design and physical construction of concrete projects, thereby providing the framework through which sovereignty and hegemony is institutionalized and internalized. Here questions of location, identity, and power have been established through the ongoing day-to-day mechanisms of confrontation, compromise and creation of consent among economic, social, and political forces. Housing, in its spatial configuration and organization, played a crucial role in making a sense of space and concretizing the sovereign territory of the nation-state. Thereby the state of Israel has always had a double agenda in its housing production: for the outside world, it was intended to establish ownership, control and sovereignty over the national territory; whereas for internal consumption it was to facilitate the creation of a living space from which residents were expected to draw their identities and life-styles, while bestowing symbolic meaning upon the daily practices embedded in that space.[38] Here architecture, especially modern architecture, played a major role, since it was an effective means by which the environment could be interpreted, modified and contested. This is, of course, not unique to Israel.

But the Israeli experience, in its vast and speedy development, enables review of the ways in which the nation-state project of housing production has been constantly assisted by modern architecture.[39] Israel provides here a unique laboratory, since it presents a valuable case study with which to examine the role of housing in the continuously shaped and reshaped urban environments in neighborhoods and cities, and the unique role played by the architectural culture in this process.

The Story of Gilo: The Construction of the Everyday

With the reunification of Jerusalem following the 1967 war, the development of Jewish settlements has become a strategic geo-political effort to eliminate the future separation of Jerusalem. This was the base for the initial decision to build the perimeter neighborhoods around Jerusalem, a strategic planning operation based in the age-old Zionist ideological assumption that Jewish settlements are the only means for Israel's territorial sovereignty.

The construction of the new neighborhoods around Jerusalem was achieved in three major steps (see Figure 6.4). The first step was meant to break the continuation of the northern Palestinian built up area from the Old City up to Ramallah, by introducing what was referred to as the 'Northern Lock', in the form of three new Jewish neighborhoods (Ramat Eshcol, The French Hill, and Givat Hamivtar) continuing the Jewish urban sprawl into the Palestinian territory. The second step was meant to expand the Jewish settlements beyond the Green Line[40] and consisted of four new Jewish neighborhoods (Ramot, Neva Yaacov, East Talpiot, and Gilo). The third step was a means of increasing the Israeli hold over Palestinian territory (Pisgat Zeev), and is continued today with further building of Israeli neighborhoods[41] in Palestinian territory[42] (Har Homa being the latest and still under construction[43]).

These strategic steps were backed up by major confiscation of land, which has been condemned by the UN Security Council. Faced with major international pressure, the Israeli government has decided to speed up the development process, thereby taking various time-saving planning and construction measures. Regular planning procedures were often over ruled, construction has often started before the final approval at all planning levels was achieved, sometimes even before the instigation of building permits, and pre-fabricated elements were used in order to speed up construction.

Despite these time-saving measures, the construction of the new neighborhoods, part of a large-scale national financial, planning and

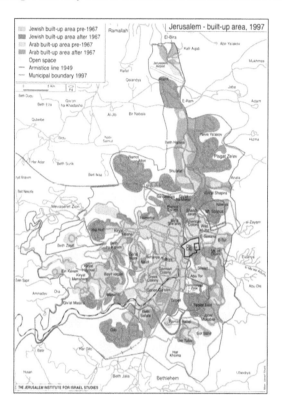

**Figure 6.4   Residential neighborhoods constructed around Jerusalem
1967–90**

construction effort, has produced residential areas far better than the housing
stock built before the city's reunification. Instead of the standardized
architecture of the 1950s and 1960s, expressed in the monotonous building
blocks and the limited replications of dwelling types, a serious effort was made
to diversify and innovate in the new areas. This is not unique to Jerusalem, but
is the result of a major shift in public policies to improve housing stock and
make both its quality and location more attractive, based on lessons learnt from
failures of earlier schemes. This represents the adaptation of the housing market
to changing socio-cultural and geo-political conditions, combined with the
architectural challenge of mass housing. It is an attempt to revise the inhuman
project of modernism, thereby fostering a new sense of community and place,
humanizing the *machine a habitar*, finding the lost city, and addressing, at
least to some extent, issues of personal and cultural identity.

By the end of the 1960s the massive construction of large-scale public housing projects in the urban and the national periphery had almost come to an end. The public housing sector had gradually adopted a variety of spatial strategies, intended to diversify its location and to develop a wider house-type choice for its changing clientele. A large stratum of middle class had been emerging in Israel in the 1970s, and the housing market was eager to serve its needs. As described by Gonen, the demand for 'villa' and 'cottage' by new middle class home seekers has produced neighborhoods of detached and semi-detached housing in all sorts of locations: very close to the inner urban areas, on the urban fringe, in outlying suburban towns or villages, and far out on the mountains of the Galilee and the hills of the newly occupied West Bank.[44] Each one of these locations represents a spatial preference for low-density residences, and many of them have been used by the government as a mean to direct residents yearning for a 'private house with a garden' to its 'high priority areas'.[45]

Furthermore, until the early 1970s population strategies were based on allocating designated areas to specific eligible groups, such as immigrants, young families, etc. From 1978 this strategy changed, attempting to create a better social integration by combining strong households with weaker ones. This has required a large variety of dwelling types, organized in different spatial combinations. It has also demanded a far more inventive and less routine architecture, which could respond to changing needs. These housing strategies, and their design implications, correspond to, and are also informed by, the break from modernism in architecture culture worldwide and the growing call for more human environments and the return to the lost city.[46] In this sense the housing schemes of the 1970s and 1980s express two competing trends. On the one hand they demonstrate a yearning for suburban low-density living, but at the same time they articulate the Israeli version of postmodern urbanism and are thus an Israeli version of new-urbanism.[47] Planned projects of residential developments, catering for middle class home-improvers, have been designed by attempting to recreate the urban environments missing in the massive modernist housing estates of the 1950s and early 1960s, thereby fostering a sense of place. This 'return of the city' approach is clearly declared by the Ministry of Housing and Construction, and is presented not only in its various manifestos and in various research publications[48] but also in the range of projects built and constructed during the 1970s and 1980s.[49]

Following these developments, Gilo, built in the late 1970s, is organized into four sub-areas of about 300 dwelling units, serving also as sub-social units with a population of about 1,000–1,200 residents each (see Figure 6.5). Of

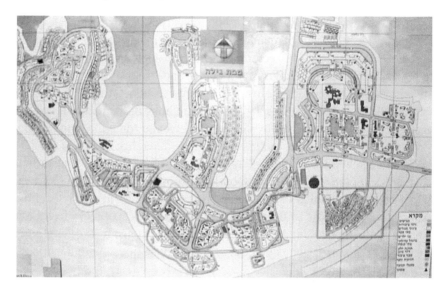

Figure 6.5   Plan of Gilo

course, the assumption that these sub-units could support community services is not new to town planning. However, the way in which this social framework becomes a base for the development of a physical spine-like architectonic structure, attempting to connect well-defined pedestrian walkways and squares to transportation and public services infrastructure is worth noting. These social attributes have been further translated into sophisticated formal categories, suggesting an intricate interplay between built and open spaces. The major attempt is to contextualize a Mediterranean setting architecturally[50] by referring the building types and to the urban patterns of the Jerusalem urbanscape, thus, making Gilo a catalogue of 1970s and 1980s typology of perimeter blocks locally styled. This politicized nostalgia for the vernacular of the colonized[51] is expressed in architectural authentication made with reference to topography, weather conditions, and local materials, all appropriating the building-types to distinct local circumstances.[52]

This cluster-based formation of the neighborhood, considered from both physical and social points of view,[53] allows for low density; while at the same time maintaining an urban character, as clearly stated by the planning team:

> Seemingly this [3 families per 1,000 square meters] is a low urban density, however, because of the concentration of the dwellings on the high areas only (with the wadies reserved for parks) the relative density of the dwelling areas is increased, insuring thus an activity space appropriate for urban life.[54]

But, vitalization of the neighborhood by giving an urban character to the clusters is not intended merely for internal communal life. It is also seen as an important means of overcoming the suburban nature of this remote neighborhood, placed away from the city center, by creating '... the feeling that it is an organic part of Jerusalem ...'.[55]

Gilo is built on the highest ridge south of Jerusalem, as a boundary against the Palestinian towns of Bethlehem and Beit Jalla (see Figure 6.4). This location is officially presented as a historical border of Jerusalem. For example, as the introduction of the Gilo project in the annual Ministry of Housing and Construction publication clearly states, Gilo is built on a '... a ridge which is a clear boundary ever since Jerusalem's early build-up'.[56] This planning rational is based on the notion that Gilo is 'clearly part of the wall enclosing Jerusalem'[57] and that 'The character of this area dictates a continuous line of buildings on top of the hill, creating a "wall" protecting the town from the outside ...'.[58]

However, the wall-like character of the neighborhood (see Figure 6.6) is also taking a full advantage of the views offered by the sloping – both southwards towards Bethlehem and Beit Jalla, and northwards towards Jerusalem. Based on this location, the neighborhood from its initial conception has been recognized as offering exceptional views:

> Outwardly there is a view of green open spaces and inwardly, a magnificent view of Jerusalem, with the old city in the midst of it, and with the crests of Mount Scopus and Augusta Victoria enclosing (*sic*) in the North.[59]

Needless to say, the architectural approach takes full advantage of this superb topographical location, treating the neighborhood's unique panoramic view with the greatest design sensitivity, both in its detailed studies of the entire neighborhood layout and in the way it is viewed from different strategic points (see Figure 6.6), as well as in the careful considerations during the development of the various building types:

> In order to give the quarter its architectonic character, four types of building and site layout have been chosen: on the mountain top, groups of buildings around courtyards with gardens; on the slopes descending towards Jerusalem, buildings in stepped parallel rows, like hillside terraces looking out toward the city; along the ridge, 7 high-rise towers giving rhythm along the extended spur and serving as landmarks for the precinct; and on the slopes descending into the wadis, detached houses that provide a transition from the closely built-up area to the open wadi.[60]

View from Bethlehem

מבט מבית-לחם

Figure 6.6   A study of Gilo as viewed from Bethlehem

The architectural design of the buildings, with its variety of dwelling units, similarly treats the views offered by this exceptional location, providing the residents with various means to enjoy the site. Many of the units are designed with balconies, terraced gardens and large panoramic windows. Large clusters of semi-detached stepped cottages, with private yards, roof gardens, or large balconies are built along the streets facing Beit Jalla and Bethlehem, with the scene of the oriental skyline as their main attraction. Many of these units are provided with living rooms having large windows facing the scenic view. Needless to say, these dwellings units are marketed on the basis of the attractive panorama they offer and are priced accordingly.

The Fortification of Gilo

In the period leading up to the second wave of the Palestinian Intifada it seemed as if Gilo had fulfilled most of its goals. Offering better housing alternatives to Jerusalem's middle-class inner-city populations, it has drawn in public-housing-eligible inhabitants, many of whom being home improvers who had willingly left the dense and poorly maintained inner-city neighborhoods for a newly developed quasi-suburban area. The improved housing units, with the promise of high level schools and communal facilities, brought in many families seeking a better environment in which to raise children. Thus, marketing of most dwelling units in the neighborhood was quite easy, although of course, the units at the south end of the neighborhood, overlooking the panoramic view of Bethlehem and Beit Jalla, by offering inner city dwellers the opportunity to consume the magnificent landscape of the hinterland, had a greater appeal.

The fortification process started at the end of 2000, when as part of the second wave of the Palestinian Intifada, Gilo came under fire from its neighboring town Beit Jalla.[61] It was based on the Prime Minister's Office decision taken in November 2000, in response to the growing demands of Gilo's residents, supported by political figures, especially from the Right Wing.[62] Initial fortification steps were meant to protect educational facilities and public outdoor spaces in the neighborhood, by constructing protective walls on the streets facing Beit Jalla (see Figure 6.2). With the continuation of the shooting, and growing public pressure, it was decided to protect the residents in their own homes as well.[63] It was initially assumed that a defendable space in the form of one sheltered room in each unit would be sufficient. However, it was soon discovered that the units' layout, their structural features, and the nature of their use could not support such a scheme. Thus, it was decided that all windows in the apartments facing Beit Jalla would be protected by replacing them with bulletproofed glass set in lead enforced frames. The fortification process was administrated through the Jerusalem municipality, which has set up a special fortification council in the neighborhood. The council's main mission was to organize the fortification procedure according to the government's direction. Thus, all decisive factors, such as determining the fortification standards, defining dwelling-units eligibility and priority, were decided by the IDF Back Command Unit and implemented by the fortification council.

The shielded windows used were based on a ballistic specification, custom manufactured and installed by a private firm. The selection of units to be fortified was based on mapping of the 'threat sources'[64] defining the extent of the danger area, and thereby the places in need of protection (see Figure 6.5 – blue line marks the areas where fortifications took place). So far about 700 units have been installed at an average cost of US$ 1,400–1,700 per square meter of window surface. Assuming windows' surface of about 10–14 square meters per unit, it is estimated that about US$ 17,000–20,000 were spent on each unit. Thus, with the government covering all expenses of this indoor fortification process, the estimated budget spent so far is at about US$ 13 million.[65]

Gilo's Everyday Lived Experience

Gilo marks not only a physical border, but also a socio-cultural one, in which the everyday becomes a setting for conflicting realities of contested identities. Thus it is not only a space defining the divide between nation-states (Israel and Palestine), but also an arena of constructed schisms within the nation-state (Israel). Despite its contested legal situation and its obvious geo-political uncertainty, Gilo has been presented from its initial planning stages as part of Jerusalem. Good road connection to the city, efficient public transportation and its municipal administration being part of Greater Jerusalem, have served to obscure its actual location between two Palestinian towns – Beit Jalla to its south and Beit Tzafafa to its north (see Figure 6.4). Public discourse has also obscured its reality; above all by referring to it as a neighborhood, insinuating it as being part of Jerusalem. Although, in fact, its location vis a vis the city, its population size and area range could easily make it a small town in its own right. Secondly, declaring Gilo as an extension of Jerusalem, conveniently avoids any reference to its location in the occupied territories beyond the Green Line. This has been supported by the unilateral act of annexation of all areas around Jerusalem to the city and the declaration of the new neighborhoods as part of the metropolitan area, thereby making Gilo 'a neighborhood in southeastern Jerusalem' and not a settlement. But of course, for the great majority of its residents, Gilo could have become a home exactly for its inherent contradictions and what they have facilitated in terms of housing. With its ideological foundation and ambivalent borderline, supported by economically-based urban policies, Gilo has drawn its initial inhabitants, by the lure of affordable housing, thereby making it what is often referred to as 'suburban colonization'[66] and its residents 'economic settlers'.[67]

Gilo's innate contradictions, as embedded in its unsettled geo-political situation, even more painfully evident since the latest Intifada, are the cause for acute anxiety in its residents. This is clear in the call to the government by many residents to 'bring back security' and the effort to see Gilo as distinguished from the occupied territories of the West Bank. But nonetheless, for some residents living in Gilo its distinctive situation has in fact been a source of self-respect and newly acquired identity. As interviews with recent immigrants from the former Soviet Union living on Anafa Street have shown, their borderline experience in Gilo has been a process whereby their Israeliness has been initiated. Women interviewed in their apartments overlooking Beit Jalla have expressed their strong sense of identity and belonging to their new place, achieved through the hardship of daily life in Gilo.[68] In fact, one of them, formally having no legal documents told me how she had acquired citizenship immediately after her apartment was set on fire in one of the shooting incidents. With very little Hebrew, but with a great deal of pride, these women are happily confronting the 'other' as part of their newly gained Israeliness, thereby re-enacting the long-lived Zionist legacy of pioneering and frontier. These women are the living evidence of the home as a national territory, owned by the social deed of national citizenship.

Examining different lived experiences in Gilo's complex reality of being a state-constructed everyday, shows the need to overcome the often used binary oppositions of 'self' and 'other' ('us' and 'them'), to enable the discovery of a multifaceted arena, which when carefully deconstructed reveals layers of internal colonization, creating a highly contested class, gender, ethnic and national identity. The women interviewed have come to live in Gilo for its affordability, probably moving to housing that became available and affordable due to Gilo's changing conditions. But, living in Gilo these women are so caught up in the confrontation with the 'other' (Palestinians) that they have no space to actually realize their own colonization acted through the domination and control of the state. In fact, their living conditions, in apartments sealed from the outside for protection and with little chance to develop social connections in free and democratic urban open spaces has left these women captured citizens. With unsafe public spaces, confined to their interior private spaces, and exposed to propaganda by the media,[69] these women have very little chance to experience any real democracy, let alone develop a sense of place of resistance.

As in the case of other fortressed and militarized urban situations the urban model of spatial segregation has transformed Gilo into what is known as 'fortified enclave'. Caldeira discussing the situation of Sao Paulo, has explained this form of urban living as privatized, enclosed, and monitored space for residential consumption, leisure and work.[70] Davis discussing Los Angeles, has also described a process of privatization of the urban space, whereby the city is divided into 'fortified cells' of affluence and 'places of terror' where police battle with the criminalized poor.[71] These grim descriptions of urban situations have shown how obsession with security has supplemented previous hopes for urban reform and social integration in the modern city. What is clear in both Sao Paulo and Los Angeles, as in many other cities and of course in Gilo, is the constant destruction of any truly democratic urban space. The deteriorated quality of urban public life has eliminated the basic principles of openness and free circulation that have been among the most significant organizing values of modern cities. Thus, especially in Gilo, where these processes are supported by public action, and in fact carried on by the government, the elimination of citizenship, based on public urban experience of democracy, is most frightening.

Housing: a Home or a National Territory?

The way in which private homes, the core of personal life, have been used to construct and defend national territory, has been in practice in Israel even before its establishment as a state. As a result, the function of public intervention in the housing market has assumed a role far beyond that of providing improved housing facilities and shelter for powerless populations who are presumed to be unable to take care of their own housing needs in the free market. A closer look at the practices and routines of housing provision in Israel reveals the dual political and social role of housing: in shaping the land, through the Zionist proclamation of a new (Jewish) territory, and in shaping identity, by determining a new (Israeli) citizen.[72] Thus, in the process of building the state, it could be said that the residents are the primary raw material. No less, and possibly more than the materials of concrete and cement, they are the resource required to fulfill the national-spatial aspirations of the state. In the context of the particular situation of Gilo, the underlying state's intentions have come to the surface, thus exposing such housing as a first-rate national instrument. The chapter has attempted to reveal the role of housing in the physical and social production of the everyday, through investigation of the formal strategies, the planning policies and the architectural practices used in creating Gilo. The apparent role of the nation-state in constructing the everyday became more obvious while examining Gilo's fortification process, where it has been an active agency of housing production, but also a leading force in everyday consumption.

While the customary meaning of the everyday usually apply to situations where the state is being confronted by its subjects,[73] the Gilo case sheds light a situation of confrontation with the exterior 'other'. In Gilo, arguably still at its early stage of territorialization, the state is using the residents to establish control over territory. Thus, the shaping of its space has been overshadowed by the attempt to ensure domination over newly created territory. Thereby, the fortification measures are openly declared as an effort to maintain day to day life. They come as a response to the residents' demand to get on with their daily activities, as well as the state's insistence of keeping up its obligation to its citizens' well being.[74] However, in the context of this massive public intervention in private properties and the intrusion of personal domain, the state's dual agenda has come to the fore. This has put in doubt the space of the residential environment as ever having been private, not being from the start a national asset. Hence the nation-state's interest in the practice of everyday life, which in times of crisis, enables it to make even a stronger claim over space and territory. Thus, in a militarized culture, where the margin between civilians and soldiers is constantly blurred,[75] this strategy has obviously served the state. But, when every house is a frontier, and every resident a warrior, put into combat in his/her most private sphere, the war takes place everywhere, including the most intimate and personal domains. This double-edged strategy has deliberately turned the environment of the everyday into a civilian battlefield where housing has become the frontline and the streets a war zone.

Notes

1   For analysis of the second Intifada see S. Tamari and R. Hammami, 'The Second Intifada', *MERIP*, 30 (4), (2000), pp. 4–10.
2   Eyala Cohen as cited in *Yidi'ot Aharonot*, 10 October 2000, p. 3.
3   Ibid.
4   See for example the work of the Dutch painter Johannes Vermeer, depicting everyday life in the city of Delft in The Netherlands during the seventeenth century.
5   H. Lefebvre, *Critique of Everyday Life*, Vol. 1, 2nd edn 1958, trans. J. Moore, London: Verso, 1991 [1947]).
6   M. de Certeau, *The Practice of Everyday Life*, trans. S. Rendall (Berkeley: University of California Press, 1984).
7   Two recent books that take this position are S. Harris and D. Berke (eds), *Architecture of the Everyday* (New York: Princeton Architectural Press, 1997) and J. Chase, M. Crawford and J. Kaliski (eds), *Everyday Urbanism* (New York: Monacelli Press, 1999).
8   S. Harris, 'Everyday Architecture', in Harris and Berke, *Architecture of the Everyday*, pp. 1–8.
9   See M. Mclaod, '"Other" Spaces and "Others"', in D. Agrest, P. Conway and K.L. Weisman (eds), *The Sex of Architecture* (New York: Harry N. Abrams, 1996); B. Colomina, 'The Split Wall: Domestic Voyeurism', in B. Colomina (ed.), *Sexuality and Space* (New York: Princeton Architectural Press, 1992).
10   For example: D. Hyden, 'What Would a Non-sexist City be Like? Speculations on Housing, Urban Design, and Human Work', in C. Stimpson et al. (eds), *Women and the American City* (Chicago: University of Chicago Press, 1980); L.K. Weisman, *Discrimination by Design: A Feminist Critique of the Man-Made Environment* (Urbana: University of Illinois Press, 1992).
11   For example see the work of Venturi such as R. Venturi, D. Scott Brown and S. Izenour, *Learning From Las Vegas* (Cambridge: MIT Press, 1977).
12   The notion of post-partition does not the deny the colonial or the postcolonial experience, but shifts attention to local power structure, spatial transformations and societal struggles, based in typical ruptures and dislocations resulting from the moment of partition.
13   There are of course serious doubts weather architecture could actually be considered a cultural production, and if it can communicate meaning, let alone politicized meaning. See for example, F. Jameson, 'Is Space Political?', in C. Davidson (ed.), *Any Place* (Cambridge: MIT Press, 1995), pp. 192–205.
14   I am referring here to recent examination of the politics of professional practice of Israeli architecture in the context of the Israeli/Palestinian conflict, i.e. R. Segal and E. Weizman (eds), *Civilian Occupation, The Politics of Israeli Architecture* (Tel Aviv: Israel Association of United Architects Israel Association of United Architects, 2002) (banned by the publisher). For suggestion of a more complex approach to the connection of architecture and politics, see D. Monk, *An Aesthetic Occupation: The Immediacy of Architecture and the Palestine Conflict* (Durham: Duke University Press, 2002).
15   On this issue see: W. van Vliet, 'The Privatization and Decentralization of Housing', in W. van Vliet and J. van Weesep (eds), *Government and Housing: Developments in Seven Countries*, *Urban Affairs Annual Reviews*, Vol. 36 (Newbury Park, CA.: Sage Publications, 1990), pp. 9–24.

16   Peter Marcuse, 'Housing Policy and the Myth of The Benevolent State', *Social Policy*, January/February, (1978), pp. 21–7.

17   For example: D.P. Salins, 'Housing is a Right? Wrong!', *Housing Policy Debate*, 9(2) (1998), pp. 259–66; David Clapham, Peter Kemp and Susan J. Smith, *Housing and Social Policy* (London: Macmillan, 1990); R. Bratt, 'Public Housing: The Controversy and Contradiction', in R. Bratt, C. Hartman and A. Meyerson (eds), *Critical Perspectives on Housing* (Philadelphia: Temple University Press, 1986), pp. 335–61.

18   On this issue, especially in the US context see: L. Vale, *From the Puritans to the Projects, Public Housing and Public Neighbors* (Cambridge: Harvard University Press, 2000).

19   B. Anderson, *Imagined Communities* (London: Verso, 1983).

20   For work on this subject in relation to architecture see: S. Bozdogan, *Modernism and Nation Building* (Seattle: University of Washington Press, 2002); R. Kallus and H. Law Yone, 'National Home/Personal Home: Public Housing and the Shaping of National Space', *European Planning Studies*, 10 (6) (2002), pp. 765–79.

21   J. Holston and A. Appadurai, 'Cities and Citizenship', in J. Holston (ed.), *Cities and Citizenship* (Durham: Duke University Press, 1999), pp. 1–18.

22   On the nation-state as a socio-construct see: K. Anderson, 'Thinking "Postnationality": Dialogue across Multicultural, Indigenous, and Settler Spaces', *Annals of the Association of American Geographers*, 90 (2) (2000), pp. 381–91.

23   On the relation of nationalism and modern architecture see: Bozdogan, 2002 (cited in note 20); L. Vale, *Architecture, Power and National Identity* (New Haven: Yale University Press, 1992).

24   Of the literature on nationalism see especially: E. Hobsbawm and T. Ranger, *The Invention of Tradition* (Cambridge: Cambridge University Press, 1983); E. Hobsbawm, *Nations and Nationalism since 1780* (Cambridge: Cambridge University Press, 1990); L. Greenfeld, *Nationalism: Five Roads to Modernity* (Cambridge: Harvard University Press, 1992).

25   E. Balibar, 'The Nation-Form: History and Ideology', in E. Balibar and I. Wallerstein, *Race, Nation, Class* (London: Verso, 1991), pp. 86–106.

26   See for example: P. Jackson and J. Penrose (eds), *Construction of Race, Place and Nation* (London: UCL Press, 1993); H. Bahabah, *Nation and Narration* (London: Routledge, 1990).

27   On nation-state building in settler societies, see: D. Stasiulis and N. Yuval-Davis, 'Introduction: Beyond Dichotomies – Gender, Race, Ethnicity and Class in Settler Societies', in: D. Stasiulis and N. Yuval-Davis (eds), *Unsettling Settler Societies* (London: Sage Publications, 1995), pp. 1–38. For discussion of the American frontier see: F.G. Turner, *The Frontier in American History* (New York: Holt, Rinehart and Winston, 1962). For discussion of Turner's thesis of the American frontier see: J.M. Faragher, *Rereading Frederick Jackson Turner: The Significance of the Frontier in American History, and Other Essays* (New York: H. Holt, 1994). For discussion of the nineteenth Canadian frontier see: N.S. Forkey, *Shaping the Upper Canadian Frontier: Environment, Society, and Culture in the Trent Valley* (Calgary: University of Calgary Press, 2003). For the Australian case see: L. Anderson and J. O'Dowd, 'Borders, Border Regions and Territoriality: Contradictory Meanings, Changing Significance', *Regional Studies*, 33 (7) (1999), pp. 593–604.

28   On Estonia see: G. Shafir, *Immigrants and Nationalists: Ethnic Conflict and Accommodation in Catalonia, the Basque Country, Latvia and Estonia* (Albany: SUNY Press, 1995). On Sri Lanka see: D. Little, *Sri Lanka: the Invention of Enmity* (Washington DC: US Institute for Peace, 1994). On Hong Kong and Singapore see: M. Castells, Lee

Goh and Y.W. Kwok, *The Shek Kip Mei Syndrome: Economic Development and Housing in Hong Kong and Singapore* (London: Pion, 1990).

29   F.W. Boal, 'Integration and Division: Sharing and Segregating in Belfast', *Planning Practice and Research*, 11(2) (1996), pp. 151–8; S.A. Bollens, 'Urban Planning Amidst Ethnic Conflict: Jerusalem and Johanesburg', *Urban Studies*, 35 (4) (1998), pp. 729–50.

30   *The Knesset* (Israel Parliament) *Chronicle*, 1949.

31   For example, N. Carmon and D. Czamanski, 'Housing in Israel, 1948–1988: From Planned Economy to Semi-free Market Management', *Housing Science*, 16 (1) (1992), pp. 47–59.

32   Settlement and other practices involving land acquisition and management have been historically in the hands of non-governmental institutions such as the Jewish National Fund and the Jewish Agency. For further discussion of these practices see: O. Yiftachel and S. Kedar, 'Landed Power: the Making of the Israeli Land Regime', *Theory and Criticism*, 16 (Spring) (2000), pp. 67–100.

33   On what is often referred to as 'the Judiation of the Galilee' see: O. Yiftachel, *Planning a Mixed Region in Israel: The Political Geography of Arab-Jewish Relations in the Galilee* (Aldershot: Avebury, 1992); N. Carmon, H. Law Yone, G. Lifswitz, S. Amir, D. Czamanski and B. Kipnis, *New Settlement in the Galilee – Evaluation Research* (Haifa: Technion Center for Urban and Regional Studies, 1990).

34   The publications of Adva Center provide valuable data on Israli government spending on infrastructure for the Jewish settlements in the West Bank. See for example: S. Swirski, E. Konor-Atias and A. Atkin, *Government Financing of Israeli Settlements in the West Bank and the Golan Heights in the 1990s: Municipalities, Housing and Road Construction* (Tel Aviv: Adva Center, 2002a).

35   S. Swirski, E. Konor-Atias, A. Atkin and B. Swirski, *A Look at the Budget Law and the Arrangement Law for 2002* (Tel Aviv: Adva Center, 2002b).

36   E. Werczberger, 'Will a Secondary Mortgage Market Solve the Housing Problems of Young Couples?', proposition paper (Tel Aviv: Adva Center, 2001) (Hebrew); Gavriel, 1985, in N. Carmon, 'Housing in Israel: The First 50 Years', D. Nachmias and G. Menahem (eds), *Public Policy in Israel* (Jerusalem: The Israel Democracy Institute, 1999), pp. 381–436 (Hebrew).

37   E.R. Alexander, R. Alterman and H. Law Yone, 'Evaluating Plan Implementation: the national statutory planning system in Israel', *Journal of the American Planning Association*, 50 (1983), pp. 314–27.

38   For further discussion of the implications of the state's double agenda through its involvement in housing, especially in Israel's formative years, see Kallus and Law Yone, 2002 (cited in note 20).

39   The example of Israel joins the growing scholarship on the employment of modern architecture during nation-state building. Most of this work focuses on the city (e.g., J. Holston, *The Modernist City, An Anthropological Critique of Brasilia* (Chicago: The University of Chicago Press, 1998)) and its national monuments (e.g., Vale, 1992, cited in note 23). Discussion of evolving national modern architectural culture tends to consider housing (e.g., Bozdogan, 2001, cited in note 20). But, the employment of modern architecture in the design of public housing is usually discussed in the realm of the colonial attempts of modernization (e.g., M. Eleb, 'An Alternative to Functionalist Universalism: Ecochard, Candilis and ATBAT-Afrique', in S. Goldhagen and R. Legault (eds), *Anxious Modernisms, Experimentations in Postwar Architectural Culture* (Cambridge: MIT Press, 2000), pp. 55–74).

40 The Green Line is the boundary that separated Israel from the West Bank between 1949 and 1967. It had resulted from the armistice which was signed in January 1949 during the Rhodes Agreements to end the 1948 Arab-Israeli war. Today it exists as an informal 35-year-old administrative line, separating civilian Israel from the occupied territories in the West Bank, and delineating between two different legal and structural systems. Although official Israeli maps do not show the Green Line boundary, Israel's municipal boundary stops at the Green Line.

41 Due to their size, some of these neighborhoods are practically cities. They are considered part of the municipal area of Jerusalem in an attempt to demographically maintain its Jewish majority.

42 According to the 1949 armistice agreement this territory was under Jordanian jurisdiction from which it was occupied by Israel during the 1967 war.

43 The location of Har Homa is between the village of Umm Tuba and the Palestinian town of Beit Sahur in the east of Jerusalem in a highly Palestinian populated area. But, officially it is often represented as seated between Ramat Rahel and Gilo attempting to make it seem as a logical extension of Jerusalem and the last link in the chain of new neighborhoods built around Jerusalem since 1967.

44 A. Gonen, *Between City and Suburb* (Aldershot and Brookfield: Avebury, 1995).

45 Areas of national priority are zones strategically defined according to the government's political goals. Development in these areas is heavily subsidized by the state, either directly through generous loans given to home buyers, or through various tax benefits and other financial adjustments.

46 For further discussion of these trends as a paradigm shift in architecture culture see: M. Hays, *Architectural Theory Since 1968* (Cambridge: MIT Press, 1998); K. Nesbitt, *Theorizing a New Agenda for Architecture, An Anthology of Architectural Theory 1965–1995* (New York: Princeton Architectural Press, 1996). For specific discussion of these trends in urban design see N. Ellin, *Postmodern Urbanism* (Cambridge: Blackwell, 1996).

47 On new urbanism see: P. Katz (ed.), *New Urbanism: Towards Architecture of Community* (New York: McGraw-Hill, 1994).

48 Such as R. Carmi, 'Human Values in Urban Architecture', in Misrad Ha-Shikun, *Israel Builds* (Jerusalem: State of Israel, Ministry of Housing and Construction, 1977), pp. 320–28; M. Wallfson, *Planning Residential Environment and Providing for Human Needs – Examples from New Neighborhoods in Jerusalem* (Jerusalem: Programming Division, Israel Ministry of Housing and Construction, 1976) (Hebrew); A. Paldi and M. Wallfson, *Searching for Neighborhood Building, Cluster, Court, Street – Planning the Everyday* (Jerusalem: Programming Division, Israel Ministry of Housing and Construction, 1989) (Hebrew).

49 For an overview of the nature of public housing projects planned and built in Israel during the 1970s and 1980s see the Housing Ministry publication *Israel Builds* (Misrad Ha-Shikun, 1973; 1977; 1988).

50 A reference is made always to Mediterranean architecture and not to the local Middle Eastern context, hinting the dismay and the repression in which the architects view this highly contested space.

51 On this issue see: A. Nitzan-Shiftan, 'Israelizing Jerusalem: The Encounter Between Architectural and National Ideologies 1967–1977', unpublished PhD dissertation (Cambridge: MIT, 2002).

52   Misrad Ha-Shikun, *Israel Builds* (Jerusalem: State of Israel, Ministry of Housing and Construction, 1977), pp. 194–209.

53   Ibid, p. 200.

54   Misrad Ha-Shikun (1973), *Israel Builds*, Jerusalem: State of Israel, Ministry of Housing and Construction, p. 153.

55   Misrad Ha-Shikun, 1977 (cited in note 52), p. 207.

56   Ibid, p. 153, althoug the translation is made from the Hebrew on the same page, which is different from the English version on this page.

57   Ibid, p. 209.

58   Misrad Ha-Shikun, 1973, (cited in note 54), p. 153.

59   Ibid. Note that this description does not account for the view of Bethlehem and Beit Jalla to the south.

60   Misrad Ha-Shikun, 1977, (cited in note 52), p. 207.

61   It has been claimed that the shooters were not residents of Beit Jalla, but members of the Tanzim, taking advantage of its strategic location, and fleeing when Israeli retaliation started (*Yide'ot Aharonot*, 18 October 2000, pp. 4–5; *Ma'ariv*, 24 October 2000, pp. 2–3).

62   The prime minster at the time was the Labor leader Ehud Barak, so 'beefing up' residents' discontent was undertaken by members of the Likud party, in an effort to built opposition from the Right. Among these much-publicized acts taken by the opposition members of the Knesset was a declaration by the Jerusalem Right Wing Mayor Ehud Olmert that he would move his office to Gilo in an act of 'solidarity with the hard-hit residents'. His feigned office was set up in the headquarters from which the bulletproofing process was managed. Another example is the on-line news from Gilo put by the extremist Right Wing (Herut) Knesset member Michael Kleiner on his website.

63   This decision was contested with heavy opposition from hard core settler leaders, demanding retaliation instead of defensive measures. Sharon Katz, editor of *Voices* said in the settlers' broadcast channel (Arutz 7): 'What about bulletproofing the caravans in Dagan? And bulletproofing every home, school and car in N'vei Dekalim? … and bulletproofing the homes on the perimeter of Pesagot? … and bulletproofing the homes in Hevron? ... Wouldn't it be smarter instead of bulletproofing the entire country, to stop the Arabs who are shooting into every neighborhood?… Instead of bulletproofing the country and letting them continue shooting at us, stop the shooting!' (published in arutzsheva.org, 13 December 2000).

64   Uri Menahem, Vice-Director of the Defense Council of the Jerusalem Municipality (interview on 12 February 2002).

65   Ibid.

66   D. Newman, 'How the Settler Suburbs Grew', *New York Times*, 21 May 2002).

67   S. Foa, 'Blame Brooklyn, Fundamentalist Settlers Have a Borough Accent', *The Village Voice*, 22–28 May 2002.

68   So far only limited number of interviews with women, recent immigrants from the former Soviet Union, living on Anafa Street, in their homes, was conducted in June 2002.

69   On ethnic and citizenship perception of immigrants from the former Soviet Union see: D. Shomski, 'Ethnicity and Citizenship in Russian Israeli Perception', *Theory and Criticism* 19 (Fall) (2001), pp. 17–44.

70   T.P.R. Caldeira, 'Fortified Enclaves: The New Urban Segragation', in J. Holston (ed.), *Cities and Citizenship* (Durham: Duke University Press, 1999), pp. 114–38.

71   M. Davis, 'Fortress Los Angeles: The Militarization of Urban Space', in M. Sorkin (ed.), *Variations on Theme Park the New American City and the End of Public Space* (New York: Hill and Wang, 1992).

72   R. Kallus and H. Law Yone, 'National Home/Personal Home: The Role of Public Housing in Shaping Space', *Theory and Criticism*, 16 (Spring) (2000), pp 157–85.

73   This view of the everyday is taken for example by Lefebvre, 1991 (cited in note 5) and de Certeau, 1984 (cited in note 6).

74   Nevertheless, as was declared by Ia'alon, the former Israeli Chief of Stuff, in a recent interview '… as much as the settlements in the occupied territories need the army, the army needs these settlements …' (*Ha'aretz*, Friday supplement, 30 August 2002).

75   Kol Ha'Am Tzava (the whole nation an army) is an expression often used in Israel during nation-building years and up to the present time, in various official propaganda, by national leaders and in popular songs.

PART III
MIXED SPACES – SEPARATED PLACES

Chapter 7

Urban Iconoclasm: The Case of the 'Mixed City' of Lod[1]

Haim Yacobi

Introduction

> The city has a symbolic dimension; monuments but also voids, squares and
> avenues, symbolyzing the cosmos, the world, society, or simply the state.
> (Lefebvre, 1996, p. 116)

The term 'mixed cities' is widely used in Israel, describing an urban situation
in which Jewish and Arab communities occupy the same urban jurisdiction.[2]
However, a critical examination questions this terminology that brings to mind
integration and mutual membership of society, while reality is controversial.
Similar to other cases of ethnic nationalism, a clear spatial and mental division
exists between Arabs and Jews in Israel,[3] and hence the occurrence of 'mixed'
spaces is both exceptional and involuntary. Rather, it has resulted from a
historical process during which the Israeli territory, including previously
Arab cities, has been profoundly Judaized. In this process, the Palestinian
community remaining in Israel following the 1948 war, has become a
marginalized and dispossessed minority. Beyond the significant effect of
the social and political processes, the ex-Palestinian urban fabric has been
dramatically transformed.

Following the above, in this article I will analyse critically the dynamics in
which the contested urban landscape in the 'mixed city' of Lod is produced,
transformed, and reproduced. My analysis will focus on: (i) the actual changes
of the built environment; and (ii) the contents and meanings embodied within
the planning discourse as expressed by planners, architects and policy makers.
Exposing the discursive meaning of the professional sphere, I will argue, is a
key to understanding the spatial dynamics of a city as well as its ideological
agenda, which are often neglected in the literature of ethnic urban relations.

I would also suggest that in these disputed cities the built environment
reflects both physical and symbolic transformations, and cannot be seen as

autonomous from the existing socio-political context. Rather, I would propose that the built environment in the Israeli 'mixed cities' accentuates the inherent nexus between the disciplines of planning and urban design and the realization of the modern Zionist utopia.

In relation to the empirical findings to be presented, I argue that similar to other studies of colonial urbanism (King, 1990; Mitchell, 1991; Celik, 2000), the production of physical and social division in the city of Lod has re-ordered perceptions of reality. This is expressed in an epistemological antinomy of 'here' and 'there'; 'we' and 'they'; 'enlightened new' and 'backward old' respectively. Following this existing body of literature, I would also claim that the roots of this division had already appeared during the British Mandate period in Palestine in general, and in the city of Lydda – the Arabic name of Lod – in particular.

This attitude was accentuated when the Israeli state was established, employing a dual mechanism that expressed both a tangible project of colonization of a settler society, and a symbolic construction of a collective national consciousness based on ethnic belonging. I would propose to define the physical act of this process as urban iconoclasm, which was rhetorically presented under a scientific planning approach, rationally and objectively concerned with the 'public interest'. Very often this approach demanded massive destruction of the existing Palestinian built environment, an act that was justified as being a hygienic necessity as well as a functional progress towards modernity.

However, a total replacement of the Arab urban landscape was impossible. This was a result of both political and demographic circumstances to be discussed in this paper, as well as from the development of a controversial approach towards the indigenous built environment. This will be presented using the orientalist discourse, which opened a new perspective for understanding culture as a product of social dominance (Said, 1978). It shows how the Jewish settler society had constructed its imagined sense of place while deforming the content and meaning of the local vernacular and transferring it into a subject of 'local' and 'authentic' but 'non-Arab' belonging.

In the first section of this paper I will present briefly the historical circumstances that transformed the Palestinian city of Lydda into the Israeli 'mixed city' of Lod. The second section will explore the beginning of the colonization of Lydda through planning and urban design, as had appeared during the British Mandate period in Palestine. The third part will examine the dramatic changes that had taken place post 1948, following the project of Judaization and modernization. The fourth section will aim to clarify whether the resulting shift in planning and architectural discourse was meaningful in relation to the contested sense of belonging in the 'mixed' urban setting. All four sections are based on discourse analysis[4] of a fieldwork findings taken from January 1999 to May 2002 which included documentation, data collection, archival research as well as interviews. Finally, in the light of the findings and the theoretical debates, I will discuss and critically formulate some conclusions.

The 'Mixing' of Lydda

Lod is located at the edge of the coastal plain of Israel and has developed around a junction of routes leading from west to east (Jaffa-Jerusalem) and from south to north (Egypt-Syria-Lebanon). There is evidences of intensive commercial activities in this area (*Cook's Tourist's Handbook*, 1876) and the first railway line to Lod was constructed as early as 1892. The British occupied the city in 1917, and invested widely in development, including the construction of a train station, the renovation and enlargement of the rail tracks and the establishment of an airport. In 1920 Lod was declared capital of its region (Vacart, 1977; Kadish et al., 2000).

In 1922 the British Mandate Department of Statistics reported 8,103 inhabitants, and in 1944 the Anglo-American committee counted 16,780 inhabitants (Vacart, 1977). Beyond the demographic and economic developments, some changes had occurred in the administrative and municipal levels since Ottoman rule. In 1934 a new law had been passed concerning municipal elections, and as a result some of the elite families had gained political positions in the city. These changes have brought about a spatial extension of the built area outside the borders of the old city, followed by new urban schemes initiated by the Mandate regime and designed by British planner Clifford Holliday, and later by Jewish architect Otto Polcheck.

1948 was a turning point in the history of Lod and other Palestinian cities and villages. Israeli armed forces occupied the city, which was to be part of an Arab State according to the 1947 UNGA. In 'operation Dani', initiated by the Israeli forces, 250 Palestinians were killed, and about 20,000 inhabitants escaped or were forced to leave the city (Morris, 2000; Kadish et al., 2000). Yet, the need for labor and specific professionals such as the railway workers, was one of the reasons for allowing 1030 Palestinians to remain in the city.[5] The establishment of the Israeli state and the 1948 war had created, indeed, a new reality in the city of Lod. As a first step, the Israeli Military Administration moved the remaining Palestinians to the Large Mosque and to the area around the church of St George, which was enclosed by a wire fence and named the 'Sakna'.[6]

The 'Sakna' was used as a means for constant surveillance through direct and indirect mechanisms of control over the remaining Palestinians that were perceived as enemies. The security forces were the main body that coordinated the relations between the Palestinians and the Israeli governance. These forces had total control over the Palestinians' conduct[7] in different aspects of their daily lives including their movements and their right to work.[8] In order to gain these rights, proper political behavior of the Palestinian individual towards the Israeli governance was necessary (Bishara, 1993). This viewpoint was expressed in Israeli public discourse: Prime Minister Ben-Gurion accused the Palestinians in Israel of supporting the surrounding Arab countries, and president Ben-Zvi claimed that the Palestinians aimed to complete Hitler's project (Benziman and Mansour, 1992, pp. 16–20).

The Military Administration regime had ended soon after the occupation of Lod, in the summer of 1949,[9] but there was still a wide agreement concerning the necessity to control the Palestinian population in the city. Thus, every aspect of this population's life was under surveillance including education, social services and spatial planning. The next period was characterized by intensive settlement of Jewish immigrants in the abandoned Arab houses of Lod;[10] some of these newcomers were Holocaust survivors.

Settling Jewish migrants in frontier cities and regions was part of a governmental scheme that was termed by McGarry (1998) as 'demographic engineering'. The massive expropriation of Palestinian land and houses, and their transformation into Jewish State property through legislation was one of the efficient means to implement this program (Kedar, 1998). In Lod, for instance, all properties and land were listed under the name of the Trustee of Absentee's Property and the Development Authorities, who financed renovation, subdivision and adjustment of the Arab houses, and rented them out very cheaply to the Jewish migrants.[11] Following McGarry, I would propose that this process reflects the social construction of both actors; the Arabs as enemies and the Jewish immigrants as agents:

> On one hand, state authorities move *agents*, that is groups which are intended to perform a function on behalf of the state. State agents are normally *settled*, that is made provision for, and they are normally moved to peripheral parts of the state occupied by minorities. On the other hand, the authorities move *enemies*, that is, groups, which in their present location pose a problem for

the authorities and an obstacle to their goals. 'Enemy' status is subjectively assigned by the authorities, and need not correspond with anti-state activity on the part of targeted groups. (McGarry, 1998, pp. 614–15)

Nevertheless, the 'enemies' in Lod as reported at that time, were a fragmented society that could not endanger the Jewish hegemony.[12] The Palestinians who remained under Israeli rule had become powerless, their urban culture had been undermined as well as their collective identity and leadership (Bishara, 1993; Zahalke, 1998).

In the first period after the war, Palestinian refugees tried to penetrate and resettle in their vacant houses in Lod. The authority's reaction included military acts against them as well as massive settlement of Jewish immigrants[13] – mainly Mizrahim.[14] Since the foundation of the Israeli state and up to 1949, 126,000 (66 per cent) of the 190,000 Jewish immigrants, who arrived in Israel, were settled in abandoned Palestinian houses in the 'mixed cities', including Lod (Morris, 2000, p. 263). Starting in the early 1950s – following a new master plan for Lod – the city had witnessed the massive construction of modernist housing blocks, infrastructure and public services for Jewish population while the Arab urban fabric became subject to intensive demolishing.

The above socio-spatial process demonstrates clearly a specific case of settler society; a model that refers to the colonial legacy in which European invaders immigrated to other territories and settled there, perceiving these places as 'terra nullius'. This model identifies social and spatial patterns that can be also recognized within the Israeli case, and three social groups can be schematically marked (Shafir, 1993; Stasiulis and Yuval-Davis, 1995; Yiftachel and Meir, 1998; Yiftachel and Kedar, 2000). The first is the founding charter group, which gains the dominant political, cultural and economic status during the first period of establishment. In Israel, this group is composed of mainly Ashkenazi Jews, the 'founders' of the state. The second group are waves of immigrants, 'homecomers', that follow the charter group, and are often ethnically different and fixed into an inferior social status. This group, in reference to Israel, are the Mizrahi Jews as well as those who immigrated in the nineties from the former Soviet Union. The indigenous people forming the third group are excluded from the process of constructing the new state, and members of this group are generally fixed into their inferior ethno-class status. In the Israeli context this group includes the Palestinian citizens that are discriminated against in different spheres of public life.

The importance of 'mixing' Lod's demographic composition – i.e. controlling the numerical ratio between Jews and Arabs – had been accentuated again in 1989 when a new flow of Jewish immigrants arrived to the city. This group came mainly from the former Soviet Union and they are currently form 25 per cent of the city's population (Lod Municipality Report, 1992), forming the second generation of 'agents'. Nonetheless, at this point it is important to note that not all new 'agents' in the city are Jewish. The mass immigration to Israel in the last decade has been characterized by having around 30 per cent non-Jewish newcomers. These immigrants were able to settle in Israel by virtue of the 'Law of Return'.[15] In this context Lustick (1999) argues that despite the contradiction between the Jewish nature of Israel and the non-Jewish immigrants-agents, this migration serves the goal of demographic engineering and hence the shaping of Israel as a 'non-Arab state'.

'Can One Name it a City?'

As already mentioned, the British occupied Lydda in 1917 and in 1920 they established a civil regime in the city. An aerial photograph taken in 1918 (see Figure 7.1) describes the spatial nature of Lydda's landscape when the British arrived. Most of the built area was concentrated in 'El Hara Esh Sherquiya' – the eastern quarter – that was surrounded by agricultural plots and olive groves. The built area, forming a triangular shape, was characterized by a dense fabric of one – and – two floor houses constructed of stone and surrounded by patios for domestic use. Similar to other traditional Arabic-Islamic cities shaped by religious law (Hakim, 1986) the commercial and manufacturing activities were attached to the dwelling environment. Nevertheless, a closer view shows that already in 1918 the city had expanded partly towards the 'Hara El Gharbiye' – the western quarter – and partly to 'Hara El Jnubie' – the southern quarter. These clusters formed an ordered structure with wider streets than the old city alleys. The new houses were one – and – two floors height and surrounded by walls and private gardens.

Some of the significant acts of the Mandate regime project of mapping Palestine as well as promoting western norms in regional and urban planning within both Jewish and Palestinian settlements. In this context, declaring Lydda as the capital regional city included massive investments in infrastructure, the establishment of the international and military airport as well as the improvement and changes in the Ottoman railway route. The latter had a significant effect on Lydda's urban landscape since 'Lydda Junction' – the new train station and its surroundings – was not merely a rail-track intersection in mandatory Palestine, but an attempt to realize colonial utopia. Lydda Junction

**Figure 7.1   An aerial photograph of Lydda, 1918. Source: The Hebrew
University Aerial Photographs Collection.**

is located two kilometers southwest of the core of the old part of the city,
isolated by a round circle drawn by the new railtracks. At first the families
of the British staff dwelled in huts, forming a camp-like layout, and in the
beginning of the 1940s the temporal structures were replaced by permanent
stone houses. The urban scheme was a typical example of British colonial
design, shaped by planning principles of the garden city such as health, light
and air, as well as by a set of social and aesthetic norms. Also, as noted in
the literature that deals with colonial urbanization (Chatterjee and Kenny,
1999; King, 1990; Rajagopalan, 2002) one of its central characteristics was

segregation on a racial basis. And indeed, mainly British workers and a few Jewish families dwelled in the new isolated quarter.

Moreover, the sketches made by the Drawing Office of the Palestine Railways Engineering Branch (see Figure 7.2) clearly represents a European vision of an architectural style: red tile roofs, brick chimneys as well as front and back gardens. In an interview with one of the oldest inhabitants of the neighborhood[16] I was also informed that the population had to follow a strict set of rules especially in relation to sanitation. The social life of the inhabitants was organized around the local club, but the recreation center and tennis courts were only accessible to British inhabitants. However, British influence did not end in the construction of the Lydda Junction district.

The earthquake of 1927 as well as population growth caused the Mandate authorities to initiate a new urban scheme for Lydda. The plan was defined as 'Earthquake Reconstruction Scheme' for constructing new housing and improving sanitation and infrastructure;[17] Clifford Holliday, a British town

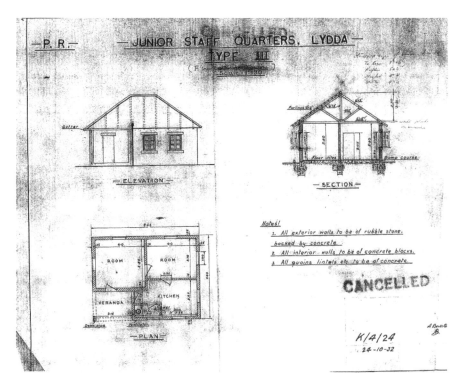

Figure 7.2   The new western design for Lydda Junction housing district. Source: The Israeli Rail Company Archive.

planner, was in charge of this project[18] that was approved in August 1929 (Hymann, 1994). Holliday had worked in Liverpool for Patrick Abercombie who had encouraged him to apply for a job as the Civic Advisor to the City of Jerusalem. Holliday was involved in urban planning in Palestine from 1922 to 1935 (Holliday, 1997; Erlich, 1984). When he returned to Britain he became preoccupied in preparing urban schemes for the colonies such as Colombo and Gibraltar, while his best known project was the design of Stevenage New Town in Britain (Holliday, 1950). This background is reflected in Holliday's scheme for Lydda as illustrated in Figure 7.3 and in his essay on town planning in Mandatory Palestine (1938). Both sources point to Holliday's attempt to re-shape Lydda's urban fabric according to western norms.[19]

An aerial photograph dated 1944 (Figure 7.4) presents the results of the implementation of Holliday's vision to the city. Mainly the southern and the northern quarters had been expanded, forming a pattern of ordered plots, blocks and roads, with the olive groves and agricultural landscape around them. Commercial buildings were constructed facing the main road as well as public buildings such as the schools at the edge of the southern quarter.

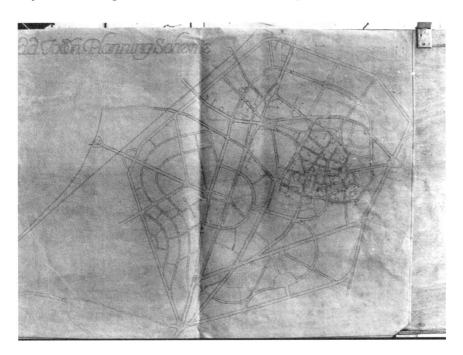

Figure 7.3   Holliday's Town Planning Scheme. Source: The Israeli Antiquities Authority Archive. In Hymann, 1994.

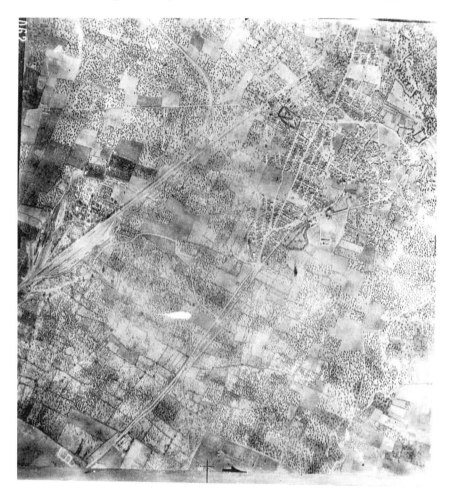

**Figure 7.4   An aerial photograph of Lydda, 1944. Source: The Hebrew
University Aerial Photographs Collection.**

The development of a road system is also significant, including a new road
within the old city fabric. This photograph also shows the growth of Lydda
Junction district. Despite the formal design of the city, Holliday advocated
the preservation of the ancient eastern quarter. This attitude is expressed in
his article 'Town Planning in Palestine', which criticizes the approach of the
British Mandate Planning Commission to cities such as Gaza and Hebron that
he thought were beautiful and romantic, and it might be better to leave them
'unplanned' (Holliday, 1938, p. 202).

In the beginning of the 1940s the local municipality, encouraged by the British officials to initiate a new plan, selecting this time a Jewish architect and town planner, Otto Polcheck, who had acquired his experience in town planning in Mandatory Palestine. The architect's voyage to Lydda gives an impression of the reality in Palestine at that time:

> … How did one arrive to Lod at that time? You had to go to Jaffa, there in King George Street there was a kiosk in front of the Anglo-Palestine Bank. There you took an Arab bus … It was in 1942 or 1943, I took the bus and I was afraid … at that time there were always riots. I had a special costume for such dangerous areas, I was dressed as a Christian pastor.[20]

Polcheck's description of Lydda during the Mandate period does not fully fit with the image of the city as appears in the aerial photograph from 1944:

> About the landscape of Lod … it was poor, quite poor; one-floor houses, maybe two floors. There was a hint of a public garden and it was very miserable … the only interesting buildings were the church and the mosque.[21]

I would propose that Polcheck's view – as an agent of a modernist-planning paradigm – was shaped and culturally constructed according to western norms:

> I studied in Czechoslovakia … I studied for one trimester in the Bauhaus School. Ernest Neufert was the bible for architects in the entire western world and he was my professor of architecture.

And indeed, when I mentioned the existence of the ancient part of the city, Polcheck's reaction was clear: 'Can one name it a city? It was not a city! It was just a local municipality.' Polcheck's definition of the urban entity contains a whole set of norms, conceptions and images that can be read in relation to the orientalist discourse. Said (1978) analyses the way in which Europeans have constructed an image of the oriental culture and people as less civilized, and thus demanding to be governed by others. This discourse is reflected in governmental texts as well as in literature constructed on dichotomies articulated into a 'natural' way of thinking, suggesting that western culture is superior, civilized and progressive while oriental culture is characterized by being barbarous and backward.

What was, then, the vision for Lydda and what was the purpose of the new plan? According to Polcheck the main expectation from his scheme was to create a town plan on the basis of which Lydda's inhabitants will be able to apply for building permits. It seems that the previous scheme was too general and thus the landowners in the city built 'wherever they wanted and the municipality was satisfied'. As the new town plan demonstrates (Figure 7.5) the concept was to make a dense core with a principal commercial road and to allocate land for other uses including less dense housing districts around the center such as industrial zones, parks and public open spaces. Beyond the area that was already built, Polcheck suggested to develop a 'garden town that was at that time in fashion'. Indeed, one of the issues raised here is the importance of newness; a theme that is a central concept of the colonial city which aims, according to King (1990), to transform the local population economically, socially, culturally and politically.

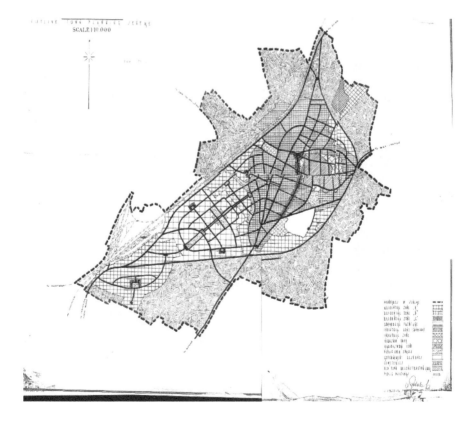

Figure 7.5   Polcheck's Town Planning Scheme. Source: Polcheck's
private archive.

Polcheck's approach was a product of the 'scientific' professional knowledge that had governed the planning discourses at that time. Formal education and conferences that became popular events of exchanging ideas had transferred this knowledge. One of the main issues that were discussed in these conferences was the professional position towards the ancient urban fabric as noted – for instance – by Professor Brown at the London Town Planning Conference who had:

> … [L]ittle sympathy for the iconoclastic city planner who yearned for a 'clean slate' that he might make design untrammeled by the past … Cities are not only made but grow … The growth is continued not only by physical but by human environment, and is closely dependent on history. To wipe out this history's evidence may be to take away more then the town planner can give. (*The Architectural Record*, 1910, p. 456)

This notion must be remembered while discussing the work of Holliday and Polcheck. On one hand the 'chaotic' texture of the city represented backwardness and underdevelopment on the part of the local population, and on the other hand it had picturesque qualities that were to be preserved:

> I did not want to touch the old city, I did not want to demolish it. I have kept it. We should have done that not only here but in Haifa and Jaffa as well, where we have destroyed too much.

Additionally, an important point that preoccupied Polcheck was that a garden city urban scheme demanded expropriation of private land for public use; a planning paradigm that had legal support from the Mandatory Planning and Construction Law in Palestine. However, the implementation of such an approach contradicted local perception of both the meaning of private property (Hakim, 1986) as well as the cultural need for garden city style open spaces. Hence, Polcheck's vision met the objections of private landowners as well as city council members who were, according to him, 'concerned mainly with their [building rights] in their own plots'. Here, we can recognize the inherent contradiction outlined by King (1990) in relation to colonial planning. Statutory control that represents the 'collective will of society' as well as assumptions concerning the cultural use of space (housing typologies and public open spaces for instance), cannot be applied to culturally different pre-capitalist societies, who become by definition non-democratically governed.

The scientification of the act of planning required definitions of rules and norms concerning the aesthetic nature of the urban:

> The city which is white has the greatest refinement and charm. Paris, of modern cities the most beautiful in the world, is a city of ivory studded with pearly gray in a setting of green. Regents Street, London, is painted in white and cream, and to this is entirely due its attractiveness to the fashionable throng ... All greater cities are either white or gray. (Prof. Adsheads, *The Architectural Record*, 1910, p. 142)

In 1945 when Polcheck's town planning scheme was finally approved, it included not only parceling regulations, but also architectural design standardization instructions that would 'whiten' the built landscape of the Arab city:

> ... The external walls of all houses including outbuildings and garages and all columns and piers shall be constructed of or faced with natural dressed stone.
> ... It shall be competent for the local commission to exercise full control over the design of any proposed building in all matters pertaining to appearance, choice of materials or manner of construction.
> ... The District Commission may require the Local Commission to prepare or may itself prepare an architectural design for any street or quarter. ...
> No corrugated or sheet iron shall be used externally on any building, part of a building, other than a door, shutter or similar fixture.[22]

Towards a 'Hebraic City'

> ... [Lod] has changed from a neglected and undeveloped Arab city into a city of 16,000 [Jewish] inhabitants ... Lod, with its clean streets and plantings and its organized management ... is a living example of the dynamic power of the Jewish people. (Lod municipal report, 1952)

Following the 1948 war Polcheck's vision of the garden city of Lydda could not be realized. A different ideology and power-relations had reshaped the urban meaning and form of Lydda, who was renamed after its biblical source: Lod. Simultaneously with the acts of expulsion of most of the Arab population and the concentration of the remaining Arabs in separated areas, there were massive acts of demolition of the built area by the Israeli army. This often occurred under direct governmental commands, and sometimes as a result of an army initiative. However, total demolition of the Palestinian city was not realistic. The flow of Jewish immigrants demanded a housing stock, and the intentions of some Palestinian refugees to penetrate back to their homes required an active reaction in the form of settling the 'agents' in those houses as defenders of the frontiers.

Indeed, similar to other frontier areas, the Judaization of Lod was a high priority project, and in 1950 only 9 per cent of the 12,100 inhabitants were Palestinians, while the rest were Jewish:

> The composition [of the Jewish inhabitants of Lod] is diverse and characterizes the kibbutz galuiot:[23] the origin of 50 per cent is from Poland, Romania and Bulgaria and the origin of the other 50 per cent is Morocco, Tunisia, Turkey and Iraq. However, seven years after it has been established as an independent Hebraic city, it is still far from being an integrated urban entity that has taken off its diasporic dress. The different [Jewish] communities preserve their customs and manners, a fact that prevents healthy and desirable development.[24]

This critic explores an essential stratum within Zionist discourse, which aims to create a modern and aggregated society in an old-new space – the 'Hebraic city'. Furthermore, it uses the indigenous architecture that remained in the Sakna as a symbol not only of the intimidating enemy landscape, but also the diasporic Mizrahi origins of the Jewish migrants:

> While one walks in the Sakna ghetto and watches the degenerated life style of its inhabitants, who live in dark houses and cellars without basic sanitation … one has the feeling that nothing had changed in the life of these people that were transferred from the dark known ghettoes of Morocco.

Paradoxically, the 'agents', mainly the oriental Jews, that were implementing the project of Judaization, were marked as the cause for the backwardness of the city. The migration policy of the Israeli authorities had 'compressed' into the city welfare and social problems, as defined by Zvi Itzkovitch, the mayor of Lod 20 years later:

> Everyone [of the Jewish migrants] occupies a house from the abandoned property… Instead of searching for solutions to rehabilitate the Arab city that was neglected, such as erecting new factories, the Sochnut [the Jewish Agency] and the government compress in the city welfare cases from the entire country.[25]

What urban landscape should be created in the city of Lod in order to replace the Arab, diasporic and backward built environment? The response for this question was clear at the time, and it had to do with the project of modernization that represented the ideological denial of the oriental past and present. In a deterministic way the Israeli public discourse tied the characteristics of the 'dark ghetto' to social maladies that could be healed if the physical conditions would improve:

> ... [T]hose who live in the new housing blocks will not live peacefully with the ghetto inhabitants ... Let's realize the experts' and engineers' decision concerning the necessity to demolish all this area in order to clean the social dirt.[26]

The 'housing blocks' mentioned above were an outcome of the first Israeli master plan initiated in 1954 and approved in 1958. The architect, Michael Bar, advocated modernistic planning principles such as zoning, public open spaces and efficient transportation systems, demonstrating total estrangement to the existing Arab landscape (Figure 7.6): 'There is a necessity to prepare a detailed urban scheme without any relation to the existing buildings' (Lod Municipality Publication, 1952). Bar's modernistic attitude was compatible with the ideology of the Jewish settler society. While Polchek's orientalist view advocated the reconstruction of the old city of Lydda, Bar's ambition was to transform Lod's Arab urban fabric in the name of modernization:

> We should gradually transfer the inhabitants of the ruins [the old city] to new housing zones. The vacant area we should use partly for new planned housing districts and mostly for construction of public buildings and public open spaces for the needs of the population that lives around it. (Lod Municipality Publication, 1952)

In this context I will mention Sandercock's claim that modernist planners were the 'thieves of memory'. Modernist planners, she suggests, have been 'embracing the ideology of development as progress' and by doing so they 'have killed whole communities, by evicting them, demolishing their houses' (Sandercock, 1998, p. 208). In the case of Lod this argument is valid, though Sandercock's focus on the project of modernization is partial, since modernization is not an independent concept; most certainly, it is strongly tied with other ideologies. As illustrated above, similar to other colonial practices, the act of planning was justified as being more civilized than the indigenous population's use of space.

Urban iconoclasm was not presented as an act of aggression, but as an expression of modernization. Hence, the modernist interpretation of the city and its inhabitants was used as a moral and ethical foundation to make them

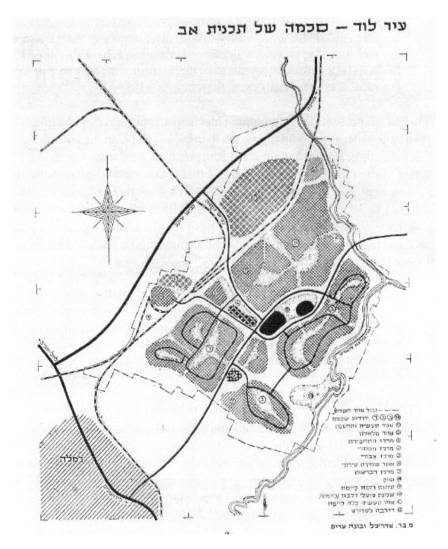

עיר לוד – סכמה של תכנית אב

Figure 7.6   Bar's Town Planning Scheme. Source: Bar, 1953.

subject to surveillance and control. For without constructing the 'other' as a less civilized enemy, there would have been no justification to control them. This approach characterized the first decade of the establishment of Israeli Lod, and was again repeated 20 years later:

> We have received the city abandoned, underdeveloped and dirty, with no electricity, sewerage system and running water, roads and sidewalks. The

> streets, especially the main street, were miserable alleys. In order to build a city, excellent city-builders are necessary. In order to rehabilitate an underdeveloped city there is a need for good will, energy, initiative and above all – money … Roads and sidewalks were paved all over the city, neglected places became well-groomed, deteriorated squares are flowering. (Zvi Itzkovitch, Lod's mayor)[27]

This citation presents the overlapping mechanisms that have produced Lod's urban landscape; a modernist approach to hygiene and planning, and a concept of Jewish ethnicity that had replaced filth and backwardness.

Bar's urban scheme took into account that some of the religious buildings in the ancient part of the city will remain. However, their original function, that used to be part of the communal Palestinian city life, was no longer valid in the planner's eyes. Here, urban design had a special role in confining space and memory into the limitations of the hegemonic Jewish, western and modern frameworks:

> The religious institutions and the special buildings will remain within the [new planned] parks, and if they will be vacant we will use them for museums and exhibitions as well as an artists' hostel where artists will find the special atmosphere inspiring for their creations. We might consider as a memory of the past the preservation of some alleys that are characterized by special buildings.[28]

Until the beginning of the seventies most of the Arab city had been demolished, leaving a void named 'Park Ha-Shalom', the Hebrew word for Peace Park. The planners – these are the 'physicians of space' (Lefebvre, 1996, p. 99) – followed their ideological commitment of healing the oriental space. Indeed, urban iconoclasm was used both as a means and as an end in itself. In the name of modernization and newness, it enabled the physical colonization of Lod, while symbolically transforming Lod's sense of place.

Urban iconoclasm in Lod had a significant role in implementing the project of demographic engineering. However, the Palestinian population in Lod has nonetheless increased from 9 per cent in 1950 to over 20 per cent in the year 2000, while the Jewish population decreased from 91 to less than 80 per cent respectively.[29] The authorities, both on local and on national levels see these changes as a problem that endangers the Judaization of Lod, as stated by Mayor Maxim Levi:

> In relation to the special demographic characteristics of the city … [it is] appropriate to consider unconventional solutions, and to act towards the

dispersal of groups out of the city as well as to entirely prevent the continuation of illegal invasion in the future. The problem of the Arab population is difficult and urgent, it demands an overall fundamental and immediate solution …[30]

Among other strategies designed to solve this 'problem' the authorities began some intensive negotiations with families who lived in the Arab district at the city center. The purpose of this was to evacuate the existing houses and to rebuild modern, dense housing projects by the private sector. The Arab population was expected to move to new housing projects out of the city center or even out of Lod. Since 1988 a massive act of evacuation began in Lod's city center, and the empty lots, which remain, are still waiting for developers.

'Children Come Again to Border'

While Lod underwent massive urban transformations, new voices began to be heard; voices that called for preserving Lod's architectural heritage. The local municipality, who organized a conference that dealt with this issue, accepted this apparent shift towards the historical importance of remnants: 'The participants of the conference expressed their impressions by stating that the historical sites in Lod are more beautiful than those they had visited abroad.'[31] Preserving Lod's indigenous built environment was seen, first and foremost, as a source for tourist attractions. This approach was mentioned before in reference to planner Michael Bar and was repeated in this conference forty years later: '… [W]e could restore this area, and it can easily compete with old Jaffa: with museum, galleries, tourists shops and art centers.'[32]

Boyer (1996) criticizes this approach stating that it represents a point of view of white, middle-class professionals who focus on depoliticizing the act of planning. This attitude that evaluates cities according to their location in the global restructuring of capital, dominates present-day urban design. However, in relation to the apparent shift towards the Arab landscape of Lod, I would propose that its commercialization must be viewed within the context of the architectural discourse in Israel during the 1960s, that was dominated by a yearning to define 'deep-rottenness' and a national local regionalism. According to Nizan-Shiftan (2000), this was a dialectical process that adopted the indigenous landscape as an object of inspiration on one hand, and excluded the 'intimidating other' on the other. Hence, my argument is that the apparent shift towards advocating for preserving the Arab urban texture that had remained in the city of Lod, is but another form of hegemonic dominance. Behind it lies the domestication of landscape, time and space as a means for constructing the hegemonic collective sense of 'primordial' belonging, referring back to an 'authentic being'. This is a central ideal of settler societies in general, and particularly of the Israeli case that emphasizes the concept of return of the Jewish people to their homeland.

This raises the question to what extent is such a conceptual discussion reflected in the built environment. Based on Bourdieu's concept of habitus, Dovey (1999) has theorized the role of architecture and urban design as an act of 'framing place'. According to him routine everyday activities in space are framed by planning, not only in an obvious sense but on a discursive level as well. The framing of places is indeed the very act of producing/reproducing meaning to a certain urban context, articulating hegemonic dominance which in turn penetrates into the 'obvious' social order and provides more ideological power to the built environment. This epistemological process requires conscripting collective imagined historical, religious and cultural interpretations that become 'facts', backed up by professional disciplines that are perceived as objective and rational such as history and archeology.

In 1990 Sa'adia Mandel, one of the leading architects in urban conservation in Israel, was hired by the Israeli Committee of Site Preservation to prepare a Conservation Master Plan for Lod. The main focus of his plan was the potential for tourism:

> The municipality of Lod aims to return the city to the route of tourism, and to present to visitors from Israel and abroad the historical sites of the city. These sites demand rehabilitation that will rediscover and transform them into attractive tourists sites (Preservation Master Plan, 1991).

However, this task required re-reading the city's history and reshaping memory as noted by the architect himself:

> ... [A]t the end of the process I had reached some conclusions ... I told Maxim Levi, the mayor, that this was conservation and tourism as well as an educational project. There are people in Lod aged of 50, and they have the right to know whether Lod was occupied or liberated ... I am not using such words 'occupation' and 'liberation' fortuitously. In this story I have a clear position. I know that Emek Izrael was occupied. The Zionists bought the land there from the Arab landowners and settled there – this is an elegant occupation if you want, I am not saying that it was fair, but elegant is the most innocent term I can use. Then, the war of liberation broke out and Lod was

occupied – this was less elegant … My thesis is that in Lod there is space for both them and us.[33]

Mandel's work was published in a detailed report presenting the history of the city chronologically, and showing each period and its central monuments that should be preserved. It is a voyage in time goes back to the Neolithic and Iron ages, the Roman, Byzantine and Early Arab periods, the Crusader, Mamluk, Ottoman and British Mandate time, and finally the period from 1948 up to the present. This document describes the characteristics and monuments of each period, though its subtext deals with a reconstruction of the historical narrative of the city. Two significant examples illustrate this. While describing the transition from the Byzantine period to the Early Arab period, the text mentions the process of urban iconoclasm that had characterized the Muslim occupation stating that the city 'was demolished by Muslims, and the stones were taken for the construction of the city of Ramla'. However, while relating to the period from 1948 onwards, the almost total destruction of Lydda is ignored, and 'rational' circumstances explain its Judaization after the 'liberation' of the city was completed:

> On July 11th 1948 Lod was occupied by the Israeli Defense Army as part of 'Operation Dani'. Its Hebraic name – Lod – was restored. New home-comers populated the abandoned dwellings, thus using efficiently the existing stock of housing.

Obviously, framing Lod's sense of place is an ideological task. The efforts to domesticate the city's memory and monuments serve the construction of an Israeli collective identity within the context of the Zionist narrative, claiming that the Jewish people have returned to 'a land without people' waiting for a 'people without land'. However, since Lod is a 'mixed city' of both Jewish and Arab citizens, relying on the concept of terra nullius is problematic; despite its iconoclastic past the city still contains both Arab architectural characteristics and Arab population. Mandel is conscious of this and though he declares his commitment to Zionism, he tries to give a wider meaning to the Master Plan:

> When I mentioned the oil-press building of the Hassona family I was asked to hush this matter up. I said that I want the Hassona family to operate their ancient oil-press. There is no reason why they shouldn't … Every Arab building we found had been destroyed, the Jews have deceived themselves – we thought that we could erase … but everything is still alive and I have tried to express this in the Master Plan.[34]

Yet, as noted by Rabinovitch (1993) any reference to the non-Jewish legacy and cultural heritage are conceived as problematic for the Jewish majority, for recognizing the 'other past' may expose the competitive identity to the Zionist one. Indeed, Mandel's plan has never been realized, and the Arab remains in the city were left to disappear 'naturally'.

One of the only projects that were realized in preserving Lod's urban fabric is the conservation of an Arab building in the city center. This building was renovated in order to serve as the offices of the mayor of Lod. Until then a modernist building housed the offices as well as the representative functions of the municipality. Nevertheless, the apparent shift towards the Arab landscape was translated into a symbolic transfer of the city leaders into an ancient house, situated at the end of a visual axis. The house, which was once part of an urban texture, had become a displaced 'monument'. Its typical symmetry, the hierarchical movement in space from private to public as well as its ceremonial entrance have been deformed. A decorated sign quoting a verse from Jeremiah (31:17) greets those who approach the building: 'Children come again to border'.

Who are the children to come again to border? It seems that the offices of Lod's mayor come within an approach discussed by the orientalist discourse in relation to the international exhibitions (Mitchell, 1996; Crinson, 1999). In these displays the artificial objects were supposed to create an impression of order, authenticity and scientific chronology; this in turn suggested a certain organization of knowledge. In the Conservation Master Plan Lod's 'historical collection' suggested by the architect was a succession of anecdotes, organized so as to present contemporary reality as a linear and natural continuation of the past. This point, in the context of Lod that was a Palestinian city, is very central for understanding the reconstruction of the historical narrative that presents a linear historical continuity of Jewish settlement in the city, as stated by the Mayor at the time of preparing the plan:

> During hundreds of years, in different époques since the Canaanite period ... Lod was the only place in the world in which a Hebraic settlement existed.[35]

Discussion

The struggle over the city's meaning and form has not ended in Lod. To a visitor coming to Lod today, the city looks unordered with many demolished houses and abandoned vacant lots hinting on a different past. In this paper I have followed the ways in which this urban landscape has been produced, reordered and transformed. Theoretically this analysis shows that the built environment is not entirely a representation of a given social context, but rather urban spatial forms, in turn, organize and construct much of our social and cultural existence, framing 'our' sense of place. Furthermore, from the presented findings I conclude that the very act of planning, in its discursive and practical dimensions, has a significant role in 'rationalizing' power relations.

Indeed, the production of urban space is an ongoing process that involves hegemonic ideology and socio-political relations, while in the case of the 'mixed city' of Lod these are based on two interrelated processes; Judaization on one hand and de-Arabization on the other. This ethnic logic requires 'scientific' disciplines to contribute to the 'struggle over geography'. This struggle, as noted by Said, is complex as 'it is not only about soldiers and cannons, but also about ideas, about forms, about images and imaginings' (Said, 1994, p.6). In this context the contribution of the spatial practices such as architecture, urban design and planning is significant; on one hand they codify within their professional discourse the ideologies they serve, while on the other they transform the built environment, urban form and space. Their power, indeed, is in their 'unquestionable' effect of framing daily experience and accumulating meaning, which can be decoded through analysing the discourses in which they are produced.

Exposing the discursive level of the spatial dynamics in Lod has leaded me to link the urbanization of the Israeli mixed city with the colonial legacy. This link offeres a different understanding of the spatial changes that have occurred in Lod which are very often described in the literature as an evolutionary urban process that responded to the need of housing the Jewish settlers (Ephrat, 1976; Vacart, 1977). These descriptions ignore the contested cultural, political and symbolic meaning of space.

Similar to other processes of urban colonization, the planning of Mandatory Lod demonstrates two strategies that had been implemented. The first was isolation, as in the railway district that was built separated from but close to the Arab city, and the second was the introduction of new western urban principles and regulations as Holliday's and Polcheck's schemes show. However, the old city, though underdeveloped, was seen as a picturesque object, reflecting the exotic distanced image of the oriental indigenous. This attitude was largely continued from 1948 onwards, yet the colonial tone was covered by the Jewish settler society nationalism that in addition to the principles of planning mentioned above introduced the concept of urban iconoclasm. As expressed in Bar's urban scheme urban iconoclasm had been 'rationalized' and in the name of westernization and modernization enabled the destruction of the existing landscape, transforming the Palestinian town into a 'Hebraic city'. Through this process, the Arab urban landscape became a signifier of the intimidating enemy as well as a symbol of the diasporic-Mizrahi culture. The conjunction of the modernist planning paradigm with the Zionist project has transformed Lod tangibly and symbolically, assuring the complete dominance of the ethnic logic; this in turn has paved the road to a post-modern shift towards preservation in architectural and urban design discourse. But as the case study demonstrates this was an apparent shift; its aim was to reproduce dominance through the domestication of time, space and memory.

Notes

1   I am grateful for the PARC (Palestinian-American Research Center) for its financial support of my project. A version of this article was published in: *Geografiska annaler*, 84 B (3–4), pp. 171–87.
2   Three main types of 'mixed cities' can be identified in Israel: a) pre-1948: cities such as Haifa where Jews and Arabs lived under the same municipality prior to 1948; b) Judaized: Palestinian cities prior to 1948, such as Ramla, Acre, Yaffa and Lydda, which became dominated by a Jewish majority; c) recently mixed: Jewish-Israeli new towns accommodating Arab migration such as Upper Nazareth and Carmiel.
3   At this early stage I should clarify my use of terminology, without entering the controversy over terms, which is beyond the scope of this paper. 'Arab' and 'Palestinian' are interchangeable terms in the paper, denoting residents of Israel/Palestine who belong to the Arab culture. There is a political distinction between Arab citizens of Israel and Arabs residing in the Occupied Territories, but the ethno-national identity of both is Palestinian-Arab. 'Israel' is the area within the internationally recognized pre-1967 borders. 'Israel/Palestine' denotes the entire area under present Israeli control (between the Jordan River and the Mediterranean Sea). 'Ashkenazi' Jews (Ashkenazim in plural) are those originating from Europe or America, while Mizrahi Jews (Mizrahim in plural) come from the Arab and Muslim countries. 'Russian' Jews are immigrants of Russian culture who have arrived in Israel since the early 1990s.
4   From a methodological perspective the interpretation and analysis of the texts and documents in this article are based on vast literature that deals with discourse analysis in general and in relation to the built environment and planning in particular. See: Markus and Cameron, 2002; Fairclough, 1992; 1995; Scollon, 1998; Dovey, 1999; Finnegan, 1998, Hastings, 1999.
5   Military Administration Report, 10 October 1948, IDF archive, 1860/50–31.
6   Israeli State archive 2401/21.
7   The Military Administration controlled the Arab areas within the borders of Israel after the 1948 war.

8   IDF archive, 1860/50-31; 1860/50–32.
9   'Denial of the Military Administration in Yaffa, and Ramlah-Lod', 23 June 1949, IDF
      archive, 1860/50–31.
10   Ben-Gurion archive, 11075–21/4/49; Military Administration Report, 2 June 1948, 23
      December 1948, IDF archive, 1860/50–31.
11   Israeli State archive, 2401–B4; 2401–B22.
12   Military Administration Report, 10 October 1948, IDF archive, 1860/50–31.
13   Military Administrator Reports, 23 December 1948, 28 December 1948, 11.1.49, IDFA,
      1860/50–31.
14   In 1969, for instance, it is reported that there were 50 per cent Jewish immigrants from
      North Africa, 18 per cent from other Middle Eastern countries, 24 per cent from Europe
      and 8 per cent Arabs (Hashimshoni, 1969).
15   The Law of Return (1950, 1954, and 1970) defines the right of the Jewish people to return
      to Israel. For a wider discussion of the social implications of this immigration see: Lustick
      1999.
16   Interview with Abu-Taufik 25 March 2001. Following this interview as well as some other
      conversations with old inhabitants in the city, it seems as some Arab-Christian families
      dwelled in Lydda Junction district as well.
17   Palestine Department of Health, 1929. In Hymann (1994, pp. 476–7).
18   'Times' Journal, 29 September 1960, RIBA archive, London.
19   Israeli State archive, Z/TP/47/39–262.
20   Interview with Otto Polcheck, 18 May 2001.
21   All the quotations of Otto Polcheck are from an interview with him, 18 May 2001.
22   Israeli State archive, Z/TP/14/48–2647; Z/TP/12/47–2237.
23   Kibbutz Galuiot is an essential concept in Zionist discourse. Its meaning is the melting
      pot of diasporic Jewish immigrants from different countries of origin.
24   Galili, 'Al Hamishmar' newspaper, 17 May 1954.
25   Lod Municipality archive, 1972.
26   Galili, Al Hamishmar newspaper, 17 May 1954.
27   Lod Municipality archive, 1972.
28   Lod Municipality Publication, 1952.
29   Statistical Abstract of Israel 1999; Census of Population and Housing Publications
      1995.
30   Lod Municipality and the Ministry of Housing and Construction Report 1987.
31   *Moked* – local newspaper 30 October 1992.
32   *Alternativa* – local newspaper, 9 October 1992.
33   Interview with Sa'adia Mandel, 20 September 2000.
34   Interview with Sa'adia Mandel, 20 September 2000.
35   Preservation Master Plan 1991.

References

Architectural Record (New York: The Architectural Record Company, 1910).

Benziman, U. and Mansour, A., *Subtenants* (Jerusalem: Keter, 1992), (Hebrew).

Bishara, A., 'On the Question of the Palestinian Minority in Israel', *Theory and Criticisms*, 3
      (1993), pp. 7–35.

Boyer, M.C., *The City of Collective Memory* (Cambridge, MA: The MIT Press, 1996).

Celik, Z., 'Colonialism, Orientalism and the Canon', in I. Borden and J. Rendell (eds), *Intersections – Architectural Histories and Critical Theories* (London and New York: Routledge, 2000), pp. 161–9.

Central Bureau of Statistics, *Statistical Abstract of Israel*, No. 50 (1999) (Hebrew).

Chatterjee, S., and Kenny J.T., 'Creating a New Capital: Colonial Discourse and the Decolonization of Delhi', *Historical Geography*, 27 (1999), pp. 73–98.

Cook's Tourists' Handbook for Palestine and Syria (Ludgate Circus, EC: Thomas Cook and Son, 1876).

Crinson, M., *Empire Building – Orientalism and Victorian Architecture* (London and New York: Routledge, 1996).

Dovey, K., *Framing Places – Mediating Power in Built Form* (London and New York: Routledge, 1999).

Efrat, E., *Cities and Urbanization in Israel* (Tel Aviv: Achiasaf, 1976) (Hebrew).

Erlich, A., 'The British Architects in Mandatory Eretz-Israel', *Tvai*, 22 (1984), pp. 48–50 (Hebrew).

Fairclough, N., *Discourse and Social Change* (Cambridge: Polity Press, 1992).

Fairclough, N., *Media Discourse* (London: Edward Arnold, 1995).

Finnegan, R., *Tales of the City – A Study of Narrative and Urban Life* (Cambridge: Cambridge University Press, 1998).

Hakim, S.B., *Arabic-Islamic Cities – Building and Planning Principles* (London, New York, Sydney and Henley: KPI).

Hashimshoni, Z., *Lod – The Old City Census* (The Evacuation and Construction Authority, 1969) (Hebrew).

Hastings, A., 'Discourse and Urban Change: Introduction to the Special Issue', *Urban Studies*, 36, 1 (1999), pp. 7–12.

Holliday, C., 'Town Planning in Palestine', *Journal of The Town Planning Institute*, London (1938), pp. 202–3.

Holliday, C., 'The New Towns', *Journal of the Town Planning Institute*, London (1950), pp. 180–82.

Holliday, E., *Letters from Jerusalem During the Palestinian Mandate*, ed. J. Holliday (London, New York: Radcliff Press, 1997).

Hymann, B., 'British Planners in Palestine' a thesis submitted for PhD (London: LSE, 1994).

Kadish, A. Sela, A. and Golan, A., *The Occupation of Lydda, July 1948* (Israel: Ministry of Defense, 2000 (Hebrew).

Kedar, S., 'Majority Time, Minority Time: Land, Nation and the Law of Adverse Possession in Israel', *Iyunay Mishpat*, 21 (3) (1998), pp. 746–65 (Hebrew).

King, D.A., *Urbanism, Colonialism and the World Economy – Cultural and Spatial Foundations of the World Urban System* (London and New York: Routledge, 1990).

Lefebvre, H., *The Production of Space* (Oxford: Blackwell, 1991).

Lefebvre, H., *Writings on Cities* (London: Blackwell, 1996).

Lod Municipality (1991), *Preservation Master Plan* (Hebrew).

Lod Municipality and the Ministry of Housing and Construction Report (1987), (Hebrew).

Lod Municipality Report (1952), *Lod – Three Years of Municipal Regime* (Hebrew).

Lod Municipality Report (2000) (Hebrew).

Lustick, I., 'Israel As A Non-Arab State: The Political Implications of Mass Immigration of Non-Jews', *Middle East Journal*, 53, 3 (1999), pp. 416–33.

Marcuse, P., 'Not Chaos, but Walls: Postmodernism and the Partitioned City', in S. Watson and K. Gibson (eds), *Postmodern Cities and Space* (Oxford: Blackwell, 1995), pp. 187–98.

Markus, T. and Cameron, D., *The Words between the Spaces – Buildings and Language* (London and New York: Routledge, 2002).

McGarry, J., 'Demographic Engineering: The State Directed Movement of Ethnic Groups as a Technique of Conflict Regulation', *Ethnic and Racial Studies*, 21 (4) (1998), pp. 613–38.

Ministry of Construction and Housing (2000), *Preliminary National Report Habitat.*

Mitchell, T., *Colonising Egypt* (Cambridge: Cambridge University Press, 1998).

Morris, B., *The Birth of the Palestinian Refugee Problem, 1947–1949* (Tel Aviv: Am Oved, 2000) (Hebrew).

Nitzan-Shiftan, A., 'Whitened Houses', *Theory and Criticism*, 16 (2000), pp. 227–32 (Hebrew).

Rabinivitch, D., 'Oriental Nostalgia: How the Palestinians Became Israeli-Arabs', *Theory and Criticism*, 4 (1993), pp. 141–51 (Hebrew).

Rajagopalan, M., 'Dismembered Geographies: The Politics of Segregation in Three Mixed Cities in Israel', *Traditional Dwellings and Settlements Review*, 13, 11 (2002), pp. 35–48.

Said, E., *Orientalism – Western Conceptions of the Orient* (New York: Penguin Books, 1978 [1995]).

Sandercock, L., *Towards Cosmopolis* (Chichester: Wiley, 1998).

Scollon, R., *Mediated Discourse* (New York: Addison Wesley Longman Inc., 1998).

Shafir, G., 'Land, Labor and Population in the Zionist Colonization', in U. Ram (ed.), *The Israeli Society – A Critical View* (Tel Aviv: Brerot, 1993), pp. 104–19 (Hebrew).

Stasiulis, D. and Yuval-Davis, N., 'Introduction: Beyond Dichotomies – Gender, Race, Ethnicity and Class in Settler Societies', in D. Stasiulis and N. Yuval-Davis (eds). *Unsettling Settler Societies* (London: Sage Publications, 1995), pp. 1–38.

Vacart, E., *Lod – A Geographical History* (Chericover, Lod, 1977) (Hebrew).

Yiftachel, O. and Kedar, S., 'Landed Power: The Making of the Israeli Land Regime', *Theory and Criticisms*, 16 (2000), pp. 67–100 (Hebrew).

Yiftachel, O. and Meir, A., 'Frontiers, Peripheries, and Ethnic Relations in Israel: An Introduction', in O. Yiftachel and A. Meir (eds) *Ethnic Frontiers and Peripheries: Landscapes of Development and Inequality in Israel* (New York: Westview Press, 1998), pp. 1–11.

Chapter 8

Planning to Conquer: Modernity and its Antinomies in the 'New-Old Jaffa'

Mark LeVine

Introduction

This article explores the role of architecture and town-planning in the struggles of territory and identity in the city turned neighborhood of Jaffa. While it was often overshadowed by its younger and larger neighbor to the north, Tel Aviv – the 'most modern city in the Middle East' – Jaffa was in fact the pre-1948 economic and cultural capital of Arab Palestine, home to upwards of seventy thousand Palestinian Arab inhabitants and thirty thousand Jews by the late 1940s. It was also a borderland, or frontier, between the two competing national movements, and thus a pivotal space for the conflicted interactions between them.

The constant interactions between Jews and Palestinian Arabs often muddied the nationally determined boundaries between the two towns, and through this the two nations; because of this the Tel Aviv municipal and Zionist leadership needed (from the start) to use every method at their disposal for creating and enforcing separation between Jews and Palestinian Arabs. Modernist architectural and town-planning discourses, which themselves evolved in good measure in colonial settings (where separation of the 'races' was a primary concern), proved to be indispensable tools in generating and sustaining a narrative of modernity and progress versus tradition and stagnation; a depiction that reflected the larger dual society paradigm that was central to the enactment of Zionist ideology and politics. Indeed, they would create a 'discursive erasure' of the territory and history of Jaffa and its Palestinian Arab residents that would make possible the literal erasure of ninety per cent of the pre-1948 population during the war.

The success of these discourses contributed to their continued saliency within the space of Jaffa in the post-1948 period, particularly beginning in the 1980s when the neighborhood became the site of gentrification increasing competition for housing between Jewish and Palestinian citizens – that is a renewed conflict over land, only this time ostensibly for reasons of the market, not of state. But before we understand how this occurred we need to discuss in a bit more detail the pre-1948 situation in the Jaffa-Tel Aviv region.

As I have elsewhere examined in great detail, the space of Jaffa-Tel Aviv constituted a particularly powerful site for the unfolding of modernity in Palestine.[1] Specifically, we can see the Jaffa-Tel Aviv region as a primary generator of the boundaries established by a mutually constitutive fourfold matrix of discourses composed of modernity, colonialism, capitalism and nationalism, and the numerous binaries they create and sustain. Together they constituted an extremely potent force that when deployed by the leaders of Tel Aviv and the larger Zionist movement made possible the 'overthrowing' of the existing geography of the region.

Modernity – specifically, European modernity – possessed a penetrating, almost pulverizing power to reshape the landscapes of Palestine, and Jaffa in particular. Yet however ultimately harmful was this Euro-modernity, the six decades leading up to the catastrophe/miracle of 1948 witnessed a much more ambivalent, if conflictual interaction between multiple modernities inspired and directed by different agendas and powers. At the same time that Tel Aviv was being established the fourfold modernity matrix would produce two powerful and contradictory discourses and landscapes of spatial production – the first a hierarchized, stratified and planned space generated by modernist/colonial ideologies of planning, the second a lived, often subaltern space generated by (post)modern Einsteinian physics, through which the concretized, hierarchized modernist spaces could be shattered.[2]

The former discourse makes possible, as a French planner of the Hausmannian period described it, a 'cleansing the large cities' that was a necessary and sometimes sufficient condition for the cleansing of the 'whole country'.[3] Such *tabulae rasa* were difficult to create in Europe, but were much easier in the 'backward' colonies, which were deemed ready for the radical (re)inscription of modern planning and development.[4] The colonial modernity that emerged in part out of this developmentalist impetus was doubly asymmetrical, designating both a break in the regular passage of time (i.e., modern versus traditional) and as important, 'a combat in which there are victors and vanquished … Other cultures "bec[o]me premodern by contrast"'.[5]

Once the colonized space was vanquished and cleansed, the plan for and founding of a new colonial (and often capital) city would be 'a civilizing event … giv[ing] form and identity to an uncivilized geography'.[6] Central to this process was a logic of 'creative destruction' whose power was especially strong in the space of the urban.[7] If the leaders of Tel Aviv believed that it was from the overthrow of geography that Tel Aviv came into the world – as the quote beginning this chapter/article suggests – then it was through the process of creative destruction that this action was realized.[8]

Thus Zionism, as a modern nationalist movement, is an inherently colonial discourse, and Tel Aviv, the 'modern capital' of Zionist Palestine and now globalized Israel, cannot be understood or examined other than as a colonial city. Once we begin to expose the cracks of this fourfold matrix of modernity-colonialism-capitalism-nationalism, the rich history of Jaffa and its struggle with modernity – long buried under the debris of Tel Aviv's perpetual 'growth' and (today) gentrification – begins to emerge.

The Discourses of Architecture and Town-planning in Mandate Palestine

To accomplish this I shall in turn discuss the architectural and planning histories of the Jaffa-Tel Aviv region in the late Ottoman and Mandate periods. To begin with, it is clear modern architecture was a crucial symbol for the successful implantation of Zionism in the soil of Palestine. As Herzl put it in *The Jewish State*, 'We shall not dwell in mud huts; we shall build new more beautiful and more modern houses', while one of the founders of Tel Aviv explained that while they didn't have money, 'a plan we do have …'.[9]

Yet as Gendolyn Wright has saliently observed, even the most aesthetic designs have political implications,[10] and in the case of Jaffa and especially Tel Aviv, the evolution of styles from garden suburb through International Style would reflect an increasing focus on defining 'modern' Tel Aviv against its (apparently non-modern) Other – although the reality is that Jaffa saw the same architectural developments as its daughter to the north. The core of the turn-of-the-twentieth century Zionist architectural and planning imagination reflected the view of these movements at large, which increasingly leaned towards creating a 'clean slate' in the (re)design of modern cities.

Specifically with regard to the Jaffa-Tel Aviv region, Meir Dizengoff, Tel Aviv's first and longest-serving mayor, decried the Ottoman governance of Jaffa, explaining that local authorities 'did not know any law and did not have any rules (professional regulation) for building the city. So there was complete anarchy.'[11] The 'filth and ugliness, from a lack of comfort and aesthetic beauty' from which Dizengoff and others felt that the Jews suffered in Jaffa were also decried, which was why building a 'clean and healthy' Jewish city prompted such joy.[12] In the Zionist imagination Tel Aviv would be for those who 'possessed energy and vigor',[13] 'beautiful, arranged according to laws, all of it Hebrew'.[14] It was therefore only natural that the youthful Jewish city should displace Jaffa and become the 'natural' center for business, commerce and industry, health and rest, sports and tourism and home industry.

How was this specifically reflected in the architecture and planning of the period? To begin with, we need to recall that in the mythology of Tel Aviv the city was literally born out of the sounds, and thus out of the 'overthrow' of the existing geography of the region.[15] That is, there was a clear need/desire to avoid building just another Jewish neighbourhood of Jaffa. Thus the goal of Tel Aviv's founders was to 'establish a Hebrew urban center in a healthy environment, planned according to the rules of aesthetics and modern hygiene in the place of the unsanitary housing conditions in Jaffa'.[16] As adjacent lands were purchased and other new neighbourhoods planned, steps were taken to ensure that 'attention will be paid to all the modern facilities of Europe'.[17]

It was this 'modern' rationality and consequent rejection of older Jewish neighbourhoods of Jaffa that has led scholars like Katz to consider Tel Aviv 'the first Zionist urban undertaking in Palestine',[18] as a modern self-identity and rejection of the existing indigenous Palestinian Arab or Jewish cultures were defining features of Zionist ideology. The most obvious way to achieve this was through physical/spatial separation from the existing Arab, and even more interestingly, *galut* (Diaspora) Jewish environment;[19] not surprisingly, the leadership of Achuzat Bayit agreed that new land purchases had to be as far away from Arabs as possible.[20]

The central planning and architectural motif of the early period of Tel Aviv was Ebenezer Howard's garden city/suburb paradigm, which was considered the most modern planning system in Europe, and which was given a specifically colonial twist through the twin ideas of spatially and ethnically segregating Tel Aviv from 'Arab' Jaffa,[21] which architecturally was reflected in debates over whether to build in 'Arabic-Yafo' versus 'European' styles.

With the end of World War I British planners, like their Jewish/Zionist counterparts, found 'the transformation of the oldest country into the newest fascinating' and believed that 'this paradox gives architects and engineers a golden opportunity'.[22] The trademark of this decade was the 'eclectic style' of architecture that had already become predominant in British and French colonies and reflected an idealized meeting of East and West, which was concretized in the town plan prepared by Patrick Geddes for Tel Aviv in 1925. Indeed, one critic revealingly labeled it an 'eastern-modern' style – a conjoining of two discourses rarely sanctioned within Zionist ideology, unless it is Zionists who are the agents of that meeting.[23] The local architecture was thus believed to be, however 'primitive', in 'harmonious union with the landscape'.[24]

While the symbolism of eclecticism suggests the attempts at conciliation by many Jews during this decade, it must also be noted that Zionist eclecticism emerged concomitant with British attempts to create a new 'colonial' style in Palestine, at least in British government or church buildings.[25] And while the development of eclecticism in French colonies – to broaden the context – did reflect an often sincere desire by administrators/planners such as Lyautey, Prost, and even Le Corbusier to achieve a harmonious admixing of East and West, such a balance did not disturb the basic hierarchy of culture and civilisation upon which colonial discourse was erected. We can therefore also understand the reasoning underlying this aesthetic as being rooted in a feeling that the local architecture was 'exotic and romantic',[26] a quaint sentiment that was part of the desire to built a 'cocoon' that would be a simulacrum of the local environment while ultimately remaining separate and European.[27]

Moreover, if developments in Zionist architecture and design in the 1920s were spawned both by the sincere desire to understand the *indigene* society shared by many colonial administrators and planners and the continued attempts to arrive at a *modus vivendi* with the Palestinian population, the style changed dramatically with the onset of the 1930s in good measure in response to the country-wide eruption of violence in 1929 (although even before this episode, by several accounts around 1927, Zionist architects had begun to 'turn back to Europe').[28]

In response to this renewed threat to the emerging Jewish hegemony in Palestine[29] and the concomitant needs for both increasing separation and justification of Zionism as uniquely civilized and modern, International Style (IS) architecture quickly became the dominant style of the 1930s, with the eclectic style coming under criticism for bad architectural form and lack of culture or planning[30] – that is, for being too Galut. The closing of the Bauhaus school – where seven Jewish/Zionist architects studied – with the onset of Nazism in Germany in 1933 also contributed to this trend.

It is thus no surprise that in the environment of the 1930s Zionist architectural discourse became ever more militantly 'modernist', just as the larger Zionist argument that the Palestinians were incapable of developing the country, and thus undeserving of ruling, or even remaining in it, became ever more vigorous. The IS/Bauhaus vernacular accorded so well with the Zionist spirit of renewal that by the early 1930s modernism became the visual mould for the Zionist project,[31] not just in Tel Aviv, which became the White City and the world center for the style, but for the agricultural *kibbutzim* (collective Jewish farms) too.

As David Ben-Gurion put it in a speech before the 22nd Zionist Congress, 'development and progress' were the keys to rejuvenating Palestine – if the push toward IS marked the conflation of economic, political and ideological/ aesthetic motives underlying 'development and progress' within Zionist built space, there was no better measurement of their abundance in the Jewish sector than the block upon block of gleaming white IS buildings being erected in Tel Aviv, which reflected a refusal of accommodation to existing urban and social conditions. Thus we see how the need for a 'clean slate' which articulated the very beginnings of Zionist urbanism was realized by the followers of Le Corbusier, who believed that 'Where one builds one plants trees. We uproot them.'[32]

The evolving Zionist architectural and planning visions were indeed given biblical justification – the entrance to the Tel Aviv Museum, located in the home of Tel Aviv's first mayor, Meir Dizengoff, is inscribed with a verse from the Book of Amos: 'I will restore the fortunes of my people Israel, and they shall rebuild the ruined cities and inhabit them.'[33] But in order to rebuild the ruined cities, the cities must first be ruined, or 'erased'.

Thus far I have discussed how this was achieved on the visual/aesthetic level primarily through architecture and the garden city paradigm. Now I will briefly discuss the role of town-planning specifically, which was equally and perhaps even more essential to the visual and discursive separation of 'modern, Jewish' Tel Aviv from 'ancient', 'Arab' Jaffa, marking the city as one of the premier symbols of Jewish/Zionist rebirth in and of Palestine and obscuring Jaffa's own experience of modernity.

Town-planning was and remains a crucial tool in the expansion of Jewish/ Zionist control of the land in and borders of the Jaffa-Tel Aviv region for many reasons. Perhaps chief among them (during the Mandate period) being the resonance that Zionist planning had with the British authorities, who enacted numerous town-planning ordinances that facilitated Tel Aviv's expansion even when other policies aimed to curtail Jewish acquisition of and control over Arab land.[34] As I have elsewhere explored, from the perspective of both Tel Aviv's leaders and the British Mandatory/colonial Government there was a strong link between the need to reform the country's land laws, and the need to increase the power of its Town-Planning Committee and size of its Town Planning Area. Thus, for example, Tel Aviv's Municipal Engineer explained that the dearth of open spaces in Tel Aviv arose because of both 'inadequate municipal town-planning area' and because of the 'system of land ownership'.[35]

The history of and strategies underlying the expansion of Tel Aviv during the Mandate period reflected the development of town-planning legislation, especially during the 1930s and 1940s. During its first 12 years Tel Aviv's leadership attempted to make the town 'completely independent' from Jaffa despite the fact that the quarter, like other recently established neighbourhoods, enjoyed no separate legal status or autonomy from Jaffa.[36] The goal, as stated by Arthur Ruppin as early as 1913, was to 'conquer Jaffa economically'.[37]

Just as important, in describing the region surrounding Tel Aviv, Ruppin explained that 'unlike other cities which are surrounded by large unbuilt-on areas and have wide town development possibilities, Tel Aviv is bounded on the west by the sea, on the south by Jaffa, on the north by the river and to the east by Sarona'.[38] Despite being written in 1942, at a moment when the Tel Aviv Municipality was engaging in a 'war' over land with the surrounding villages, there is no mention here of the six villages that also surrounded Tel Aviv, several which would soon see large parts of their lands annexed to the city. Such lacunae typified the process of discursive, and then physical, erasure or disappearance of these villages discussed in the introduction to this essay.

As important, during the 1930s, that is the period of increased ideological/discursive separation reflected in the predominance of IS architecture, a new planning administrative structure was established that centralized authority and power for planning and development that had previously been under the jurisdiction of municipal councils such as the Jaffa Municipality, which had enacted its own town-planning legislation as early as 1923.[39] This was also the period when the importance of 'land settlement' grew significantly, particularly because the cadastral survey and newly established land registry offices mandated by it facilitated the purchase of land at the same time a recession was making it harder for the often debt-ridden Palestinian smallholders to resist the speculative prices offered for their lands.[40]

The Palestinian Arab Revolt that began in Jaffa in 1936 led to closer cooperation between the British and Zionists than had been seen since the 1929 uprising, and the shared focus on 'security' issues – the citrus groves of the surrounding villages were frequently the staging ground of attacks on Jews – and town planning facilitated the annexation of village lands during the remainder of the Mandate period, especially after the 1939 White Paper limited the ability of Jews to purchase agricultural land. In fact, the British provided an excellent example of the conflation of security and planning when they destroyed large swaths of the old city of Jaffa in June 1936 in retaliation for the Arab Revolt, but justified their actions as an act of town-planning, 'renewal' and 'improvement' of the old city.[41]

To provide an example of how this discourse worked in practice, during the 1940s Summel, along with parts of the villages of Jammasin el-Gharbi and Jerisha, were in the sights of British and local Zionist town planners. Such lands were still considered 'practically undeveloped', and it was considered vital to bring them under 'complete municipal authority' because only the 'legal and administrative machinery of a municipal corporation' would have the power to draw up a 'creative or positive machinery of development' through which Tel Aviv could 'redeem some of these defects which have deformed and stunted its past growth and to prepare for a better planned and more spacious urban future'.[42] The government was not of one mind as to the merits of annexing these lands. While the Attorney General and Solicitor General opposed it, the District Commissioner (the senior official of the regional Town-Planning Commissions) felt that 'the concentration of industry in a suitable place ... is of the greatest importance both to Tel Aviv and to Palestine as a whole'.[43] The next year a similar divergence of views occurred when the Tel Aviv Municipality requested permission to annex land from Salama and Sarona for the erection of cheap dwellings to relieve the acute housing shortage in Tel Aviv.[44] The Colonial Office officials was more skeptical of Tel Aviv's intentions while town-planning and local officials believed that to build it within the current borders of Tel Aviv would 'completely spoil the whole layout of the area which is intended for residential houses of a more expensive type'.[45]

A similar process occurred in the village of Sheikh Muwannis to the north of Tel Aviv (today's Ramat Aviv) as the Jewish city needed land for numerous 'workers neighbourhoods'.[46] First the government during the 1920s attempted to gain title to the sandy lands lying to the west of Sheikh Muwannis until the seashore by claiming that it was 'waste and uncultivated' and re/mis-interpreting Ottoman land laws to claim that 'nobody has a right to waste of uncultivated land'.[47] Ultimately, the land was incorporated into Tel Aviv, which was made possible by the government's long-standing concern to fix permanently the status of the land based on its present condition in order to facilitate both land purchases by Jews and urbanization.

There is not space here to engage in a detailed discussion of how the Palestinian Arab residents of Jaffa participated in and responded to the modernization of architecture and town-planning in the region. However, it should be pointed out Jaffa too was engaged in a long process of evolution in both street design and architecture that actually mirrored closely developments in Tel Aviv.[48] This included, on the one hand, the same movement through neo-Oriental, eclectic and then IS style architecture – which was recognized by even the most conservative members of the Palestinian Arab elite of the city as an attractive symbol of modernity – and the development of town-planning in the city, where from the late Ottoman period through the last years of the Mandate city leaders sponsored modern, European-style planning improvements (including detailed city-wide development plans) that sought to place Jaffa in the same position vis-à-vis Arab Palestine as Tel Aviv symbolized for Zionism.

As important, the Palestinian Arab residents of the Jaffa-Tel Aviv region – indeed, often the least educated and poorest among them in the surrounding villages – well understood the implications of the Zionist-cum-British discourses of development generally, and their implementation through town-planning schemes, perhaps even better than the Zionist and British planners and political leaders. To take just one example from the Jaffa-Tel Aviv region, when the Tel Aviv Municipality attempted to build a road on Arab land the Jaffa-based religious daily *al-Jam'iah al-Islamiyyah* complained:

> [I]n reality the plan in the Town-Planning Commission now including Sheikh Muwannis is not really a 'plan', but rather a plan to take the land out of the hands of the owners … We have farmed land north of the Auja for a long time and then Jews came and wanted to buy it because it is close to Tel Aviv, and we said no, and they tried to get it through various means, including using the Government to push a plan to open a road through our farmland … after it proved incapable of gaining ownership through [other] means. [We declare] that this project has no benefit returning to the village, either from a planning or moral perspective.' Moreover, the project 'has no benefit returning to the village, either for planning or from moral perspective.'[49]

The reality of Palestinian Arab experiences and perceptions of, and responses to Zionist architectural and planning discourses are important to bear in mind as we move to the contemporary period. This is because while the vast majority of the pre-1948 residents no longer live in Jaffa, there is a communal memory of the period before the war and the events that led not just to the exile of so many citizens but of the various mechanisms involved in the competition over the territory of Jaffa between the Jaffa Municipality and Tel Aviv Municipality.

What this means is that while Palestinian Arab citizens of Jaffa no longer have municipal and/or administrative control over their city-turned-neighbourhood, of the ability to mirror or even anticipate the architectural developments in their neighbour to the north, they still have a living understanding of how they lost power, kept alive in good measure because the Israeli state and Tel Aviv Municipality continued to deploy these discourses. This then produced an 'architecture of resistance', that has been the source of both communal solidarity and continued political and communal struggles with the government. In the second half of this chapter I will explore how and why this imagined architecture is so crucial to the maintenance of the community in the face of continuing efforts to 'judaize' Jaffa under conditions of globalization.

Architecture and Planning in Contemporary Jaffa

History records that in May 1948 the city of Jaffa was conquered by a combination of Hagana/IDF and Jewish irregular units after fierce fighting throughout the Jaffa-Tel Aviv region. Soon after the war the two towns were unified, with 'Yafo' been attached to the name Tel Aviv in order to preserved the historical name, while 'Tel Aviv' symbolized the Jewish settlement renewing itself in Israel.[50]

With only a few thousand native Palestinian inhabitants remaining and tens of thousands of Jewish immigrants needing housing, over the next decade the largely emptied town quickly saw new waves of Jewish immigrants, many from North Africa or the Middle East, replace the former Palestinian Arab population. Jaffa was thus turned from the economic center of Arab, and in many ways, Jewish Palestine to a marginal and poor neighbourhood of Tel Aviv. This situation did not change until the late 1980s and 1990s, which began a period in which the neighbourhood has become an object of 'development' – as both a site for tourism and as a new and chic neighbourhood for the burgeoning Jewish elite of 'global Tel Aviv.'

More specifically, in the face of creeping dislocation, accompanied (and supported) by daily media and television portrayals of Jaffa as both poor and crime-ridden *and* chic, exotic and romantic (and thus the ideal tourist site). Palestinian residents have attempted re-imagine their 'city' and open up new spaces for agency and empowerment through which they can articulate a more autochthonous synthesis of the city's history and its architectural traditions – one that will allow them to remain on the land and develop Jaffa for the benefit of the local, as well as international community.

The post-1948 remnants of the Palestinian Arab community of Jaffa were the poorer Arabs from the surrounding villages and a few Jaffans who remained. Jewish immigrants, mainly from the Balkans, were settled in empty Palestinian properties in the early 1950s. Later, when many of them moved to newer neighbourhoods in the Tel Aviv region, Palestinians resumed renting and buying properties, moving back to live in Jaffa.

After a precipitous drop in the Tel Aviv metropolitan region's population during the 1972–83 period, the 1980s began a period of transformation of Tel Aviv into a 'post-industrial era' which saw the relocation of most of the major financial and industrial corporations of Israel to the city, and with it, numerous 'yuppy/dinkie' couples.[51] This immigration was augmented by the wave of Soviet Jewish immigration that began in 1989. Within Jaffa, the Palestinian population almost trebled since 1972; on the other hand, the Jewish population of the two predominantly Palestinian Arab neighbourhoods of Jaffa – Ajami and Lev Yafo – fell dramatically, down to less than 3 per cent in the case of Ajami.

Despite the population increase discrimination has continually played a central role in the social life of Jaffa's Palestinian residents. It is a major factor in wage differentials, access to jobs, and educational level between Palestinians and Jews,[52] all of which has been exacerbated by the large increase in its Palestinian population and the influx of Russian immigrants which took away jobs and housing from the local population after 1989. In fact, the situation in Jaffa 'meaningfully deteriorated' in recent years and especially since the outbreak of the al-Aqsa intifada and boycotts by Jews of Palestinian businesses, to the point where Palestinian Jaffa has become the most depressed and disadvantaged community in the entire country – an important cause of the renewed 'intifada' in Jaffa in the fall of 2000.

The Symbolic Functions of Tel Aviv and Jaffa

The symbolic and discursive functions of Tel Aviv and Jaffa within the Zionist enterprise have always been as important as their economic and political functions. Consequently, they have exercised a determinative influence on the current political-economic situation in Jaffa described above. On the one hand, 'modern', 'clean', and 'well-planned' Tel Aviv was from the start contrasted with 'backward', 'dirty', and 'unplanned' Jaffa. At the same time the 'first modern Hebrew city in the world' was also contrasted with Jerusalem, the religious and cultural capital of pre-Zionist Jewish Palestine.

In the post-1948 period this dichotomy has continued to be a major theme in the Israeli and Western imagination of both cities, in no small part due to the trend during the last two decades (the era of postmodernist planning and architecture) for cities to seek to distinguish or differentiate themselves through their architecture, particularly through the selling of their image.[53] Some American and European publications have also contrasted 'secular', 'normal', 'cosmopolitan', 'unabashedly sybaritic', and most important, 'modern' Tel Aviv with 'holy', abnormal Jerusalem:[54] 'a visitor wanting to see what the 50-year old Jewish state is really all about would do well to plunge into the casual, self-consciously secular and thoroughly modern metropolis on the sea back where the dunes used to be' – as if Jerusalem and the seemingly interminable conflict it symbolizes are in fact a mirage on the 'Sahara Desert' upon which Tel Aviv was imagined, then built.[55] In a similar vein, the Chief

Architect of Tel Aviv titled her recent book on the history of International Style architecture in Tel Aviv *Houses from the Sands*.

As I explained earlier, this 'discourse of the sands' is intimately connected to an aesthetic of erasure and reinscription which itself lies at the heart of most modernist planning ideologies, particularly Zionist/Israeli planning. Not surprisingly, the discursive erasure epitomized by the symbolism of sands and the changing of streets names has lasted till today: as the *Economist* explained in comparing Tel Aviv and Jerusalem, 'Unlike Jerusalem, Tel Aviv contains hardly any Arabs. It has swallowed the old Arab port of Jaffa, but in the main it was built by Jews, for Jews, on top of sand dunes, not on top of anybody else's home'.[56] The purported absence of 'Arabs' from the land on which Tel Aviv was built is an important reason why Tel Aviv is not considered a 'national' space in the way that the *Times* conceives of Jerusalem; an ironic development considering that Tel Aviv was created as the living embodiment of a Zionist – that is, Jewish national – utopia.[57] It was also created, as I would suggest as the living embodiment of the *modern* project, in which dreams of utopia all too often turn dystopic, particularly for those considered 'other' than the group for whom the utopia is being created.

The Contemporary Tel Avivan Imagination of Jaffa

Such precise renditions of Tel Aviv's creation mythology by the Western media have had a profound impact on the way Jaffa has been imagined by Israelis and foreign writers during the past 50 years, because the landscape of Jaffa has remained central to the Tel Avivan self – and thus 'Other' – definition. If Palestinian Arabs were discursively (and ultimately physically) erased from Tel Aviv, the process was even more determined in Jaffa. Two contemporary depictions of Jaffa, one negative and one quaint and 'aggressively restored'[58] have framed its envisioning.

On the one hand, Jaffa has been and continues to be visualized as poor and crime-infested. The neighbourhood is the site of many crime or war movies and television shows since the 1960s[59] because 'it resembles Beirut after the bombardments – dilapidated streets, fallen houses, dirty and neglected streets, smashed cars'.[60] This image is reinforced by the media and government depictions, and to a lesser extent reality, of the neighbourhood as being a major center for drug-dealing in the Tel Aviv metropolitan region.

The other image of Jaffa, specifically designed for tourist consumption, is based on its being 'ancient', 'romantic', 'exotic', and 'quaint': 'Old Jaffa … is the jewel of Tel Aviv', is how an official brochure described it.[61] Such depictions of Jaffa are linked to its re-imagining as a historically Jewish space, one that was 'liberated from Arab hands' as the museums and tourist brochures inform visitors.[62] These visions of Jaffa are connected to Jaffa's place as a historic, archeological, and thus tourist site within Tel Aviv: 'A port city for over 4,000 years and one of the world's most ancient towns, Jaffa is a major tourist attraction, with an exciting combination of old and new, art galleries and great shopping ... Great care has been given to developing Old Jaffa as a cultural and historical center...'[63]

In fact, the 'city of the sands', imagined without a past or history, has always required Jaffa to complete its identity:

> Once Tel Aviv became Tel Aviv-Yafo the young city all at once acquired itself a past – the 3000 years of ancient Yafo ... [and] was ready for the great leap forward which transformed it into a metropolis. Yafo ... one of the oldest cities in the world, acquired a future and renewed youth, with widespread progress streaming its way from its youthful neighbor.[64]

Not surprisingly, contemporary Jaffans protest how their city has become little more than 'a margin on the name of Tel Aviv' since 1948.[65] One reason, of course, is that pre-1948 Jaffa was considered the 'jewel' of Arab Palestine, and was continually depicted in the Palestinian press as the country's most beautiful and important Arab city; as *Falastin* described it in 1946 – 'No one doubts that Jaffa is the greatest Arab city in Palestine, and it is inevitable that visitors to Palestine will stop by to see the model of Palestine's cities'[66] – that is, Jaffa was a symbol and epitome of Palestine's modern urban landscape.

Yet the erasure of Jaffa has come to be accepted by many Diaspora Jaffans, particularly those returning to visit the city in recent years, who have come to regard present-day Jaffa as a 'figment of the imagination' or at best an object of critical nostalgia that borders on cynisim.[67] In many ways Tel Aviv has displaced Jaffa in the Palestinian imagination – when the facilitator of a peace mission in Palestinian-controlled Nablus asked people what their vision of peace was, a Palestinian artist replied 'visiting Tel Aviv and watching the sun set'[68] a sentiment whose echoes return to the pre-1948 period.

On the other hand, the attachment of the remaining Palestinian population to Jaffa has grown significantly during the past two decades; in part in line with the larger trend toward increasing 'Palestinianization' of 'Israeli Arabs' in the wake of the reuniting of all of Mandatory Palestine after the Six Day War, and then the outbreak of the Intifada in 1987. Yet this nationalistic re-imagining of Palestinian-Israeli identity also added greater relevance to the question of territoriality. In fact there have been several violent protests in Jaffa during the 1990s – most recently during the Al-Aksa Intifada that began (and within Israel, ended) in October 2000 – and leaders of the Palestinian community have called for Jaffa's municipal independence.[69]

Indeed, more than four years before the latest Intifada, in response to continued attempts by the Tel Aviv Municipality to evict long-time Palestinian residents of Jaffa the community's leadership threatened a 'housing *Intifada* in the streets ... declaring with a loud voice that we are planted here and that they will not be able to uproot us from our homes the way the uproots the orange and olive trees'.[70] Both Intifadas shared a focus on rootedness that is deeply imbedded in the Jaffan, and the Palestinian, psyche, as evidenced by the following painting done by a young Jaffan artist, Suheir Riffi, who depicted a mother, nursing her child, rooted into the earth and connected through it to her dilapidated home.

Globalization, Architecture, and Planning in Tel Aviv-Yafo

The specificities of contemporary Jewish and Palestinian imaginings of Jaffa have influenced the way the Jaffa-Tel Aviv region has experienced the impact of globalization and the attempts by Israel's leadership to make Tel Aviv into a 'global' or 'world' city. The drive to 'globalize' Tel Aviv by the city's leaders is understood as being part of their increased desire to shape and deploy a unique identity, separate from the rest of the country, especially from Jerusalem. Such an identity leaves planners, architects, and commentators to wonder 'what to do with a world city that is so different from the rest of the country in which it is located'.[71]

Naturally, Israeli social scientists have begun to analyze Tel Aviv as a global city, focusing on its entrance into the international market,[72] the increasing disparities between rich and poor, the 'marketization' of social services such as the education system,[73] and the influx of increasingly illegal migrant guest workers.

Most architects working in Tel Aviv have refused to criticize the Municipality's planning policies for 'Global Tel Aviv', which call for building numerous high-rise projects throughout the city to maximize market value of the land. One fantastical project, called 'Tel Aviv on the sea' which would build several islands off the coast connected by bridges that would each contain several 'millennium towers' of up to 170 floors, was developed by researchers from the Technion in Haifa. Indeed, at least one architect, Massimiliano

Fuksas, the designer of the as yet unbuilt Peres Center for Peace (which I discuss below), sees such towering buildings as creating order out of urban 'chaos'.[74]

On the other hand, the eminent architect Peter Kook, who has worked in Tel Aviv, has described the present Tel Aviv 'style' as 'paranoia on the one hand, and the world wide trend of the worship of money on the other:

> The paranoia is reflected in the fact Israeli architects are closed to any outside styles, they only see what the Housing Ministry does, and not what's going on in the wider world. The power of money rules here in a dominant way on both aesthetics and on urban planning ... Also, there is a psychological factor. Israeli architects take the fortress as their model ... the security room in their apartments.[75] They are afraid to do more elegant architecture here, with more feeling, because maybe something will [destroy] the building.[76]

What is interesting is that in striving for 'elegance' and 'order', as we will see Fuksas has attained in his design for the Peres Center, both his design and the Center it will house play into a century-long Zionist imagination of Jaffa as the antithesis of both.

Architecture and Planning in Tel Aviv and Jaffa during the 1980s and 1990s

If it is clear that Tel Aviv used town planning as a tool of the 'war over land' in the Jaffa-Tel Aviv region during the Mandate period, it is not surprising that much of post-1948 Israeli planning literature, particularly when dealing with Tel Aviv, has avoided any discussions of the Palestinian minority that would disturb the apolitical suppositions upon which it is based, focusing instead on planning as 'change-oriented activity' in order to 'shift attention away from the document – the plan – to the political process whereby intentions are translated into action'.[77] Thus in a 1997 edited volume on planning in Tel Aviv, a chapter on 'Conflict Management in Urban Planning in Tel Aviv-Yafo' consisted of a case study on underground parking in stores in central Tel Aviv.[78]

Within frontier regions, spatial policies can be used as a powerful tool to exert territorial control over minorities, while on an urban scale, majority-controlled authorities can exercise more subtle forms of spatial control through land use and housing policies, and in so doing create segregation between social groups.[79] This is particularly true when the government takes almost all the power out of the hands of local Palestinian communities to plan their own development.

The discussion thus far suggests that the object of analyses of planning in Tel Aviv and Jaffa should be to clarify the complex web of relations between governmental, semi-governmental, pseudo-governmental organizations and institutions that control the planning system in Israel. The number of institutions involved and the complexity of their relation indicate that despite claims to the contrary, planning is highly politicized and ideological. What is new in this equation is the increasingly prominent role of private interests in planning and development in Israel, and in Jaffa in particular, and how this development is shifting the internal boundaries within the land and planning system while maintaining the traditional focus on permanent Jewish ownership of as much land as possible. Thus, for example, the pseudo-governmental Jewish National Fund[80] announced in November, 1998 that it was severing ties with the Israel Land Authority, the semi-governmental agency that administers both state and JNF-owned lands (and which heretofore has been composed of both government and JNF representatives), precisely because by going 'private' it could buck the legal trend toward equality between Jews and Palestinians in the government sector and ensure that its huge reserves of land remained 'in the hands of the Jewish people'.[81] As I explain below, one can understand the transfer of the land on which the Peres Center for Peace is supposed to be built as serving a similar goal.

Fueled by the larger discursive, even epistemological, shift towards privatization in Israeli society, the strategic shift toward privatization in city planning has led to is a situation in which planners chart a course of development focused on middle and upper class Israelis and implemented through private developers, that pit Jews against their Palestinian co-citizens. Thus Palestinian land is expropriated, the construction of new, privately-developed Jewish housing on that land is subsequently approved, and the new Jewish 'owners' – often self-described liberals and supporters of Palestinian rights – naturally take the lead in fighting against the claims the previous (or now, 'illegal') Palestinian inhabitants, since by then they have invested time and money into their new homes. This is how the government, working through private developers, brings the economic interests of liberal Israelis in line with its perceived 'national' interests vis-à-vis increasing Jewish ownership, control and presence on the land.

By the early 1980s a new generation of 'renewal' efforts began in the older neighbourhoods of Neve Tzedek and Lev Tel Aviv. The renewed activity was prompted by the structural reorganization of the city's economy which began in the previous decade, and sought to 'reviv[e] the region as a space for living in the center of the city by drawing a mainly young population to it'.[82] Both neighbourhoods featured architecture that made them attractive for gentrification. Lev Tel Aviv, having already undergone extensive reconstruction in the 1930s, featured the International Style buildings that put Tel Aviv on the architectural map. Neve Tzedek featured much older buildings and attracted a Bohemian crowd trying to escape both the austere, International Style architecture that dominated the city from the 1930s through 1970s, and the more recent postmodern fetishization of consumption that took the ironic form of an easily identifiable uniform 'postmodern style'.[83]

If the gentrification of Tel Aviv's older neighbourhoods generates and reflects contradictory impulses and desires, the process is much more complicated in Jaffa, which is officially part of Tel Aviv, yet is heavily invested with symbolism that portrays it as Tel Aviv's alter ego. How is this separation mediated? The answer becomes clear when we understand that through the various Zionist/Israeli visions of 'ancient' Jaffa the neighbourhood becomes 'a discursive object created by Israelis as part of turning Israel ... into particular socio-political spaces'.[84] If we view Jaffa as a frontier region – indeed, as a frontier region of a frontier region; that is, as Tel Aviv first saw itself[85] – it further becomes clear how the spatial policies of the Municipality are used as a powerful tool to exert territorial control over and physically shape this discursive yet material space.[86]

In the resulting process of cognitive and physical boundary demarcation between Tel Aviv and Jaffa, the Jews yuppies moving to Jaffa 'see residential exclusivity and the redeeming modernizing impact of Zionism as simply engendering a demarcation between two types of territory'.[87] In this vision Jaffa is the onto-historical 'Other' of Tel Aviv, against which Tel Aviv defines itself. At the same time, having been 'liberated' from its Arab identity and 'united' with its daughter city, Jaffa is continuously undergoing a process of 'renewed youth and progress', the life blood of which is the architectural and planning policies of the Municipality. Yet the neighbourhood's renewal is dependent upon its permanent fixture in time and space as 'ancient' or 'quaint' – the ideal site for tourist and elite development.

In fact, if architecturally Tel Aviv has become a 'tragedy'[88] because architects are afraid to build imaginatively there, Jaffa has become the space where the imagination, although remaining under Government supervision, has had freer reign. That is, as 'picturesque' has become the architectural fashion, the Government realized that 'old, dilapidated Arab neighbourhoods have an 'oriental' potential'.[89] Thus the function of the numerous rehabilitation projects of the past two decades has been to expand commerce, tourism and hotels in line with the 'specific character' of the area.[90] More specifically, 'today the slogan is, "Gentrify!" As land becomes available, it is sold on stringent conditions that only the wealthy can meet'.[91] Thus the current style among the Jewish architects practising in Jaffa is to build with arches, 'thousands of arches, wholesale', as one architectural critic put it.[92]

The end result of this process has been expressed in 'the systematic erasure of the identity of the city of Jaffa as a Palestinian Arab city'.[93] This may seem ironic given the 'oriental' feel of current building styles, but in fact Jaffa has had to be emptied of its Arab past, and Arab inhabitants, in order for architects to be able to re-envision the region as a 'typical Middle Eastern city' and construct new buildings based on this imagined space, unaware that such a city only ever existed in the world view of the architects.[94]

The Role of Tourism and the Discourse of the Market

As the world economy and the peace process have faltered, Israel's economic leaders believe that the key to continued economic growth in the country was the real estate market, of which the Tel Aviv metropolitan region is the center. This view has clear implications for the current 'renewal' efforts in Jaffa, as the Municipality has even less freedom or incentive to commit scarce resources to rehabilitating a poor minority community that is sitting on valuable land whose marketization is seen as essential for the country's economic health.[95]

The influence of the market discourse is readily apparent in current planning in Jaffa and should be used to contextualize statements such as Mayor Ron Holdai's campaign promise that 'the time has come to lower the walls between Jaffa and Tel Aviv'.[96] Thus while current policy guidelines declare that the new regional plan for Tel Aviv must involve residents in planning and work to increase housing for young couples,[97] when Palestinian community leaders have complained that most young Palestinian couples can not afford to live in Jaffa officials have responded by explaining that 'the market is the market'[98] and that 'selling some apartments more cheaply would hurt profits'.[99]

The most important impact of the marketization of planning in Jaffa has been the partial or total privatization of several of the bodies directly responsible for the rehabilitation of the quarter since the mid-1990s. Until then, as much as 90 per cent of the housing units in Jaffa were partly owned by the government,[100] and a large part of the real estate in Jaffa was in the hands of quasi-governmental companies such as Amidar and Halmish. The transfer of the development projects to private developers was described Jaffa's Palestinian councilman (in the same language, it is worth noting, used by the

Jaffa newspaper *al-Jam'iah al-Islamiyyah* in 1932 to describe the burgeoning land conflicts in Jaffa-Tel Aviv and Palestine as a whole) as a major turning point for the quarter.[101]

One of the projects partially transferred to private developers involves the redevelopment of 'old Jaffa's' port, home to a fishing industry supporting 250 families, which is to be 'resurrected' and 'developed' as a tourist spot.[102] Yet the symbolism surrounding the port give a clue to how such the project will be realized. Thus the official *Tel Aviv-Jaffa* guide of the Ministry of Tourism explains:

> [T]he old city today is alive, her buildings and alleys restored amidst cobbled streets and green parks as a thriving artist's colony ... great care has been given to developing Old Jaffa as a cultural and historical center while preserving its Mediterranean flavor ... Jaffa Marina [part of the development project – M.L] has been established in the heart of the ancient port ... and offers all a sailor could desire ... Visit Old Jaffa anytime. By sunlight and starlight, it is the 'jewel' of Tel Aviv.[103]

'Project Shikum (Rehabilitation)' is another project ostensibly designed 'develop and rehabilitate Jaffa.' It was turned over by the Municipality to a private developer, Yoram Gadish, in 1996, and when mismanagement and concerted local opposition led the government and Municipality to terminate Gadish's contract, a new private company headed by former Tel Aviv mayor Shlomo Lahat was awarded the contract to continue the neighborhood's gentrification.[104] The goal of the project was to develop Jaffa since:

> Jaffa is not developed ... that is, develop infrastructure, sewers, streets, schools, etc., and to develop the empty lands in Jaffa. We want to revolutionize Jaffa, to change Jaffa from a neighbourhood with so many problems to a tourist city – there's lots of potential for development into a tourist city ... But you need to have a *plan*, and like New York or anywhere, sometimes you have to destroy a building as part of development for public needs, and we're working with a committee of architects and the Municipality ... However, there residents want to keep the status quo because development increases prices, and their children won't be able to live and buy apartments there; also Arabs won't go to other cities like Bat Yam, Herzliyya because there are no services for them. They *can* go to Lod and Ramle, but they're not *ready* to go and don't want to develop ... but with Jews [Jaffa] becomes more beautiful and develops.[105]

Today, a more official quasi-governmental body, 'Mishlama leyafo' has now assumed responsibility for administering Jaffa, from providing municipal services to residents to supervising the quarter's gentrification. Its goal, in the words of Tel Aviv's Mayor, is nothing less than to 'to build Jaffa, to revolutionize Jaffa into a dream space'.[106] Here we must recall the almost identical word used in describing the birth of Tel Aviv in 1933, as an 'overthrowing of the existing geography.[107]

Here we see how the importance of having a 'plan', which we can recall was fundamental to the genesis of Tel Aviv,[108] has today been transferred to Jaffa, but the implication that Jaffa requires transformation under Zionist/Israeli supervision is a longstanding discourse in Zionist and Israeli writing about and policies toward the city.[109] The issue is whether the Jaffa-as-'dream space' will answer the dreams of the local Palestinian (and working class Jewish) populations, or the wealthy Jewish and foreign populations bent on remaking Jaffa in their more 'Disneyfied' image.

The Andromeda Hill Project

Perhaps the paradigmatic example of the intersection of new global, market-based postmodern architectural discourse in Jaffa with almost century-long Zionist/Israeli imagination of the city is the Andromeda Hill project, where units start at well-over US$300,000. Constructed on property at the top of the Ajami Hill, with a commanding view of the port and ocean below, Andromeda Hill bills itself as 'the incomparable Jaffa ... the New-Old Jaffa.'

To help orient prospective customers on its website, the Andromeda Hill virtual brochure explains that 'historic Jaffa' lies to the north of the development, the 'picturesque fishermen's wharf of Jaffa' to the west, and the 'renewed Ajami district, where the rich and famous come to live' to the south. The Hebrew version stresses the architecture of the place even more, in line with the greater importance of architectural discourse in Israeli culture.[110] Moreover, the section of the website entitled 'The Legendary Jaffa' recounts the Greek legend of Andromeda, which was set on a large rock facing the city outside of Jaffa's port, and explains how 'Andromeda became a symbol of awakening and renewal, and it is not by chance that the project was named 'Andromeda Hill', expressing the rebirth of old Jaffa.'

When asked why and how the architectural design and advertising campaign was chosen for Andromeda Hill, one former employee explained:

> [T]he Municipality decided on the style – the windows, the columns, the materials – after going around Jaffa and looking at the buildings ... The style was very eclectic – Arabic from the beginning of the century influenced by

> European (specifically Italian) architecture ... Arches were a main symbol in
> a project of this size... We didn't use real stone [except in a few places] but
> rather a man-made material called 'GRC', which is fake stone. In terms of
> the ads, you have to think about who's going to buy there... they expected
> people from abroad to buy it. Jaffa today is not a nice place, you have to think
> about the future, what will be attractive. People aren't living there because of
> the sea, because there's sea all over Israel, they're living there because of the
> nostalgia, the atmosphere.

The Andromeda Hill discourse, like that of Gadish, exemplifies the
conflation of architecture and planning, market forces and government control,
that comprise the forces at play in the continuing 'war over land' in Ajami.[111]
In fact visitors to the complex are shown a short video before their tour whose
narration concludes by explaining how 'Andromeda Hill is, in essence, a city
within a city' within Jaffa. This is almost identical to the language used by the
founders of Tel Aviv to describe the Jewish position in Jaffa almost 100 years
ago, when they celebrated having created 'a state within a state in Jaffa'. The
social, political, and spatial implications of such a discourse are also identical;
that is, in each case, Jaffa is the object of 'economic conquest' – as Arthur
Ruppin described it 90 years ago – by Jewish residents from Tel Aviv.

The Peres Center for Peace

The ill-fated Peres Center for Peace is another example of the interrelationship
of architecture, planning and discourses of identity and control in Jaffa,
precisely because the center was dedicated to Shimon Peres' vision a 'new
Middle East' with Israel as its engine in for the global age.[112] What was to be
on of the most innovative architectural designs in the world was designed by
the well-known Italian architect Massimiliano Fuksas, and was described by
Israeli architecture critics as 'compact and urban' and 'simpl[y] sophisticat[ed]'
– that is, a local architecture without 'local features'.[113] The building will be
a return, for the first time in decades, to a 'straightforward Israeli architecture
... without borrowed elements, an architecture that attempts ... to be here
without 'blending in' yet at the same time help continue the 're-awakening
of Jaffa.[114]

The Center officials have promised 'that the center will also be available
to Jaffa residents and will not close itself off ... Peres Center officials say
the plan to erect the center was worked out together with Jaffa community
leaders and senior Palestinian Authority officials'. Yet no community 'leader'
with whom I spoke had been contacted by the Peres Center for Peace in any

capacity, and none – from establishment school principals to young activists – believed that in the end the community would have any real access to the site apart from visiting it like any other tourist, or that the projects the Center hopes to develop will involve or impact them in any significant manner.[115]

Their view of the prospects for any real relationship between the Center and the local community were in fact seconded by my interviews with the Center's staff. Thus when I called to arrange an interview to discuss its relationship with the local community a staff member with whom I spoke immediately became agitated and said, 'what do you want to talk about that for, we have nothing to do with Jaffa or Tel Aviv', a point she went on to stress again later on in our conversation. Similarly, the International Operations officer of the Center also admonished me not to focus on the potential projects or relationships with the locals, explaining that at least until the center is built in Ajami, such coordination and cooperation is not on the agenda.[116]

Finally, for Fuksas the Peres Center for Peace has been designed:

> to shed some light on the main principle of architecture, that is Ethics – and to caution that it is going to be trapped in the bounds of aesthetics. In the dawning of the new millennium, where the world stands before harsh problems of over-population, architecture can not stand solely on aesthetics, but must develop a sense of responsibility, and distinguish between good and bad ... Once in a while, light beams out of the [building], like a bit communication center. Light... will shine from with in – this is the first symbol.[117]

This 'light-house image' is central to Tel Aviv's history, as it long was an official symbol of the city. Yet while Fuksas focuses on 'levels of history' in his multi-leveled structure, not once in the lengthy interview are the local Palestinian residents mentioned, their history seemingly forgotten, although Fuksas did state, in response to a question regarding the urgency of this building in light of the major infrastructural, education and housing problems in the quarter, that 'the peace center will be a home for all ... It will develop the Jaffa economy. I totally agree with the Peres ideology that economic prosperity reinforces Democracy ... Israel is the most culturally diverse and authentic place I know, and it works here better than most places in Europe, believe me'.[118]

The Peres 'ideology' as I explore in below, is one firmly grounded in the market, tourism, and other discourses that few of Jaffa's non-elite residents have found to be living up to their promise. But here let us complete the analysis of the style of the building, which Fuksas explains was allowed to come to fruition after he persuaded the Tel Aviv Municipality to agree to a 'totally un-ethnic building with no arches, stone face bricks, but with bare concrete' in Jaffa based on a 'willingness to regulate Israel's architectural past, of which concrete plays an essential role'.[119]

What is most surprising about this comment is that one of the most outstanding examples of Bauhaus architecture in the whole country is a former – and self-consciously modern – Arab house (now the French Embassy), located only a few hundred meters from the site of the new Center and which was designed by an Italian architect. Not to mention the many other IS buildings – concrete, of course – located in the direct vicinity of the site.[120] Indeed, the reason for these elisions is clear by the manner in which the interview progresses. About mid-way through Fuksas stops talking about Jaffa and begins referring to Tel Aviv:

> Tel Aviv interests me so much more than Jerusalem … You can only ruin Jerusalem …Your opportunity is here, in Tel Aviv. So many meeting places and young people, and vitality you don't find in most European cities. Not it is time to do good things in Tel Aviv. I dream of planning a skyscraper in Tel Aviv. It is so beautiful, the buildings, the towers in Tel Aviv.[121]

Here again he repeats the traditional view of Tel Aviv as a 'young vibrant' city where one can do 'good things.' That is perhaps why the Tulkarm/ Palestinian version of the Center is viewed as out of 'context' – perhaps a bit too 'Disney', without the modern surroundings of Tel Aviv.

In fact, the reality of the Peres vision for a New Middle East is evident – ironically – by the presence of (usually) 'illegal' Palestinians from Gaza early each morning on Jaffa's Gaza Street looking, hoping to get a day's work from local contractors. These workers are part of the tens of thousands of foreign workers – many of their main competitors are similarly illegal Arab migrants from Egypt and Jordan (a resumption of at least a century's-long pattern of 'foreign' Arab migration to the Jaffa-Tel Aviv region) who are following employment trajectories similar to their counterparts living in the vicinity of most of the 'world cities'.[122]

Conclusion: Spatializing Palestinian Jaffa

More than a century ago Theodor Herzl explained what was necessary to create a Jewish state in Palestine: 'If I wish to substitute a new building for an old one, I must demolish before I construct'[123] Five decades later, at the height of the era of modernist planning, the French architect and city planner Le Corbusier – several of whose disciples become prominent Zionist planners and architects – quoted a famous Turkish proverb to epitomize the modernist ethic: 'Where one builds one plants trees. We root them up.'[124] From a similar but more critical perspective, Henri Lefebvre has explained how the 'plan' does not remain innocently on paper. On the ground, the bulldozer realizes 'plans'.[125]

The evidence presented in this article demonstrates that Jaffa can be understood as both a space of negation and identification for Tel Aviv. This ambivalence towards Jaffa mirrors the larger relationship of the Israeli state towards the Palestinian communities living within its pre-1967 borders. The entry of the Palestinian into the Israeli 'national self', or the self-definition of the Israeli state, is both ambivalent and paradoxical. It is ambivalent in that postmodernist architectural sensitivity towards Jaffa's Palestinian Arab heritage has remained 'superficial' and economic in orientation. The place-oriented postmodern architecture is used to catch a 'global' – and implicitly, non-Arab – elite, and disallow potentially political identification from Jaffa's Palestinian community. The double economy of fixing Jaffa for Orientalist gaze, on the one hand, and developing it along the line of a changing market economy on the other, represents both the economization and depoliticization of the Palestinian community.

The contested space of Jaffa and Tel Aviv further epitomizes the complex manner in which architectural and planning movements are inscribed in the politics of national identity in Israel: erasing 'tradition' (through IS architecture) and reclaiming/re-inscribing it (through discourses of heritage promoted by postmodernist architecture). As such, both are expressed in the economic as well as political idioms that in turn are central to the process of ethno-political identity formation in Israel as a modern nation-state. This is the dynamic governing the politics of urban design in contemporary Jaffa.

Notes

1   See Mark LeVine, 2004, Introduction and Chapter 1.
2   Lefebvre, 1992, p. 25.
3   Quoted in Wright, 1991, p. 16; cf. p. 319, note 8.
4   Holston, 1989, p. 83. The planners possessed a teleological view of history in which capitalist modernity signified an advance in human civilization that began in Europe and would spread to the 'backward' regions.

5   Latour, 1993, p. 39. He continued, 'Native Americans were not mistaken when they accused the Whites of having forked tongues. By separating the relations of political power from the relations of scientific reasoning while continuing to shore up power with reason and reason with power, the moderns have always had two irons in the fire. They have become invincible ...' (p. 38).

6   Holston, 1989, p. 68. The modern nation-state process, Nico Poulantzas has observed, has always involved the eradication of the traditions, histories, and memories of the nations dominated as part of its self-creation and expression. This process is even more pronounced when the state in the making is a colonial-settler state (cf. Swedenburg, 1995, pp. 3, 8).

7   Harvey, 1989, 16. Cf. Schumpeter, 1976. Mitchell (1988, p. 165) explains why the city magnifies the power of the creative destruction at the heart of the project of modernity: 'The identity of the modern city is in fact created by what it keeps out. Its modernity is something contingent upon the exclusion of its own opposite. In order to determine itself as the place of order reason, propriety, cleanliness, civilization and power, it must represent outside itself what is irrational, disordered, dirty, libidinous, barbarian and cowed.'

8   This creative destruction was helped by the emerging disciplines of geography, sociology and political 'science' which were in the process of overturning existing ways of seeing, mapping, governing, and living in the world.

9   See LeVine, 2004, Chapter 6.

10 Wright, 1991, p. 8.

11 *Yediot Tel Aviv*, 1925, #2, p. 3, 15 September 1925 meeting of the Tel Aviv Town Council.

12 *Yediot Tel Aviv*, 1934, #11–12, p. 43.

13 Dizengoff, undated, p. 2.

14 Droyanov, 1935, p. 188. As opposed to the Ottoman governance of Jaffa, which 'did not know any law and did not have any rules (professional regulations) for building the city. So there was complete anarchy' (*Yediot Tel Aviv*, 1925, #2, p. 3, 15 September 1925; meeting of Town Council).

15 For a detailed discussion of the mythology of the sands, see LeVine, 2004.

16 Quoted in Meroz, 1978, p. 35.

17 CZA, L51/71, Bye-laws of the Nahalat Binyamin society; quoted in Kark, 1990, p. 121.

18 Katz, 1986, p. 405.

19 Thus the leaders of Tel Aviv would often speak of even the Jewish neighborhoods of Jaffa as being '*galut*' (cf. CZA S25/5936, meeting 16 May 1940. Although the title of the protocol says '... Jewish neighborhoods within the borders of *Tel Aviv*', that is clearly a mistake).

20 TAMA, Protocols of Achuzat Bayit, 1908, p. 21.

21 Thus it would have to be situated at a distance from the city in order to maximize its autonomy, and only Jews could live in the neighbourhood (the by-laws prohibited the sale or renting of houses to Arabs; CZA, L18/105/4, L51/52, L2/578; TAMA, Protocols of Achuzat Bayit, 6 June and 31 July 1907). Apparently, as Katz points out (1993, p. 284), in purchasing the land at Kerem Jebali the members of Achuzat Bayit felt that they were 'beyond the city limits of Jaffa.' The Jaffan authorities, not surprisingly, considered the area within their jurisdiction.

22 Abercrombie, quoted in Herbert and Sosnovsky, 1993, p. 189.

23 Nathan Harpaz, 'Ir "Ivrit" – Adriculut "Ivrit"', article found at TAMA, pp. 279, 281.

24   Levin, 1982, p. 233.

25   TAMA library, article by Harlap, title page missing, p. 45.

26   Arliak, 1987, 116.

27   Cf. Troen, 1991, p. 18. This simulation of the indigenous architecture would seem to involve an epistemological operation similar to that described by Timothy Mitchell when Europeans constructed the Middle Eastern exhibitions at the great world exhibitions of the latter nineteenth century; that is, the construction of a 'peculiar distinction between the simulated and the real' (1988, p. 7).

28   Irlak, 1987, p. 11, Zalmona, 1998, p. xii.

29   By this I mean that 1929 violence and many official investigations it prompted (such as the Shaw Commission and Passfield White Paper of 1930, and the Hope-Simpson and French reports of 1931) constituted the first real threat to British support for Zionist policies in the country (cf. LeVine, 1995).

30   *Mishar ve-Te'asia,* ibid., p. 100. Similarly, Arieh Sharon describes how when he first returned to Tel Aviv after studying at the Bauhaus, 'I was very depressed by the architecture' (Sharon, 1976, p. 46), while Dizengoff lamented the 'unsuitability of earlier buildings to the conditions of the place and the goals for which they were built' ('The Question of Housing in Tel Aviv', *Yediot Tel Aviv,* 1937, #6, p. 178).

31   Nitzan-Shiftan, 1996, p. 151; cf. Monk, 1997, p. 96.

32   Le Corbusier, 1947, p. 82.

33   Amos, 9:14.

34   For a detailed discussion of this issue see LeVine, 1998.

35   Y. Shiffman, 'Tel Aviv: Today and Tomorrow', in *Palestine and Near East* magazine, September 1942, p. 167.

36   TAMA, 4/2667b, undated memo in Hebrew on the history of Tel Aviv's borders and development. One way of both achieving and demonstrating this 'independence' was through enacting specific town-planning regulations intended to differentiate Tel Aviv from the Arab quarters of the town. These regulations were put into place during the years between 1909–21 when several nearby Jewish neighborhoods joined the new quarter and demonstrate how town-planning was already an important component of the Tel Avivan self-identity. The new neighborhoods thus had to agree to follows Tel Aviv's by-laws in terms of town planning, architecture and hygiene, and its prohibition against selling land to Arabs.

37   Ruppin, 1971, p. 121.

38   Ibid.

39   *Town-Planning Handbook of Palestine, 1930,* New York Public Library, SERA, p. 5. See *Falastin,* 27 February 1923, p. 3 for article on the new land ownership law enacted by the Jaffa Municipality.

40   Owen, 1991, p. 7.

41   For an extensive review of the events surrounding the destruction of the Old City see Gavish, 1994.

42   Interview with Y. Shiffman, in *Palestine and Near East,* January 1943, p. 12.

43   ISA, M/46/L176316, note dated 6 June 1947.

44   Thus one colonial office official who reviewed the file wondered why the schemes contemplated could not be carried out by the utilization of undeveloped land lying within the existing municipal boundary (particularly the larger area lying in the north of Tel Aviv).

45   ISA, M/44/89/L/308; unsigned note dated 21 June 1945.

46   Cf. Graicer, 1982.

47   In fact, preference could have been given to the villagers to obtain rights to the land, as opposed to either leaving it uncultivated state land or transferring it to Tel Aviv for urban and industrial development, which the government considered doing. Instead, the Director of Lands noted that the land was 'a very valuable site for urban development purposes', as they claimed in their defense against the villagers' suit. Barbara Smith writes that 'Not only did British officials misread the previous land laws, but they failed fully to understand local agricultural practices and appreciate the unusual post-war conditions' (1993, p. 99). ISA, M/18/LD8/3411, letter to Department of Agriculture dated 7 February 1928.

48   In fact, at the turn of the century, Ajami was thought of by Jews as the best and most modern place to live in Jaffa. Thus, when one long-time Jewish resident of Jaffa was told about the coming establishment of Tel Aviv, he exclaimed 'Go live in Ajami and enjoy exemplary cleanliness. Move into an attractive, roomy house and stop wasting your time with foolishness' (quoted in Kark, 1990, p. 116).

49   *al-Jam'iah al-Islamiyyah*, 21 December 1937, p. 3. The consolidated town-planning regulations of 1930 gave a local Town Planning Commission the right to build roads as a first step toward executing a town plan.

50   Tel Aviv Municipality, 1974, pp. 5–6.

51   Schnell, 1993, p. 41.

52   Semyonov and Hadas, 1997, p. 195.

53   Lefebvre, 1996, p. 51.

54   Corby Kummer, 'Tel Aviv: Secular City, Where Israel Meets the Modern World', *The Atlantic Monthly*, December 1995 issue; cf. 'Survey of Israel at 50', *Economist*, 25 April 1998, p. 18, and Serge Schmemann, 'What's Doing in Tel Aviv', *New York Times*, 21 December 1997. Kummer continues, 'If you want to stroll down an Israeli street lined with quirky, forward-looking shops, or sit back and enjoy a relaxed meal in a restaurant that cares about elegance and service, Tel Aviv is the place.' Cf. *Economist*, 25 April 1998, 'Survey of Israel at 50', p. 18; *Le Monde*, 25 April 1998.

55   Cf. LeVine, 1998 and forthcoming, Chapter 5, note 24, TAMA, 4/3565, August 1941 article by L.V. Beltner entitled 'City of the Jews'.

56   'Survey of Israel at 50', *Economist*, 25 April 1998, p. 18.

57   Cf. Gorny, 1984, pp. 19–27.

58   Corby Kummer, *The Atlantic Monthly*, December 1995 issue.

59   Including the popular 'Jaffa Portraits' crime series, which aired on the 'family channel' (Helen Kaye, 'A 'Portrait' of Aviva Marks', the *Jerusalem Post*, 25 March 1997, p. 7). More recently, in a TV documentary about Jaffa and its redevelopment called 'Things in their Context' (produced by a European Channel and aired on Israeli TV) the Arab population and history of the city was completely left out (cf. Nir Nader, 'Things Out of Context', *Challenge*, July–August, 2000, p. 17).

60   Andre Mazawi, 'Film Production and Jaffa's Predicament', *Jaffa Diaries*, 9 February 1998, at www.jaffacity.com.

61   Official 'Tel Aviv-Yafo' guide of the Ministry of Tourism and Tel Aviv Hotel Association, 1997.

62   State of Israel, Ministry of Defense, Museums Unit, *Brochure of the Museum of the I.Z.L;* Eretz Israel Museum, *Guide to Yafo, for Self-Touring*, Tel Aviv, 1988.

63   Ministry of Tourism, *Official Guide to Tel Aviv*, 1997, pp. 23, 32. This theme of the beauty and quaintness of Jaffa at night was already being used in articles in the local Hebrew Jaffa paper, Yediot Yafo, as far back as 1963 (see 'There's Nothing Like Yafo at Night', *Yediot Yafo*, July 1963, p. 4).

64   Tel Aviv Municipality, *Tel Aviv: People and their City*, Tel Aviv, 1974, pp. 3, 5–6.

65   *Al-Ayyam*, 19 May 1997. Story on the Jaffa Internet Discussion group.

66   *Falastin*, 9 May 1946, p. 2.

67   Tamari and Hammami, 1998, p. 73. Tamari, forthcoming, p. 8.

68   April 1998 meeting sponsored by 'Face to Face.' Other bourgeois Palestinians with the ability to travel freely in Israel also have told me of their fondness for visiting Tel Aviv.

69   This demand had, during my fieldwork in 1996–98, won some support among Jewish residents of Jaffa who also see themselves as excluded from the Municipality's plans for the neighborhood.

70   RA, public declaration of al-Rabita, 24 June 1997 entitled 'We will not Leave Because We are Planted Here' (lan nuraha innna huna manzura'una).

71   Talia Margolit, 'Cities of the World, Twin Identity', *Ha'aretz*, 2 May 1997, p. B6.

72   Nachmias and Menahem, 1993, p. 294. Shachar, 1997, p. 312.

73   *Ha'ir*, 28 February 1997, p. 31; 26 September 1997, p. 37; 6 June 1997, p. 30.

74   Esther Zandberg, Interview with Massimiliano Fuksas, *Ha'aretz*, 14 September 2000, web edition.

75   Most new Israeli apartments, especially luxury models, have special 'security rooms' that are bomb proof and are provisioned in case of chemical attacks.

76   'Yafo is Disneyland, the North is a Tragedy' (Hebres), *Ha'ir*, 12 July 1997, p. 24.

77   Bilski, 1980, p. 259.

78   Churchman and Alterman, 1997. The best example of this lacuna is the two volume edited collection by Nachmias and Menahem, *Social Processes and Public Policy in Tel Aviv-Yafo*, specifically Arieh Shachar's 'The Planning of Urbanized Areas: A Metropolitan Approach and the Case of the Tel Aviv Region' (Hebrew), in Nachmias and Menahem, 1997. Similarly, a 1996 article titled 'The Revitalization of Tel Aviv' basically excluded Jaffa – and along with it any hint of the political/national implications of the trend toward 'postmodern' architecture and town-planning in Tel Aviv – from its analysis (cf. Schnell and Graicer, 1996, especially pp. 114–15).

79   Yiftachel, 1995, p. 498.

80   An agency that since 1901 has used donations from Jews around the world to purchase land in Palestine/Israel that, once in its possession, can never be sold to non-Jews.

81   *Ha'aretz*, 6 November 1998, p. A1.

82   Schnell and Graicer, 1996, p. 106. Specifically, Tel Aviv during that time became the center for the management of the large manufacturing enterprises in the country, two-thirds of which manufactured their goods outside the Tel Aviv metropolitan periphery.

83   Ginsberg, 1993, p. 151; Portugali, 1994, p. 312. Yet as Portugali points out, the words 'uniform' and 'style' represent the very opposite of postmodernism.

84   Rabinowitz, 1997, p. 15. Also see Rabinowitz, 1992.

85   Cf. LeVine, forthcoming, Chapter 1.

86   Yiftachel, 1995, pp. 494, 496, 498. For a more detailed analysis of this dynamic in the country at large, see Yiftachel and Meir, 1998.

87   Portugali, 1994, p. 312. This remark was made regarding the Occupied Territories, but it is equally relevant in this context.

88   'Yafo is Disneyland, the North is a Tragedy' (Hebrew), *Ha'ir*, 12 July 1997, p. 24.

89   *Challenge*, May–June 1998, pp. 12–13, 18.

90   Mazawi, 1988.

91   *Challenge*, May–June 1998, pp. 12–13, 18.

92   *Ha'ir*, 20 June 1997, p. 32.

93   Architect Yitzchak Leer, quoted in Mazawi, 1988, p. 4.
94   Andre Mazawi and Makram Khoury Machool, 'Spatial Policies in Jaffa, 1948–1990', (Heb.) in Liski, 1991, p. 66.
95   Architect Yosi Tager, quoted in *Ha'ir*, 18 April 1997, p. 43.
96   *Ha'ir*, 3 December 1999, p. 24.
97   'Regional Descriptive Plan for the Tel Aviv region', (Heb.), Interim Report #3, 16 March 1998, published by Hebrew University under direction of Prof. Arieh Shachar, pp. 2–4.
98   Quoted in Ali Waqed, 'Place for Worry', *Ha'ir*, 20 September 1996, p. 1(9?). A public brochure from the Tel Aviv Municipality dated November 2000 – that is, after the outbreak of the intifada – offered residents homes at low prices, but required extensive rebuilding, which few can afford (Hamishlamah Leyafo, 'Ila Sukan Medinat Yafa al-Karam').
99   *Ha'ir*, 15 August 1997, p. 34.
100   Shachar, 1997, p. 37. Often, when new building is allowed, it is only on the roofs of existing structures, in order to keep as much vacant land as possible available for development.
101   *al-Jam'iah al-Islamiyyah*, 18 Decemmber 1932, p. 7. The phrase was 'a fork in the road', ('muftarik fi al-turuq'). Cf. Nissim Shachar, 1997.
102   Cf. *al-Sabar*, 24 Decemmber 1997, p. 8. For more information on the continuing battle between the fishermen and the Municipality, see *al-Sabar*, 7 January 1998, p. 4; 12 January 1998, p. 9. For another description of the situation see 'The Fishermen are furious at the closure of the port', *al-Sabar*, 18 March 1997, p. 9. The project would link the port directly to the lived area of the old city through the construction of up to four thousand elite residence and hotel units. See 17 April 1997 letter to Schmuel Laskar from Dan Darin of Tel Aviv Planning Commission with attached plan. Copy given to author by local journalist. Also see *Ha'ir*, 15 August 1997, p. 22, for reporting on the plan.
103   Ministry of Tourism, *Official Guide to Tel Aviv*, 1997, pp. 23, 32. As important, opposition to building a hotel/tourist complex in the port by other officials are being made in the name of 'declar[ing] Jaffa Port a *national* site', which in the eyes of Jaffa's Palestinian residents, would still exclude them (cf. comments by Tel Aviv Deputy Mayor Michael Ro'ah in *Ha'ir*, 13 August 1999, p. 33).
104   Lahat was a vocal proponent during his mayoralty of continued 'Judaization' in Jaffa as well as vigilance against attempts by the local Arab community to gain more control over the neighborhood. For a description of the declared goals of the company, named Ariel Real Estate-Yafo, see their November 1998 publication in Hebrew and Arabic, 'Yediot Yafo'.
105   Author's interview with several representatives of Gadish, 13 July 1997.
106   *Ha'ir*, 7 May 1999, p. 28.
107   *Ha'ir*, 7 May 1999, p. 28. 'Makom halomi' could be translated as place or space, but within the context of the meaning of 'place' as related to experiences of its inhabitants I have chosen 'space' (cf. Harvey, 1996).
108   Cf. LeVine, 1998.
109   Cf. LeVine, forthcoming, Chapters 1, 2, 5, and 6.
110   All quotes from Andromeda Hill website: www.andromeda.co.il/home.html.
111   As the new Arab councilman from Jaffa described it (Ali Waqed and Ronen Zartzki, 'I, Rifa'at Turk', *Ha'ir*, 9 May 1997, p. 14).
112   The Peres Center for Peace, *Developing Tomorrow's Peace*, 2000–2001 Annual Report.
113   Esther Zandberg, 'Simple sophistication', *Ha'aretz*, 7 September 2000. When Shimon Peres first saw the site he exclaimed that the land 'would make you want to cry', which coincides in an ironic way with Palestinian views of the fate of their erstwhile hospital, and

of Jaffa as a whole (see Esther Zandberg, Interview with Massimiliano Fuksas, *Ha'aretz*, 14 September 2000, web edition).

114 Esther Zandberg, 'Simple sophistication', *Ha'aretz*, 7 September 2000. The four-storey building, which will tower 16m above the street and 26m above the seafront, will contain 2,500m² of floor space. Its prominent presence on the shoreline seems a fair price to pay, thanks to both its precise appearance and the large, open public space that it leaves surrounding it. The area will be developed as a 'dry' garden, as Messer calls it – a park without grass expanses – for gatherings and use by the general public.

115 Interviews with community leaders in Jaffa, December 2000, including members of al-Rabita, an Palestinian principal of a Jewish-Arab primary school located a few hundred meters from the proposed site of the Peres Center for Peace, a Palestinian school teacher, and a Palestinian official of Mishlama.

116 Interview, Peres Center for Peace, 10 December 2000.

117 Esther Zandberg, Interview with Massimiliano Fuksas, *Ha'aretz*, 14 September 2000, web edition.

118 Ibid.

119 Esther Zandberg, Interview with Massimiliano Fuksas, *Ha'aretz*, 14 September 2000, web edition. Fuksas explained that he regards arhces as 'cheap architectural props. A modern time architectural stepping stone is missing. Architecture is uch an impotant thing. It draws together people and is good for the economy.'

120 For analyses of Ajami's architectural heritage, including IS, see Tzafrir, 1995, LeVine, 1998 and forthcoming, where I discuss how the owner/builder of the house, who was personally very 'conservative', nevertheless wanted to build the 'most modern house' in Palestine.

121 Esther Zandberg, Interview with Massimiliano Fuksas, *Ha'aretz*, 14 September 2000, web edition.

122 Cf. Sasken, 1998.

123 Herzl, 1941, p. 84.

124 LeCorbusier, 1947, p. 82.

125 Lefebvre, 1991, p. 191.

References

Arlik, Avraham, 'Architecture in Tel Aviv' (Hebrew) *Ariel*, 48–49 (1987), pp. 111–20.

Bilski, Raphaella (ed.), *Can Planning Replace Politics? The Israeli Experience* (Boston, Martinus Mijhoff Publishers, 1980).

Churchman, Arza and Alterman, Rachelle, 'Conflict Management in Urban Planning in Tel Aviv-Yafo', in David Nachmias and Gila Menahem (eds), *Mechkarei Tel Aviv-Yafo: Tahalikhim Chevratim ve-Mediniot Tzeiburit* (*Social Processes and Public Policy in Tel Aviv-Yafo*) (Tel Aviv: Ramot Publishing House, 1997 and 1997), pp. 39–64.

Dizengoff, Meir, *Renaissance of Tel Aviv* (Hebrew) (Tel Aviv: Eitan Shashoni Press, undated).

Droyanov, A. et. al., *Sefer Tel Aviv* (*Tel Aviv Book*) (Tel Aviv, 1953).

Gavish, Dov, *Operation Jaffa 1936 – A Colonial Improvement of the Inside of a City* (*Mivtzah Yafo 1936 Shipur Coloniali shel Pnei Ir, Tadfis Mtokh Eretz Ysrael*) (Jerusalem: Mehkarim be-Yediot Haaretz Veatikoteha, 1994).

Ginsberg, Yona, 'Revitalization of Two Urban Neighborhoods in Tel Aviv: Neve Tzedek and Lev Tel Aviv' (Hebrew), in David Nachmias and Gila Menahem (eds), *Mechkarei Tel Aviv-Yafo: Tahalikhim Chevratim ve-Mediniot Tzeiburit* (*Social Processes and Public Policy in Tel Aviv-Yafo*) (Tel Aviv: Ramot Publishing House, 1993 and 1997), pp. 147–66.

Gorny, Joseph, 'Utopian Elements in Zionist Thought', *Studies in Zionism*, 5 (1) (1984), pp. 19–27.

Graicer, Iris, 'Workers Neighborhoods: Attempts at Shaping an Urban Landscape through Social Ideologies in Eretz Israel during the Mandate Period' (Hebrew), PhD dissertation (Hebrew University, 1982).

Herbert, Gilbert, and Sosnovsky, Silvina, *Bauhaus on the Carmel and the Crossroads of Empire* (Jerusalem: Yad Ben-Zvi, 1993).

Herzl, Theodor, *Old-New Land*, trans. Lotta Levensohn (New York: Bloch Publishing Company, 1941 [1902]).

Holston, James, *The Modernist City: An Anthropological Critique of Brasilia* (Chicago: University of Chicago Press, 1989).

Kark, Ruth, Jaffa, *A City in Evolution, 1799–1917* (Jerusalem: Yad Ben Zvi Press, 1990).

Katz, Yossi, 'Ideology and Urban Development: Zionism and the Origins of Tel Aviv, 1906–14', *Journal of Historical Geography*, 14 (4) (1986), pp. 402–24.

Latour, Bruno, *The Pasteurization of France*, trans. Alan Sheridan and John Law (Cambridge, MA: Harvard University Press, 1988).

Le Corbusier, *The City of Tomorrow and its Planning*, trans. F. Etchells (London: The Architectural Press, 1947).

Lefebvre, Henri, *The Production of Space*, trans. Donald Nicholson-Smith (Oxford: Blackwell, 1991).

Lefebvre, Henri, 'Philosophy of the City and Planning Ideology', in *Writings on Cities*, ed. and trans. Eleanor Kofman and Elizabeth Lebas (Oxford: Blackwell Publishers, 1996).

Levin, Michael, 'East or West: Architecture in Israel, 1920–33', in Mark Scheps (ed.), *The Twenties in Israeli Art* (Tel Aviv: Ariel Press, 1982), pp. 223–40.

LeVine, Mark, 'The Discourse of Development in Mandate Palestine', *Arab Studies Quarterly*, Winter (1995), pp. 95–124.

LeVine, Mark, 'Conquest Through Town-Planning: The Case of Tel Aviv', *Journal of Palestine Studies*, 17 (1–2) (1998), pp. 36–52.

LeVine, Mark, *Ovethrowing Geography: Jaffa, Tel Aviv and the Struggle for Palestine* (Berkeley: University of California Press, in press).

Liski, Chaim (ed.), *City and Utopia: A Collection of Essays for 80 years of Tel Aviv* (Tel Aviv: The Israeli Society for Publishing, 1989).

Mazawi, Andre, 'Spatial Expansion and Building Styles in Jaffa: Past and Present' (Hebrew), in Mazawi, A. (ed.), *Art and Building in the View of the Paintbrush* (Hebrew) (Jaffa: The Center for Arabic Culture, 1988).

Meroz, Tamar, 'Tel-Aviv-Jaffa: A Story of a City' (Dr Ben-Zion Kaduri Fund, 1978).

Mitchell, Timothy, *Colonising Egypt* (Berkeley: University of California Press, 1988).

Monk, Daniel, 'Autonomy Agreements: Zionism, Modernism and the Myth of a Bauhaus Vernacular', AA Files, 1997.

Nachmias, David, and Menahem, Gila (eds), *Mechkarei Tel Aviv-Yafo: Tahalikhim Chevratim ve-Mediniot Tzeiburit* (*Social Processes and Public Policy in Tel Aviv-Yafo*), Vols I and II (Tel Aviv: Ramot Publishing House, 1993 and 1997).

Nitzan-Shiftan, Alona, 'Contested Zionism – Alternative Modernism: Erich Mendelsohn and the Tel Aviv Chug in Mandate Palestine', *Architectural History*, 39 (1996), pp. 147–80.

Owen, Roger, 'The Role of Mushaa (Co-ownership) in the Politico-Legal History of Mandatory Palestine', SSRC Workshop on *Law, Property and State Power*, Buyukada, 1991.

Portugali, Juval, 'The Taming of the Shrew Environment', *Science in Context*, 7 (2) (1994), pp. 307–26.

Rabinowitz, Dan, 'An Acre is an Acre is an Acre? Differentiated Attitudes to Social Space and territory on the Jewish-Arab Urban Frontier in Israel', *Urban Anthropology*, 21 (1) (1992), pp. 67–89.

Rabinowitz, Dan, *Overlooking Nazareth: The Ethnography of Exclusion in Galilee* (Boston: Cambridge University Press, 1997).

Ruppin, Arthur, *Memoirs, Diaries, Letters* (London: Wiedenfeld and Nicolson, 1971).

Sassen, Saskia, *Globalization and Its Discontents: Essays on the New Mobility of People and Money* (New York: W.W. Norton, 1998).

Schnell, Yitzhak, 1993, 'The Formation of an Urbanite Life Style in Central Tel Aviv' (Hebrew), in David Nachmias and Gila Menahem (eds), *Mechkarei Tel Aviv-Yafo: Tahalikhim Chevratim ve-Mediniot Tzeiburit* (*Social Processes and Public Policy in Tel Aviv-Yafo*) (Tel Aviv: Ramot Publishing House, 1993 and 1997), pp. 41–60.

Semyonov, Moshe, Lewin-Epstein, Noah and Mendel, Hadas, 1997, 'The Labor Market Position of Arab Residents of Tel Aviv-Yafo: A Comparative Perspective' (Hebrew), in David Nachmias and Gila Menahem (eds), *Mechkarei Tel Aviv-Yafo: Tahalikhim Chevratim ve-Mediniot Tzeiburit* (*Social Processes and Public Policy in Tel Aviv-Yafo*) (Tel Aviv: Ramot Publishing House, 1993 and 1997), pp. 191–210.

Shachar, Arieh, 'The Planning of Urbanized Areas: A Metropolitan Approach and the Case of the Tel Aviv Region' (Hebrew), in David Nachmias and Gila Menahem (eds), *Mechkarei Tel Aviv-Yafo: Tahalikhim Chevratim ve-Mediniot Tzeiburit* (*Social Processes and Public Policy in Tel Aviv-Yafo*) (Tel Aviv: Ramot Publishing House, 1993 and 1997), pp. 305–20.

Sharon, Arieh, *Kibbutz + Bauhaus: An Architects Way in a New Land* (Tel Aviv: Massada Publishing, 1976).

Smith, Barbara, *The Roots of Separatism in Palestine* (Syracuse: Syracuse University Press, 1993).

Swedenburg, Ted, *Memories of Revolt: The 1936–1939 Rebellion and the Palestinian National Past* (Minneapolis: University of Minnesota Press, 1995).

Tamari, Salim and Rema Hammami, 'Virtual Returns to Jaffa', *Journal of Palestine Studies*, 27 (4) (1998), pp. 64–79.

Tamari, Salim, with Ruba Qanaan (eds), 'Jaffa: The Modernity of an East Mediterranean City', manuscript under review (forthcoming).

Tel Aviv Municipality, *Tel Aviv: People and Their City* (Tel Aviv, 1974).

Troen, S. Ilan, 'Establishing a Zionist Metropolis: Alternative Approaches to Building Tel Aviv', *Journal of Urban History*, 11 (1991), pp. 10–36.

Tzafrir, Daron, *Yafo: Mavet Leajami* (*Jaffa: A Glance at Adjami, An Architectural Profile*) (Tel Aviv: City Engineers Department, 1995).

Wright, Gwendolyn, *The Politics of Design in French Colonial Urbanism* (Chicago: University of Chicago Press, 1991).

Yiftachel, Oren, 'The Dark Side of Modernism: Planning as Control of an Ethnic Minority' in Watson, S. and Gibson, K. (eds), *Postmodern Cities and Spaces?* (Oxford: Basil Blackwell, 1994), pp. 216–34.

Yiftachel, Oren, and Meir, Avinoam (eds), *Ethnic Frontiers and Peripheries: Landscapes of Development and Inequality in Israel* (Boulder, CO: Westview Press, 1998).

Zalmona, Yigal, 'To the East?', in Yigal Zalmona and Tamar Manor-Friedman (curators), *To the East – Orientalism in the Arts in Israel* (Jerusalem: The Israel Museum, 1998), pp. ix–xv.

PART IV
LANDMARKS OF IDENTITY

Chapter 9

Academia and Spatial Control: The Case of the Hebrew University Campus on Mount Scopus, Jerusalem[1]

Diana Dolev

Introduction

Two extremely powerful forces have been instrumental in determining the architectural designs and the significance of the Mount Scopus Hebrew University campuses.[2] One was the institution's central value for the Jewish national revival and the other its association to the traditional concept of the Dome of the Rock's image as the Holy Temple (see Figure 9.1).[3] When one bears in mind that as a means of connecting the Jewish national revival with a glorious past, Zionist rhetoric often made use of the symbolic Third Temple (the future Temple that would replace the ancient one; a symbol of the renewal of the Jewish settlement in Eretz-Israel), the identification of the actual Dome of the Rock with the vision of the Holy Temple becomes particularly charged. The idea of a Jewish University goes centuries back, yet the emergence of the two forces into a dialectic process parallels the pre-Zionist national revival period in east Europe. The outcome of the process had been given distinctive shape in the proposed design for the Hebrew University's first master plan and in the present campus.

Once chosen as a scene of these forces, Mount Scopus became a central protagonist in developing the architecture of the campus into a forceful nationalist tool. The occasion of the laying of the cornerstones for the Hebrew University in 1918 marked a merge of the institute and the site into a unified entity in the minds of Zionists and the Jewish community in Palestine and the Diaspora. An interrelationship developed between site and institute that led to unification which is still valid today.

Figure 9.1   The Dome of the Rock viewed through the Hebrew University Synagogue glass wall, Mount Scopus, Jerusalem

The Zionist Movement and the Establishment of the Hebrew University

At its earliest stages the Zionist Organization recognized the idea of the University as being central on its agenda; it was spread out by Zionist propaganda in communities around the world, discussions were held among Zionist activists everywhere on the different issues concerning it. The issue of the Jewish/Hebrew[4] University appears almost at every crossroad of Zionist activity accounts. Among other things, the importance of the University is indicated by the line of central Zionist leaders who promoted it; Chaim Weizmann, Naum Sokolow, Menachem Ussishkin, Menachem Sheinkin, Arthur Rupin and others.[5] Their actions toward founding the University involved political manipulations as well as attempts to define the University as an academic institution for the Jewish people. Those were often intertwined – as the history of the Zionist struggle for the erection of the University conveys.

The 11th Zionist Congress in Vienna in 1913 was a turning point for the advancement of the University plans. Weizmann devoted most of his address to the advantageous value a Hebrew University would have for world

Jewry.[6] Ussishkin advocated an approach that linked the University with the sacredness of Jerusalem. Thus the future University was presented as part of the colonization plan and of national identity, and required the Congress's resolution that the Hebrew University was of first rate political and national importance.[7]

Were the inhabitants of Jerusalem in need of a University? That question was not brought up at all. In fact the location issue was one of prestige rather than a true need. One approving delegate at the Congress was Heinrich Loewe, whose words provide a background that explains prevailing contemporary positions. He said that 'Universities are the birthplace of culture and *Bildung*: the European states have understood their value. Now Central Europe is celebrating the hundredth anniversary of the War of Liberation, in which the universities played such an important role. From where was the liberation of Prussia led? From the founding of the University of Berlin!'[8] Loewe's words echo the Zionist conception of the future University as leading the way towards creating a separate Jewish national entity and a New Jewish Man. The term *Bildung* was often used in this context by Zionist ideologists. The comparison with Prussia and the University of Berlin provided the Congress delegates with a familiar historical and cultural foundation for the idea. They unanimously resolved that the University would be erected in Jerusalem.

The Significance of the Choice of Site

At the early stages of the emergence of the idea of the University there had been uncertainty whether it should be located in Europe or in Palestine. Later, as the role of the University shifted from serving the needs of Jewish students in east Europe into a central propaganda tool, it became clear that the University should be erected in Palestine. This was followed by a debate as to the appropriate location in Palestine – whether it should be placed close to the *moshavot* (Hebrew for colonies), which were at the core of Zionist ideology and action at the time, or in a town. A transformation in the attitude of the Zionist movement towards Jerusalem was at the source of the Congress resolution mentioned above. Taking into account that the Zionist leaders who suggested Jerusalem as the future University locus were familiar with their local European trends, it would not be too farfetched to connect this choice with the international status of Jerusalem in those days.

Along the second half of the nineteenth century, prominent projects were undertaken outside the walls of the Old City by British, German and Russian official or religious organizations. As it was highly congested, every extramural addition around the Old City was most conspicuous and imposing. The fact that the Great Powers chose to build costly and prominent buildings in and around Jerusalem indicates, that beyond their primary functions as churches, hospitals and schools those were actually unofficial representatives of their governing authorities (Crinson, 1996). In the special context of Jerusalem of mid-nineteenth century and owing to its extraordinary religious and symbolic significance for Jews, Christians and Moslems all over the world, those buildings were actually meant to accumulate prestige and a demonstration of presence for their countries. Constructing the Hebrew University in Jerusalem can be interpreted as another such declaration of presence and power.

On 24 July 1918, the foundation stones for the Hebrew University were laid on the eastern slope of the Sir John and Caroline Gray Hill estate on Mount Scopus, which had been purchased for the construction of the University. As there were no academic activities that required a dwelling at that time, the ceremony was planned as a central political event.[9] The old convention of referring to the proposed University as the Third Temple gained reinforcement by the physical proximity to the Temple Mount, and the traditional association of the Mount Scopus site with the act of viewing it.[10] The image was taken up abundantly later at University convocations. Another indication to ancient roots and sacred connotations applied to the future University was the use of a quotation from Isaiah's prophecy (2:3): 'for out of Zion shall go forth the law and word of the Lord from Jerusalem' as an inscription on documents and banners on occasions of University promotion and fund raising meetings or conferences.

One celebration especially stands out when in February 1923 Albert Einstein was invited to give a lecture in the former Gray Hill residence. A large number of dignitaries attended the occasion, which was obviously organized to promote the interests of the future University.[11] Introducing Einstein, Ussishkin said:

> ... three thousand years ago on Mount Moriah opposite this site ... King Solomon built a temple to the Universal God and dedicated it as a house of prayer for all nations. We pray that this temple, the home of the Hebrew University, will be a temple of science to all nations.[12]

Ussishkin then turned to Einstein and invited him to 'mount the platform which has been waiting for you two thousand years'. Hence the sanctification of the Hebrew University had begun prior to the materialization of the institute's academic vocation.

The sanctified image of the University as a Third Temple in its chosen location on Mount Scopus had been even further fortified by the generally accepted image of the Dome of the Rock as the ancient Temple. Situated in the middle of the *Haram-a Shariff*, the plateau known also as the Temple Mount, the ancient Islamic building was perceived since the time of the Crusaders, by Christians and Jews, not for what is really was. Numerous manuscript illustrations depicting the form of the Dome of the Rock entitled The Holy Temple, demonstrate this phenomenon. It is therefore not altogether surprising that at the turn of the nineteenth century visitors to Jerusalem were able to observe the Dome of the Rock and yet 'see' the Temple. After all orientalist painters depicted sites of the Holy Land as Biblical scenes, regardless of the fact that those were built much later and that the artists-travelers themselves belonged to another era. Therefore, looking at the Dome of the Rock and yet 'seeing' a vision of the Holy Temple, was in accord with contemporary conventions. Benefiting from a glorious view of the ancient 'Holy Temple', the peak of Mount Scopus, which hosted no religious sites, offered a perfectly conspicuous site for the construction of the Hebrew University as the Third Temple, a monument to the Jewish national revival.

The Patrick Geddes Master Plan for the Hebrew University

Patrick Geddes (1854–1932), the first assigned Hebrew University planner, was a Scottish botanist, biologist, sociologist, educator and world known town-planner.[13] As early as the 1880s Geddes supported the idea that the Irish, the Welsh and the Scots should search for their separate ancient cultural roots. Consequently he developed great interest in cultural aspects of the links between nationalism, cultural identity and social enterprise. Geddes also became interested in the relationship between nationalism and the demand for higher education in the Celtic countries. It is most revealing in the specific context of the Hebrew University that Aberytwyth, the first Welsh college, established by Welsh scholars of nationalistic inclinations, was built in those years (late 1870s and early 1880s). Hence the choice of Patrick Geddes to design the Hebrew University, and his nomination in 1919, definitely had a thorough foundation.

Furthermore, Geddes's Scottish upbringing included a close study of the Old Testament. Six years before his nomination and first visit to Jerusalem, Geddes had expressed the intense emotions and stimulation the legendary city had aroused in him:

> But the best example, the classic instance of city renewal (beyond even those of Ancient Rome and Ancient Athens) is that of the rebuilding of Jerusalem; and my particular civic interests owe more to my boyish familiarity with the building of Solomon's temple, and with the books of Ezra and Nehemiah, than to anything else in literature. Jews probably know more or less how the Old Testament has dominated Scottish education and religion for centuries; these were above all the stories which fascinated me as a youngster ... The improvising and renewal of cities might, and should once more, find an initiative, an example, even a world-impulse, at Jerusalem.[14]

Geddes also recalled how as a child he had listened to the tale of the rebuilding of Jerusalem.[15] This latter idea was in harmony with the Zionist idea of building the New Jerusalem.

Geddes and his associate (and son-in-law) Frank Mears stayed in Palestine from September to November, and submitted the Geddes master plan in December 1919, accompanied by Geddes's written presentation. There had been no academic plan or development program Geddes and Mears could work by for their plans. Geddes was therefore free to conceive the University as:

> ... a hill-top meeting-place where Sciences, Arts and Humanities may increasingly work together, in mutual respect and stimulus, towards a unity of Culture in its fullest sense; and with this monumentally expressed, in the comprehensiveness and harmony of architectural design.[16]

The basis for the design was first of all the natural and topographical surroundings and the relation of the campus to the Old City. Mears illustrated the latter quite expressively when he described the proposed building layout as 'reach[ing] forwards on either hand towards Jerusalem. Thus the student standing at the porch of the Great Hall (see Figure 9.2) may feel the intimate relation of the new City of Learning to the ancient City of Ideals'.[17] Geddes combined the buildings in a hierarchic setting of the academic faculties, supported by the differing heights of the topography. The nucleus of the complex, the Great Hall, was designed to contain 2,500–3,000 auditors, and serve for ceremonies, meetings, addresses and music, not merely for university use, but for the entire population of Jerusalem.[18] Geddes referred to the Great Hall as 'The Dome of Synthesis',[19] and its centrality would have been accentuated by a floating dome. It would have thus become:

> [S]omething more than a meeting place for occasional academic functions'; it would symbolically express 'the unity of purpose lying behind the many studies of the University, and [become] ... the focus of its daily life ...[20]

Figure 9.2 Drawing of the proposed west front of the Hebrew University, Frank Mears, 1919

It is not accidental that the most striking characteristic of the Great Hall is its clear resemblance to the Dome of the Rock. The minor differences only emphasize the likeness; the octagonal shape of the ambulatory of the Dome of the Rock is replaced by a hexagonal ambulatory in the Great Hall; the dome of the Dome of the Rock is borne by a taller drum than that in the proposed Great Hall. Mears designed semi-domes above the Great Hall ambulatory roof, which add a Byzantine-like look to the building, yet do not exist in the genuinely Byzantine Dome of the Rock. Furthermore, Geddes seems to have deliberately juxtaposed the Great Hall and the Dome of the Rock. He specifically referred to his scale considerations for the Great Hall as related to the Dome of the Rock, and also confirmed that he had the juxtaposition of the two buildings in mind:

> It is moreover already larger than the Dome of the Rock, and this both as regards main building and stretch of Dome: so it is perhaps well not to exceed this further: though the distance and perspective will not render this too obvious.[21]

The oriental style of the proposed architecture is quite noticeable in the general model as well as in Mears's various drawings. The layout of the buildings in the model was designed to resemble local Arab buildings in the way they 'grow' naturally out of the soil like organically united beings. Yet more than anything else, it presents a beautiful, romantic and dignified campus that would exhibit a magnificent composition to be observed by the onlooker in Jerusalem. In the midst of the Orient, Geddes and Mears attempted to construct an improved version of oriental architecture, as Geddes declared:

> ... But let us ... try to make it the very culmination of Palestine and the Orient! How? By crystallising anew its old and simple, useful and practical, economical and homely way and style of building into their fullest and highest expression. So pray clearly understand that it is out of the *old* Jerusalem, with its broken yet surviving beauty, that we have each, and together, got our vision of this New Jerusalem upon the hill.[22]

Indeed it seems that Geddes's conception of the Orient as reflected in his University plan had not been affected so much by his first hand impressions as by conventional orientalist depictions of the Orient. A drawing by Mears of a south view of the proposed University illustrates Geddes's and Mears's image of the University as an oriental walled town, mirroring familiar visual images of the Jaffa Gate and the Tower of David that have always been popular representations of Jerusalem. Yet it is most probable that the topographical drawing was not made directly from the actual view of the site itself, rather Mears must have used as his source a depiction typical of mid-nineteenth century orientalist style. A print by W.H. Bartlett, on the cover page of his book *Jerusalem Revisited* (London, 1855), depicts the wall of the Old City with the Jaffa Gate and the Tower of David. Striking similarities between Mears's drawing and the Bartlett print lead to an assumption that even if Mears did not use the Bartlett print as a model, it is just as much characteristic of nineteenth century romantic portrayals of Jerusalem. Furthermore, this kind of depiction followed an aesthetic convention that dictated a type of ideal landscape from which all signs of indigenous real life had been excluded. Both the Bartlett print and the Mears drawing suggest an imaginary Biblical scene rather than a contemporary one. In Mears's eyes the image of the future University had been depicted as a Biblical scene of the past.

Hence the Geddes and Mears plan must have conformed perfectly to the Zionist sanctifying approach towards the site and the institution. Apparently there seems to be a similarity between the Zionist sanctification of the *idea* of the Hebrew University and Geddes's grand architectural plan with the duplicate Holy Temple at its centre. Both parties bestowed upon the University significance and magnitude that are alien to its customary functions. Geddes went as far as describing his design as: '… a version of Revelation XXI–2':[23] 'And I John saw the holy city, new Jerusalem, coming down from God out of heaven, prepared as a bride adorned for her husband.' Mears's drawing of the proposed University west façade seems to illuminate this vision of St John's.

Not only was Geddes's plan much too presumptuous to be implemented, it also comprised of qualities that made it actually unrealizable. The planners' ambition, implied in the plan's scope, symmetrical layout, and association with the Holy Temple, to create a 'Heavenly Jerusalem', procured a conceptual and transcendental quality that discards implementation. Rather than an applicable plan, it is a *disegno* of an *idea* for a university that exists in the realm of theory and symbolism. It was totally severed from the reality of local needs and capabilities. Geddes's own categorizing his design as inspired by the Revelation reveals exactly that quality of a 'Heavenly Jerusalem' that made the plan so popular in Zionist propaganda pamphlets. The model and drawings described a legendary place, just as the image of the Third Temple would have been conceived, and the sort of topographical views mid-nineteenth century orientalist artists presented. Mount Scopus was diverted in the Geddes plan from a site for observation upon Jerusalem into an object for observers in Jerusalem. Hence the perception of their plan as a beautiful picture acquired also a quality of a contemplation piece rather than a realistic plan.

Geddes's Hebrew University master plan had been embraced by University promoters in Palestine and around the world. It presents the kind of delicious nostalgia for a lost Biblical past that makes the David Roberts' and other orientalists' depictions of Holy Land sites so popular. If only Jerusalem could transform into what Roberts, Bartlett, Geddes and others envisioned, but alas it could not, and reality, which included a large population of Arabs, could not disappear. But it could be ignored, by enforcing western values and trends upon indigenous heritage and influences.

The Present Hebrew University Campus on Mount Scopus

After the 1967 war and its aftermath the site and the visual aspect of the Hebrew University transformed again into their former role as a means for the promotion of national interests. Once more the Hebrew University was summoned to enhance political and national interests. This time the realized architectural design determined the image of the Hebrew University as a political statement. The present Hebrew University campus on Mount Scopus

(Figure 9.3) is a visual statement of its renewed national significance. It was a political Government decision that the University should return to Mount Scopus and consequently a new large campus was constructed, with most of the old buildings included in it. An anonymous contributor to an official Hebrew University publication, expressed the depths of emotions connected to Mount Scopus shortly after the war:

> Scopus – site of the dream come true, proud home of the national university of the Jewish people, Scopus – the campus in exile, the vision cherished throughout nineteen long years as we turned our eyes to its distant prospect. Scopus – the measure of our growth and development and now – the challenge of our future.[24]

The following brief account of University and government officials' actions concerning Mount Scopus provide evidence of the unique swiftness with which the matter was treated, in a definite attempt to serve national interests. Immediately after east Jerusalem was captured, four days after the war had begun, the possibility of rehabilitating the Mount Scopus campus

Figure 9.3   An aerial view of the Hebrew University, 1974, Jerusalem

was considered. One day only after the war ended, on 12 June 1967, a special ad hoc Mount Scopus Rehabilitation Planning Committee, appointed by the University Senate, held a meeting to discuss a suggestion to erect a second campus on Mount Scopus.[25] The members of the Committee unanimously expressed their wish for the University's 'return' to Mount Scopus, notwithstanding the obstacles. They agreed, therefore, that since the restoration of the Mount Scopus campus would obtain enormous public interest and resources, the University should prepare its plans. On 18 June 1967 the government decided to take measures in order to legally unite the two parts of Jerusalem. The formal announcement of the union was delayed until 27 June for reasons of political convenience. But as the University administrators and faculty did not wait for the formal announcement, they immediately took further action to assure the University's repossession of Mount Scopus; a survey of the condition of the property on Mount Scopus was prepared.[26] In a Standing Committee meeting on 30 June, the head of the University Maintenance and Development Department reported the survey conclusions. He elaborated on the subject of rehabilitation, stating that there would be a need for a new master plan.

The special role of the University and the Mount Scopus campus as promoters of the political endeavors was demonstrated when on 28 June a ceremony took place in the open-air theatre of the old Mount Scopus campus; the Hebrew University granted an Honourary Doctorate for his achievement to Yizhak Rabin, then the military chief of staff, and the hero of the Six Day War victory.

Construction works on Mount Scopus created a *fait accompli*. Again, through its location, the Hebrew University was at the political front and an important factor in the creation of a major shift in nationalist and cultural identity. Although the University 'returned' to its original place on Mount Scopus, both site and institute did not retain their original 'place' in Jerusalem. Geographically nothing has changed, but social and political transformations created a new relationship between campus and city, the Old City in particular.

Since 1948 the myth of the 'abandonment' of the Old City and the road to Jericho (as in the lyrics of the popular song 'Jerusalem of Gold') related to places only. In orientalist depictions, the native inhabitants of those places, who were rendered non-existent, were replaced by legendary biblical figures. In 1967, Israelis 'returned' merely to places; however, the Arab inhabitants could not be replaced anymore by romantic depictions of the biblical forefathers. Yet since the native inhabitants were quite redundant, new means had to be invented in order to ignore them. Therefore, the Mount Scopus location had actually transformed and the University has been located on a new social and political frontier.[27]

An official publication of the Hebrew University presenting the first stages of the master plan for the new campus promised '... a campus combining both grace and functionalism ...'.[28] Unfortunately the realization of the plan achieved neither. The campus is a large scale compact complex, held together by a wall and an impenetrable effect of the tightly held together buildings. It is alienated from its environment, and once penetrated, it causes disorientation and confusion. The exterior does not give away the identity of the complex or the different functions of the buildings. The watchers' tower in the centre could have manifested the site's original significance as the mountain of watchers, but as it is not accessible (it is restricted military area) it had lost its potential as a belvedere, but not its domineering effect, which has actually increased by its function. Pedestrians and motor passengers 'penetrate' the building complex through underground passages and into the all-campus pedestrian passage, which connects the main buildings. This inter-campus passage is mostly closed to the outside, yet in certain sections large windows open to internal closed gardens. A set of disorientating, narrow, labyrinth-like corridors and stairs, lead to classrooms, auditoriums and offices. Occasionally spaces open up along the pedestrian street with grand staircases and monumental pillars that would indicate places of special significance, yet they lead to more offices or classrooms. The pedestrian passages along the Social Sciences, the Humanities and the Education complexes meet at a junction of assembled central functions – the Administration, the Library and the Forum. The latter is a hall designed for students' extra-curricular activities, but the breaking up of its space into differing levels does not allow communal activity. Perhaps the most striking peculiarity of the campus is that very few places allow a view of the Old City or the Judaean Wilderness. All windows and openings were designed so that they do not open to the view. The only unit that purposely opens up to the view of the Temple Mount is the synagogue (designed for orthodox worshipers), situated at the bend of the Humanities Building.[29]

In giving the campus its present form three people were probably more instrumental than others; architects David Reznick and Ram Karmi, and Yoseph Harpaz, the newly nominated General Director. Harpaz, a former army officer, was employed to set the faltering university administration in order. He soon became a dominant figure in the University and also took over the Mount Scopus development planning. It is claimed that it was he who pushed towards building a large-scale campus on Mount Scopus, which would eventually replace the Giv'at Ram campus.[30] Harpaz also initiated the establishment of an administrative supervising body that would serve under him and would be in charge of the development of the new campus. Administrative aspects of the University may have improved, but what is more significant is that faculty members were powerless vis-à-vis Harpaz's interference in academic matters.[31] This description of power struggles within the University at a time of drastic changes is relevant here, because it shows how important aspects of responsibility toward values such as freedom and independence which are essential for democratic life as a whole, and university endeavors in particular, have been abandoned in the Hebrew University. Those link together not only with the way architectural issues were dealt with, but also with the general trends that developed in Israeli society after 1967.

Reznick's preliminary plan and Harpaz's intentions for the future campus were kept secret until March 1968, even though the Mount Scopus Programme Committee was set up in January 1968, with the participation of all the heads of University departments. Associate architects were involved in the preparation of the master plan – Reznick's partner Shmuel Shakked and later also Ram Karmi and Chaim Katseff. Years after the completion of the campus Reznick explained that he intended the silhouette of the University complex to 'echo the walls of the Old City'.[32] He further elaborated on the visual and symbolic ideas, which guided him:

> I was thinking in a visionary way – the revival of the Jewish culture is symbolized by Mount Scopus. Even the location of the site in itself is symbolic – it stands between east and west, between wilderness and civilization. It is a lighthouse of the revival of the Jewish nation. When I was a child in Brazil, we knew about Mount Scopus and the Hebrew University through the *Keren Kayemet* (National Land Fund) postcards. For us it was a symbol. I came to Israel in 1949 and settled in Jerusalem in 1955. I used to watch Mount Scopus from Abu-Tor and wish some day my grandchildren will be able to go there. So for me to build on Mount Scopus was a sort of a peak. I was thinking of Jewish students coming from the Diaspora to the Hebrew University on Mount Scopus, and through Mount Scopus they will absorb the spirit of the whole land. This was part of my programme.[33]

Reznick admitted that although he and his colleagues may not have emphasized the nationalistic ideology in their declarations, it was actually an essential motivation behind the plan:

> It was my wish to create a forceful presence. Here I am and nobody will move me from here. I agree that there is in the plan a political statement. Even though I do not believe in political architecture.[34]

However, the idea of the campus as a megastructure developed as work advanced and more architects joined the original team.

When actual work toward realizing the plan began, Ram Karmi joined Reznick and Shakked on the master plan team. They divided the planning of the different parts of the campus among themselves and other architects[35] who were chosen for their status among architects in Israel, because the University authorities wanted the best architects to design the new campus.[36] As in many other cases the planning of a prestigious project remained in the hands of a traditionally elite group. Karmi's presence became extremely dominant, to the point that he dictated to a great extent the architectural concepts of the entire layout.

Karmi's Humanities building was to be the most conspicuous for onlookers in Jerusalem, and therefore – highly prestigious. The design of the synagogue which was to be included, being assigned to Karmi, determined the concept of the whole. Never before was a synagogue included in a Hebrew University master plan. Much consideration was devoted to its location; there was a demand that it would have a view of the Temple Mount, that the building would not be isolated, and that it would be singled out by its architectural design.[37] Eventually it was decided that it would be located at the west angle of the Humanities building. The large glass wall of the synagogue facing the Old City and the Temple Mount replaces the traditional Torah shrine. Traditionally a schematic design symbolising the Holy Temple decorates the Torah shrine. In Karmi's synagogue the 'true' image of the Holy Temple appears in all its glory through the glass wall in the form of the Dome of the Rock.

The synagogue, therefore, is one of the few places in the entire University that provides a panorama of the Old City; it determined the introversion concept for the entire University building complex. Karmi thus used his power as an architect to conceal the view of the Old City from the entire community of University users in order to achieve this effect. In Karmi's own words:

> The synagogue is the 'window' of the Hebrew University on Mount Scopus to the Temple Mount in Jerusalem, and is at the apex of the faculty's internal movement and orientation. Here is where the ancient pilgrims to Jerusalem, upon first sighting the Temple Mount, held ritual ceremonies. For a synagogue to represent a window means that a building that is fundamentally introverted

is required to have an extroverted character. This tension between internal and external motivation influenced the design of the synagogue on Mount Scopus.[38]

Through the Torah shrine of the synagogue, the seventh century Islamic Dome of the Rock can be nothing but the Holy Temple.

A comprehension of the national and social trends in post-1967 Israel provides a better understanding of the campus's architecture significance. The aftermath of the 1967 war and the control over Jewish sacred sites in Judaea and Samaria opened up new possibilities for the Israeli state, which actually became more of a control system[39] than before. The mountain regions, the site of the Biblical Kingdoms of Judaea and Israel, which for centuries have been densely populated by Arabs, became accessible to Israelis. As a consequence of the 1947/48 war Jews were completely barred from establishing settlements or even visiting in those regions. The expansion of the state's boundaries brought about by the 1967 occupation created an 'overlap between the boundaries of the Israeli control system and the theological "Land of the Bible"'.[40] Combining religious emotions with political actions was accelerated and its effects are a threat to democracy in Israel to this day. Locating the synagogue on the Hebrew University's 'prow', and manipulating it to 'occupy' the Dome of the Rock, was indicative of the prevailing new state of mind.

The 'return' to Mount Scopus meant far more than a political wish to occupy a conquered territory; it was also a return to the one Jewish hold of territory in the Land of the Patriarchs (between 1948 and 1967 it was an Israeli enclave). In 1967 the symbolism attributed to the combined entity of the Hebrew University and Mount Scopus lost its old Third Temple associations. Instead, years of nurturing the myth of Mount Scopus as a temporarily lost national asset and of yearning for its retrieval, placed it together with other sacred sites. The situation after 1967 was therefore unique and unprecedented. For the first time since the destruction of the Temple by the Roman legions lead by Titus, Mount Scopus could become again *the* lookout on a unified Jerusalem under Jewish rule. Yet was it really? As mentioned above, the occupation also created a dual attitude of the Israeli state toward the territories; while the occupied land was desirable its inhabitants were not.[41] The Arab inhabitants of the Old City remained there and the Temple Mount is controlled by the *Wakff* (the independent Islamic authority responsible for the *Haram-a Sheriff*). Hence it was not possible any more to look at the Dome of the Rock from the summit of Mount Scopus and 'see' a vision of Solomon's Temple unless one joins the worshippers in the synagogue. That vision had disappeared not only with the heroic accounts of the seizure of the Old City by Israeli military forces, but also through a Zionist rhetoric that had changed since those first days of Zionism.

Furthermore, since the 1930s and especially since the evacuation of the Mount Scopus campus in 1948 the Hebrew University has become in the mind of the public an institution of higher education, not a substitute Temple. This imagery, that could probably persist as long as the University was no more than an unrealized idea and site, could not be effectively used any more, it had become obsolete. Consequently a shift was inevitable in the way the Mount Scopus site was supposed to relate to the Temple Mount. On the one hand post-1967 Mount Scopus has become a new frontier; Baruch Kimmerling (1989) used the term to define the occupied land as bearing settlement potential, from the point of view of the Israeli 'control system'. The Palestinian inhabitants of the territories, though, were conceived as a threat to the definition of the boundaries of Israeli collectivity and to the state's safety in case of possible military campaign.[42] The Old City of the post-1967 era was therefore not the legendary Biblical site any more, it was seen for what it really was, a religious Islamic centre, and a town inhabited by people with an identity that belongs to the present. Physically the Old City and the Temple Mount have not changed; their image in the eyes of their Israeli observers has changed. This concept can offer an interpretation for the concealment of the Old City by the campus' architects. Hence the new campus cannot create a dialogue with its surrounding, its fortress like presence sends out one way messages of power.

Reiner Banham's description of the megastructure building type as 'above all a monumental order of heroic scale',[43] may serve to describe the campus, but it is the Mount Scopus location that provides its special political significance. Besides, a fortress also affects its inhabitants; indeed Kimmerling's use of the term 'control system' can be applied to the interior of the campus as well. Walking through the campus one looses one's sense of potency and capability, to the extent that one constantly needs orientation aids such as signs and maps, which cannot provide sufficient assistance. The architecture imposes a sense of helplessness and dependency which calls for a leading hand to direct the way.

The Hebrew University came into the world as an objective devoid of a visual image. Now, the present campus provides it a visual image that is the objective. It is the formal aspect of the Hebrew University in the context of its geographical and political location that delivers its special ideological and national message of the new Zionism. While there is nothing uniquely Israeli,

Hebrew or Jewish about the Hebrew University, its expression of dominant, forceful and oppressive physical presence, is an ultimate figuration of present Israeli nationalism.

The undercover imperialist conception behind the Geddes plan finally achieved its direct and conspicuous realization in the form of the present Mount Scopus campus. Yet while the Geddes master plan manifested an imaginary 'Ideal Jerusalem' or 'Heavenly Jerusalem', Reznick and his associates applied fashionable architectural vernacular, accepted and familiar in western architecture, to conceal the campus's true purpose. It heralded the appearance of post-1967 Zionism, which clings to a contested right over the Land of the Patriarchs and its sacred sites. Quite contrary to the early Zionist secular ideology of working the land, the occupation generated a culture that is a combination of power and self-glorification with a reverence toward sacred sites.

Notes

1   A major part of this article is based on my research for a PhD thesis: 'Architecture and National Identity; The Case of the Architectural Master Plans for the Hebrew University in Jerusalem (1919–1974) and Their Connections with Nationalist Ideology', supervised by Professor Adrian Forty, The Bartlett, UCL, 2000.

2   The Hebrew University dwelled in three different campuses in Jerusalem along the years of its existence: construction of the first campus on Mount Scopus started in 1923. It was evacuated during the 1948/49 war, and a second campus was later built on Giv'at Ram. It was replaced by a new campus on Mount Scopus after the occupation of east Jerusalem in 1967.

3   The Dome of the Rock is a seventh century Islamic building, placed in the centre of the ancient plateau in the Old City of Jerusalem, known as the Temple Mount or Haram-a-Shariff (Arabic for The Sacred Place).

4   Both terms had been in use at the time.

5   Chaim Weizmann was the president of the world Zionist movement for many years and became first president of the state of Israel. All other persons mentioned were leading figures in the Zionist movement and instrumental for the Jewish colonization of Palestine.

6   See *Stenographisches Protokol der Verhandlungen des XI Zionisten Kongresses in Wien vom 2 bis 9 September 1913* (Berlin and Leipzig, 1914), pp. 294–345, 362–5.

7   Kolatt (1997), pp. 54, 55.

8   Heinrich Loewe , *The Jewish Chronicle*, 19 September 1913, p. 7.

9   Kedar, p. 90.

10   The theme appeared in Herman Shapiro's early 1880s articles but has since been magnified. Ahad Ha'Am referred to the proposed Hebrew University as the Third Temple. While preparing for the11th Zionist Congress Weizmann wrote to his wife of his determination to achieve a significant progress for establishing the University; he enthusiastically referred

to the University as 'The Hebrew University on Mount Zion – the Third Temple!', and he added: 'To my way of thinking this is the one slogan that can evoke a response just now'. Ussishkin too, in his address to the 11th Congress said that building the University would be a compensation for the desolation of the Temple. See: Paz, pp. 286, 287.

11   Goren, pp. 368, 369; Segev, p. 167.

12   Cited in Goren, p. 369.

13   For specific information on Geddes's architectural plan for the Hebrew University's, see Geddes (1919); D. Dolev, 'The Architecture of the Hebrew University 1918–1948', unpublished MA thesis, supervised by Dr. Adina Meir (Tel Aviv University, 1990); D. Dolev,, 'The Hebrew University Master Plans 1918–1948' (Hebrew), in S. Katz and M. Heyd (1997), pp. 257–80; M. Shapiro, (1997), 'The University and the City: Patrick Geddes and the First Master Plan of the Hebrew University, 1919', in S. Katz and M. Heyd (1997), pp. 202–35; D. Dolev, 'Architectural Orientalism in the Hebrew University – The Patrick Geddes and Frank Mears Master-Plan', *Assaph*, 3, Section B (1998), Tel Aviv, pp. 217–34.

14   Letter to Amelia Defries, 1913, in A. Defries, *The Interpreter Geddes: The Man and His Gospel* (London, 1927), p. 260.

15   Mairet, p. 184 and Novak, p. 55 and n. 2.

16   Geddes and Mears, *Comments on the Romberg plan for the Hebrew University*, Central Zionist Archive, file L/12 39, 1924.

17   Mears, in *University of Jerusalem; Notes on Scheme, by F.C. Mears, To be Embodied in Dr. Weizmann's Booklet*, in Central Zionist Archive, file Z4/2790 (no date, 1919?).

18   Geddes, p. 31.

19   Boardman, p. 317.

20   Mears, see note 17.

21   Geddes, Memo I, enclosed to a letter to Dr. Magnes, 11 April 1925, Hebrew University Archive, file 31. Mears (see note 17) also referred to the juxtaposition of the Great Hall and the buildings surrounding it with the Old City and the Temple Mount: 'Just north of the site chosen for this Hall a spur runs out westwards from Mount Scopas [*sic*], enclosing a shallow valley which falls steeply towards the Kidron and the N. E. corner of the City wall. The main axis of the Hall and of the central group of buildings and terraces is directed down this valley and towards the centre of the ancient City.'

22   Letter from Geddes to Mrs Fels, July 1920, in Mairet, pp. 186, 187.

23   Letter to Louis Mumford, 14 November 1924, in Novak, p. 54.

24   *Mount Scopus Campus Master Plan*, a publication of the Building and Development Department of the Hebrew University (no date).

25   'Meeting of the Mount Scopus Rehabilitation Committee on 12. 6. 67', Hebrew University Archive, file 0801/1967.

26   'Mount Scopus', prepared by Skotnitsky and presented to Cherrick, 21 June 1967, Hebrew University Archive, file 091/1967.

27   Kimmerling, pp. 276–7.

28   *Mount Scopus Campus Master Plan*, a publication of the Hebrew University, Jerusalem.

29   The veranda of the Maiersdorf faculty club also opens to the view of Jerusalem, but it is restricted to faculty members, staff and special guests of the University.

30   The information on Harpaz's role is based on U. Benziman, 'The Impression of the Director General Joseph Harpaz', *Haaretz* (daily newspaper), 12 December 1969; on an interview with architect A. Yaski, 8 August 1994 and on an interview with architect Ch. Katseff, 18 August 1997.

31 Ibid. Benziman assessed that those faculty members who did not approve of Harpaz, gave up on opposing him because they believed that after all maintaining good order was important.
32 Interview with Reznick, 28 July 1994.
33 Ibid.
34 Ibid.
35 Reznick designed the Education building, Dan Eitan – the Social Sciences building, Karmi – the Humanities and the Synagogue buildings, Avraham Yaski – the complex of the Student Centre, the Buber-Rouseau building, Beit-Hill and the Archaeology buildings, Katseff – the Forum, Y. Rechter – the Library, Z. Ravina – the Administration building.
36 Interview with Reznick, 5 July 1994 and with Ch. Katseff, 18 August 1997, who also said that Harpaz was most active in choosing the architects.
37 Ch. Katseff, in minutes of meeting no. 34 of the Programme Committee, 1 December 1971, file 0801, 1970–71, Hebrew University Archive.
38 Ram Karmi, *Build ye Cities*, Catalogue of an exhibition of Israeli architecture, November 1985, London, p. 118.
39 Baruch Kimmerling has been the first to use the term to define the state after the 1967 war.
40 Kimmerling, p. 276.
41 Kimmerling, p. 278.
42 Ibid., pp. 274, 278.
43 Banham, p. 148.

References

Banham, R., *Megastructure, Urban Future of the Recent Past* (New York: Harper and Row, 1976).
Boardman, P., *The Worlds of Patrick Geddes: Biologist, Town-Planner, Re-educator, Peace-warrior* (London: Routledge and Kegan Paul, 1978).
Crinson, M., *Empire Building: Orientalism and Victorian Architecture* (London and New York: Routledge, 1996).
Geddes, P. (assisted by F. Mears), *The Proposed Hebrew University in Jerusalem*, unpublished (1919).
Goren, A., 'Judah L. Magnes and the Early Years of the University' (Hebrew), in S. Katz and M. Heyd (eds), *The History of the Hebrew University in Jerusalem: Origins and Beginnings* (Jerusalem: Magnes Press, 1997), pp. 363–85.
Goren, Arieh, 'Sanctifying Scopus: Locating the Hebrew University on Mount Scopus', in E. Carlebach, J.M. Efron and D.N. Myers (eds), *Jewish History and Jewish Memory* (Hanover and London: Brandeis University Press/University Press of New England, 1998), pp. 330–47.
Katz, S. and Heyd, M. (eds), *The History of the Hebrew University in Jerusalem: Origins and Beginnings* (Hebrew) (Jerusalem: Magnes Press, 1997).
Kedar, B.Z., 'Laying the Foundation Stones of the Hebrew University of Jerusalem, 24th July, 1918' (Hebrew), in S. Katz and M. Heyd. (eds), *The History of the Hebrew University in Jerusalem: Origins and Beginnings* (Jerusalem: Magnes Press, 1997), pp. 90–119.

Kimmerling, B., 'Boundaries and Frontiers of the Israeli Control System: Analytical Conclusions', in B. Kimmerling (ed.), *The Israeli State and Society; Boundaries and Frontiers* (Albany, NY: SUNY Press, 1989), pp. 265–83.

Mairet, P., *Pioneer of Sociology – The Life and Letters of Patrick Geddes* (London: Lund Humphries, 1957).

Meller, H., *Patrick Geddes: Social Evolutionist and City Planner* (London and New York: Routledge, 1990).

Novak, F.G. (ed.), *Lewis Mumford and Patrick Geddes: The Correspondence* (London and New York: Routledge, 1995).

Paz, Y., 'The Hebrew University on Mount Scopus as a Secular Temple' (Hebrew), in S. Katz and M. Heyd (eds), *The History of the Hebrew University in Jerusalem: Origins and Beginnings* (Jerusalem: Magnes Press, 1997), pp. 281–308.

Segev, T., *One Palestine, Complete: Jews and Arabs Under the British* (New York: Metropolitan Books, 2000).

Chapter 10

Re-Placing Memory

Yael Padan

Introduction

Preserving the memory of the Holocaust is becoming a more and more important issue as distance grows from the event itself, and Holocaust survivors and witnesses are aging. What has in the past been part of everyday life in Israel, where people were constantly made conscious of the Holocaust by the presence of tens of thousands of survivors amongst them, needs now to become an organized collective experience if it is to be remembered. Hence the current increase in buildings and monuments devoted to the memory of the Holocaust, commissioned by Yad Vashem – the Holocaust Martyrs' and Heroes' Remembrance Authority, as well as by survivors' organizations.

The new buildings are intended to act as 'places of memory' (*lieux de memoire*), defined by Pierre Nora as the sites in which memory turns into history. These sites appear when spontaneous memory no longer exists, and the need arises for organized places and activities to replace it. The replacements are elements of collective memory, such as monuments, archives, memorials and ceremonies, to be passed on to the next generations. Memory is defined by Nora as life itself, 'in permanent evolution, open to the dialectics of remembering and forgetting, unconscious of its successive deformations, vulnerable to manipulation and appropriation, susceptible to being dormant and periodically revived'.[1] The act of creating a 'place of memory' transforms memory into history, defined as an intellectual, secular act that invites analysis as well as critical discourse. According to Nora, memory places remembrance in the realm of the sacred, whereas history is a problematic and partial restoration of things gone by. While memory is always relevant and experienced in the present, history is a mere representation of the past. 'Places of memory' express the importance of memory in building national identity, but they also demonstrate changes in the national attitude towards its past.[2]

'Places of memory' act on three different levels, as noted by Nora: they give memory a *material* shape, they have a *symbolic* role in representing memory and its significance, and they act as *functional* places for the passing on of memory.[3] The process of transforming memory into history by creating such places of material existence raises some interesting questions regarding the role of architecture. The first concerns the institutions that commission the buildings, declaring themselves to be the custodians of memory. The second examines the choice of sites for the 'places of memory', which are especially interesting since they are very distant from the sites where the actual events of remembrance had occurred. The third question is about the capacity and the limits of architecture to convey the abstract messages of memory in physical form.

In this chapter I will look at current representations of the Holocaust memory in the Israeli collective narrative, as expressed in two recently completed Holocaust memorials. The first is the Valley of the Communities by Dan Zur and Lippa Yahalom, commissioned by Yad Vashem, Israel's main official commemoration site and a place of national importance. Yad Vashem is located on Har Hazikaron ('Mount of Memory') in Jerusalem, adjacent to the city's military cemetery. I will discuss the ways in which this proximity suggests a linkage between the Holocaust and the state, and how the different museums and memorials within Yad Vashem create a physical context for the memory of martyrdom, which is connected with the memory of national heroism.

I will argue that the Valley of the Communities reflects an approach to commemoration that invites the visitors to experience a leap into another reality through landscape architecture. This is achieved by the creation of a physical representation of the political geography of Europe, which resembles a condensed visit or pilgrimage to the actual sites of remembrance. I will refer to state- organized pilgrimages of students to the death camps, which are often used to stress national messages and conclusions. I will suggest that the symbolic geography of The Valley of the Communities, set in its wider context of Yad Vashem and the Mount of Remembrance, recreates a similar experience for the visitors in Jerusalem.

The second monument that will be discussed in this chapter is 'Yad Layeled' – Children's Memorial Museum by Ram Karmi and Partners. This museum is part of The Ghetto Fighters' House – Holocaust and Jewish Resistance Heritage Museum. It is located at Kibbutz Lohmei Hagetaot, founded by a community of Holocaust survivors. I will discuss the choice of its site, between an archaeological ruin and agricultural fields, giving the museum a local historical and ideological context as both a 'place of memory' and a timeless edifice. I will argue that this approach is further evident in the building's logic, organized around a descending spiral exhibition ramp that leads the visitor down a symbolic route to hell. This route creates an organized and didactic experience of remembrance. Furthermore, it serves to isolate the visitor from external space and time, recalling an idea of the Holocaust as 'a different planet', a singular event that can never happen again.

Finally, I will point out the problematic nature of constructing 'places of memory' for the Holocaust. The idea of shifting the crimes from their historical place to symbolic sites of remembrance and interpretation carries a danger of devaluating the terrible events and giving them banal material expressions. In addition to generating recollection, Holocaust memorials also serve to integrate contemporary visitors into past events. In spite of the limitations of architecture in conveying abstract messages, these monuments are experienced within their built context as well as within contemporary civil religion. Hence, to the uncritical visitor, these memorials are in danger of becoming didactic tools for an established experience of collective memory.

Official Memory

The memory of the Holocaust in Israel forms a fundamental part of Israeli national identity and civil religion.[4] Its formal representations and ceremonies have been shaped by official agencies throughout the state's history.[5] Their importance lies far beyond remembrance of the genocide of the Jewish people that took place during World War II. Memory itself, as noted by Idith Zertal in her book *Death and the Nation*, had become a central ideology of the state, an important national political-secular resource.[6]

According to Liebman and Don Yehiya, statism was the dominant civil religion in the early state of Israel from 1948 to the late 1950s.[7] Statism 'gives rise to values and symbols that point to the state, legitimate it, and mobilize the population to serve its goals. In its more extreme formulation statism cultivates an attitude of sanctity toward the state, affirming it as an ultimate value'.[8] Statism's approach to Holocaust memory was characterized by little emphasis on the creation of official ceremonies and patterns of collective remembrance. Israel's first Prime Minister Ben-Gurion was at first reluctant about the establishment of Yad Vashem – the Holocaust Martyrs' and Heroes' Remembrance Authority. According to Don Yehiya, this reluctance was due to Ben-Gurion's view that historical events were important for Israeli collective memory inasmuch as they contributed to the unity of the people and deepened their commitment to the Zionist state-building project.[9]

The early state, therefore, faced a simultaneous need to remember and to forget.[10] The need to forget, claims Zertal, was an attempt to restart history with the new state, erasing the memory of the shame and dishonor that were experienced by the Jews for centuries and culminated in the Holocaust. In this sense, the act of forgetting would serve as an act towards redemption.[11] Additionally, in the eyes of statism, part of the tragedy of the Jews was a result of the fact that most of the victims had failed to join the Zionist movement, which offered an alternative to Jewish life in the Diaspora. Don Yehiya and Liebman point out that statism was interested in symbols of victory and achievement, and in basing Israel's relations with other nations on terms of equality and mutual interests. Statism was therefore not interested in symbols of the Jews as defenseless victims.[12]

With the changes in statism, its approach towards the memory of the Holocaust changed as well. A distinction was made between two simultaneous narratives, that of the helplessness of the victims and that of the fighters' resistance.[13] The fighting and resistance of Jews in the Warsaw Ghetto uprising and in other places were assimilated into the official Israeli narrative, which stressed the themes of physical courage and armed rebellion.[14] The fighters were exemplary figures for the new state, and their uprising and struggle were seen as linked with the Israeli fight for independence, as well as with Israel's later wars. James Young points out that the early Israeli political leaders perceived the Holocaust as memorable 'by little more than instances of heroism and the Jewish courage it evoked in some of its victims, the hopelessness of Jewish life in exile, and the proven need for a state to defend Jews everywhere'.[15] Hence, the story of the surviving rebels who had come to live in Israel was readily incorporated into the advocated collective vision of the past as well as the future. The change in the Israeli official attitude to the memory of the Holocaust culminated in the Law for the Day of Remembrance of Holocaust and Heroism passed by the Israeli Knesset on 7 April 1959.[16]

The decline of statism in the late 1950s brought about a change in the collective attitude towards the memory of the Holocaust. According to Liebman and Don Yehiya, statism declined since it was not generated by the civil society 'but developed as an effort of the political elite to overcome the legacy of conflict among the subcommunities of the yishuv and to socialize the new immigrants to its perceptions of Zionism and the needs of a modern state'. It was replaced by a new civil religion, which was a response to the failure of statism to provide a meaningful symbol system.[17] While statism disclosed a negative attitude towards religion, Diaspora and the Holocaust, the new civil religion expresses more sympathy and identification with Jewish suffering and the indifference of the world, and the association of the Diaspora with religion has become more positive.[18]

However, some aspects of Holocaust memory are still neglected in contemporary civil religion, especially those that have a universal rather than national relevance. For example, Zertal points out that the early Holocaust trials, which took place in Israel during the 1950s, are rarely mentioned and have never been studied in Israeli schools. In these cases Jews who were accused of collaboration with the Nazis were brought to trial before Israeli courts according to a special law passed in 1950. None of these Jews were found guilty of causing death, since the judges in most of these cases (nearly 40 trials took place during the 1950s and early 1960s) were very careful in judging the Holocaust survivors and many of them had reservations concerning the law.[19] Zertal suggests that this Holocaust literature bears witness to the complexity of human existence and its incompetence in the extreme conditions of the death camps. By demonstrating the triviality of crimes performed by Jews, ordinary people who had crossed the invisible border between decency and maliciousness as may happen to any human being, it poses a threat to the concept of Israel and the Jewish people as representatives of the good in face of the ultimate evil of the Holocaust. [20]

Architectural Memory

The recently completed Holocaust memorials in Israel are part of an ongoing remembrance project, both national and private. The memory of events of communal importance is often associated with the places where they occurred, linking collective memory with specific spatial frameworks. However, in the case of Holocaust memorials in Israel, temporal distance from the events is coupled with physical distance from the places of occurrence, leaving memory in the realm of narrative, with no specific location. In order to solve this problem, the 'place of memory' is intended to create a setting for the projection of memory onto built form, providing a new linkage between memory and space.

Frances Yates has explored this relationship in her book *The Art of Memory*, a study of a system first developed by the ancient Greeks. The art of memory was based on the assumption that visual images were most easily remembered, and therefore the subjects of remembrance were memorized together with familiar architectural spaces. These places could later be revisited in the mind, evoking recollection. However, the choice of architecture did not assume a connection between the contents of memory and the buildings. On the contrary, the same set of spaces could later be reused for remembering different materials, because the contents of memory were less fixed in the mind than the visual image of places.[21] The contemporary Holocaust memorial can be viewed as a built container for memories, an architectural attempt to create a device meant to generate recollection. However, the link between spatial experiences in the building and memory is not intended to be merely coincidental or functional. It tries to give memory a distinctive shape, one that cannot or should not be reused for remembering other materials.

In addition to generating recollection, the Holocaust memorial also serves to integrate contemporary visitors into past events. Historical museums, according to Ariella Azoulay, are more than built containers for exhibits – they are also initiators and creators of meaning. They not only have the authority to represent past events, but also to generate new relations between individuals and the community. In this way, the individual's past is directly inserted into the collective, often national, narrative.[22] Israeli national narrative in the early years of the state (statism) as defined by Don Yehiya, outlined national culture as part of the political religion in which the state has a vision and a mission. It strives to educate its citizens to fulfill its objectives, thus investing their lives with content and meaning.[23] I would suggest that some aspects of this approach still exist. Among its other expressions, 'political religion' needs 'holy sites' to serve as its physical manifestations and be visited by 'pilgrims'. Holocaust memorials in Israel provide such sites or 'places of memory' as defined by Nora. These sites are described by James Young as 'insensate stone and concrete, lost to meaning. But as part of a nation's memorial ritual, they are invested with national soul and significance'.[24]

Since 1988 the Israeli Ministry of Education organizes and encourages Israeli high school group visits to Holocaust sites in Poland. In his research, social scientist Jackie Feldman who has taken part in seven such visits during the years 1992–97, notes that they are in fact 'pilgrimages' performed by observers of civil religion. Feldman points out that these pilgrimages are primarily emotional experiences, where the students face the demonic 'other' in a ritual encounter, in order to conquer it with symbols of Israel. This experience raises in the students an awareness of the constant threat to their own world, giving their everyday life in Israel new meaning. The students discover that Auschwitz is not just a state of mind, but a place (which Feldman defines as a meeting point of space and time) of evil that can be overcome only at an enormous price. The victory of the students over the 'other' using symbols of the state makes the death camps into the birthplace of the state. The students' awareness of the existential threat to their lives and its overcoming thanks to the state's existence creates and deepens their commitment to the Israeli basic cultural and national values.[25] I would suggest that Holocaust memorials in

Israel reproduce such an experience of collective identity for Israelis who have never visited the actual sites, as well as for the next generations.

Another objective of national monuments in Israel, as elsewhere, is to create and intensify the population's knowledge and relationship to the land. As noted by Young, man-made monuments and memorials gradually come to be perceived as extensions of the land. Although Holocaust memorials signify events that have taken place elsewhere, these buildings become landmarks in the Israeli national landscape, which is subsequently associated with their contents. Furthermore, both ancient biblical sites and monuments such as Holocaust memorials are then perceived as landmarks related both to the land of Israel and to the history of the Israelis.[26] The importance of this link is evident in the ongoing process of constructing new Holocaust memorials in Israel. On the other hand, Israeli officials show less interest in donating money for international remembrance projects such as the physical preservation of Auschwitz-Birkenau, itself a site of immense importance (as noted by Feldman in his article). Concerning the physical deterioration of the death camps, it seems that the official Israeli approach is almost one of indifference. 'The matter is not the responsibility of the Israeli government', says Avner Shalev, chairman of Yad Vashem and a member of the International Auschwitz Council that was established by the Polish government in the early 1990s, in an interview. 'Israel has its own crises and problems,' he observes.[27]

Yad Vashem

Yad Vashem, the Holocaust Martyrs' and Heroes' Remembrance Authority, was established in 1953 by a special act of the Israeli parliament called the 'Law of Remembrance of Shoah and Heroism – Yad Vashem'. This law was made in order to organize formally the Israeli official national memory of the murder of the Jews of Europe during World War II. Ben Zion Dinur, Israel's Minister of Education and Culture who had proposed this law, stated that one of its objectives was to assist the melting pot of immigrants from many different places into a national collective, motivated by a shared memory as well as by a common vision of the present and the future.[28]

The Site

Yad Vashem was allocated a 45-acre site in Jerusalem called Har Hazikaron (the Mount of Remembrance), adjacent to Har Herzl (Herzl Mount), where the state's ideological founder is buried, and to the city's Military Cemetery. This proximity creates an image that combines martyrdom with national heroism, suggesting a link between the state, the Jews murdered in the Holocaust and the soldiers who were killed in Israel's wars. The Holocaust victims were thus symbolically assimilated by the state, a view evident in a law suggested in 1950 whereby Israeli 'memorial citizenship' be given to them posthumously.[29] In Young's words, 'Memory of the Holocaust is brought into present moment: that which sustained the Jews during the Holocaust now sustains the current generation. Israel's defeat here would not be another Holocaust so much as an extension of the first one. In this way, Holocaust remembrance fosters a united identity between martyrs and a new generation of Israelis. The martyrs are not forgotten but recollected heroically as the first to fall in defense of the new state'.[30]

On the site of Yad Vashem were built over the years a historical museum, an art museum, different outdoor monuments, research and educational institutes. The site's amalgamated character has changed and grown since 1957, as buildings and memorials were added, reflecting the state's attitude towards its memory of the Holocaust. It is currently undergoing an expansion plan known as the 'Yad Vashem 2001' masterplan. Several projects are in different stages of planning and construction: an Entrance Plaza, a Visitors' Center, a New Historical Museum, a new Holocaust Art Museum, a Hall for Temporary Exhibitions, a Learning Center and a Holocaust Visual Center.[31] James Young notes that 'the construction of memory at Yad Vashem has spanned the entire history of the state itself, paralleling the state's self-construction. For this reason, it seems clear that the building of memorials and new spaces will never be officially completed, that as the state grows, so too will its memorial undergirding'.[32] All official guests of the State of Israel visit Yad Vashem, and the annual Holocaust Martyrs' and Heroes' Remembrance Day is opened by an official ceremony at Yad Vashem, followed by a special educational day in all the country's schools.

The Valley of the Communities by Architects Dan Zur and Lippa Yahalom

Completed in 1992, this is a memorial to every Jewish community that was destroyed during the Holocaust. Its labyrinthine shape is dug into the ground and open to the sky, with walls of rock almost 9m high bearing the name of each community. Its plan roughly resembles a map of Central and Eastern Europe, with over 5,000 names carved more or less according to their geographical location, and different fonts indicate the size of each community.

Figure 10.1 The Valley of the Communities by architects Dan Zur and Lippa Yahalom

The visitor wanders in and out of spaces in a strange landscape, surrounded by the massive walls that block all views except the sky, and by thousands of place names (Figure 10.1).

The configuration of this monument refers directly to geography, giving an almost literal expression to the idea of creating a 'place for memory'. The walls of rock enclose the visitor on the floor of a symbolic excavation, creating an artificial site of archaeology. The inscriptions, which seem partly epitaphs and partly ancient engravings, give an almost primeval shape to memory. The site is disconnected from the outside world by its huge rock walls, but the seemingly abstract setting actually represents real places that have been scaled down and condensed. Hence the visitor is lost in a sunken labyrinth, but at the same time some sense of orientation is caused by the possibility to locate geographically a specific community name in relation to others. The combination of an abstracted pit with an actual landscape is effective on two levels. One is the immediate sense of dislocation and insecurity, experienced whether or not the visitor is familiar with the map of Europe. The other is based on knowledge and recognition of the logic of movement inside the maze.

The approach to commemoration in this structure does not appeal to a sense of identification with the story of an individual or a community. Rather, it deals with the aspect of numbers. One of the difficulties in representing the memory of the Holocaust is its scope, since the immense numbers of victims and sites make them impossible to grasp. In the Valley of the Communities the architects used scale as a memory device, giving a notion of the relative location and size of the Jewish communities as an indication to the number of victims. At the same time, the size of this scaled-down space itself creates a new imagined landscape, on which memory is literally imprinted.

Memory is connected with sites, as history is connected with events, wrote Pierre Nora. Monuments derive their meaning from their own characteristics, independent of their surroundings. Even if their location were not at all accidental, it would be possible to change it without altering their meaning.[33] The Valley of Communities can be viewed as such a site, in which the map of Europe is reduced to a sunken pit placed on a Jerusalem hillside. Its national context as a 'place of memory' has been firmly secured, however, by its specific urban location at Yad Vashem and the ever-expanding projects on the Mount of Remembrance around it.

The Ghetto Fighters' House Museum

Yad Layeled was opened in 1995. It is located at the Ghetto Fighters' House Holocaust and Jewish Resistance Heritage Museum, named after the poet Yitzhak Katzenelson, at Kibbutz Lohamei Hagetaot in the north of Israel. The kibbutz was founded by a community of Holocaust survivors, former members of the Jewish underground in the ghettos and former partisans in 1949. The museum was begun the same year, and the founders stress that it is the first Holocaust museum in the world.[34]

The Site

The building stands in an open field, at the edge of a Roman aqueduct. Next to it are a large open-air amphitheater and the Ghetto Fighters' House Museum. The entire compound suggests an idea of historical continuity, from the Roman ruin to contemporary buildings and landscape elements. It is interesting to read the description of the site on the museum's website. 'The built compound of the Ghetto Fighters' House is made up of four buildings – the aqueduct, the amphitheater, the Ghetto Fighters' House Museum and Yad Layeled – which differ in their design language and in their sources of inspiration. This fact creates in the entire site a resonance of timelessness, of a dialogue between generations and of representations of universal values'.[35] In this context, Nora has remarked that the main function of the 'place of memory' is to stop time, to halt the act of forgetting, to materialize the immaterial, and to capture as much meaning in a minimum of signs.[36] The special location of Yad Layeled between the kibbutz and the Roman aqueduct has created both a sense of place as well as a sense of timelessness.

Yad Layeled, The Children's Memorial Museum by Ram Karmi and Partners

Yad Layeled (Figure 10.2) is dedicated to the memory of the one and a half million Jewish children who perished in the Holocaust. The building is entered from the top level, where a spiral exhibition ramp descends around a central cone on which names of children are inscribed. The external wall has no openings and daylight enters the building from a skylight around the cone. Hence the exhibition ramp seems to take the visitor into the earth, further continuing the idea of disconnection from place and time. At the end of this descent there is a metaphorical leap back to the present – the visitor exits the

**Figure 10.2 Yad Layeled, The Children's Memorial Museum by Ram
Karmi and Partners (on the left)**

building moving from its lowest point, dimly lit by a candle, to the contrasting
brightness of the landscape outside.

The memory device used by the architect is a symbolic route, in which
recollection is introduced to the body through the spiral movement downwards.
The idea of descending into the earth serves to isolate the visitor from the world
outside, evoking an uneasy feeling of being buried alive. This is an architectural
interpretation of a journey to another world, and hence the world outside
becomes irrelevant for the process of remembrance. Circulation inside the
building is therefore of utmost importance, determined by a strict configuration
that leads the visitor down in a single one-way route. Its linearity is further
emphasized by the symbolic exit from darkness into the light of the external
world. This approach brings to mind a very powerful definition of the Holocaust
by the writer Yehiel Dinur, known as Ka-Tzetnik 135633 ('Concentration
Camp Inmate No.135633'). He described Auschwitz as a different planet, the
Planet of Ashes, where there were different laws of nature and a different time
prevailed. But the Planet of Ashes still faces and affects Planet Earth.[37]

In an article about Ka-Tzetnik's life, Tom Segev claims that his books
express an early phase of Holocaust testimonies in Israel, characterized by
an assumption that human beings, as well as the world, are basically moral
and positive. Nazi crimes, as described by Ka-Tzetnik and other survivors,
were therefore understood as singular intolerable and abnormal phenomena
rather than an event that could possibly take place in some way or another
under different circumstances. Such an event would necessarily place moral
responsibility on every human being and society, regardless of place and
time. Israelis, concludes Segev, had read and heard the horrible descriptions

of Nazi cruelty, and had readily accepted Ka-Tzetnik's thesis of 'a different planet', which meant that the Holocaust took place outside the boundaries of human responsibility.[38] Another consequence of this approach would be that the Jews are the ultimate victims, an idea which either excludes or justifies the possibility that they would ever be responsible for the suffering of others.

Dinur later changed his views, and told Segev that one of his most disturbing insights in the following years was that humanity was responsible for the extermination of the Jews. As a human being, he too was partly responsible: the SS man who sent him to his death could be Dinur, but Dinur can also be the SS man. Segev notes that a similar change had taken place in many Israelis' perception of the Holocaust, making the culture of memory more complex, more universal.[39] However, Ka-Tzetnik's earlier theory of 'a different planet' remains embedded in many Israeli collective notions about the Holocaust.

Perhaps because Yad Layeled is intended for children visitors, it lends a very simple architectural shape to collective memory. This shape guides the visitor into a predetermined architectural experience of movement and light that has a strong perceptual impact on the senses. Moreover, by its distinct geometry the building conducts the visitor to a definite conclusion. The contents of this conclusion are, however, didactic and therefore beyond the realm of architecture. 'The historical memory of Yad Layeled has left a mark on the Israeli identity. The greatest challenge of the post-Holocaust generation is to preserve and rebuild our Jewish identity', wrote architect Ram Karmi about his museum.[40] This approach views the building narrowly, as a tool for reinforcing the Israeli narrative. On the other hand, the curator Miri Kedem expresses a more universal view of the dangers of racism. She writes of the human capacity to cause suffering, regardless of race and nation, when she describes the Holocaust as a period 'in which one part of mankind was losing its humanity, while the other was paying the unbearable price, yet remaining human'.[41]

Conclusions

The two sites reviewed here are 'places of memory', commissioned and built in Israel as part of an ongoing remembrance project that will eventually replace spontaneous memory. The buildings were commissioned by both an official remembrance authority, representing the voice of contemporary civil religion, and a private survivors' organization, with a slightly different agenda.

In order to understand the buildings as works of architecture, it is important to know the social and political background of their creation. The memory of the Holocaust that they are intended to preserve forms a fundamental part of Israeli national identity and civil religion. In the early years of the state the official attitude towards the memory of the Holocaust was torn between the need to remember and to forget. Statism, the civil religion of the early state, made a distinction between symbols of the helplessness of the victims that were not viewed as supportive of the state's goals, and symbols of the resistance of the fighters that were emphasized.

The sites chosen for the new buildings have been carefully selected to give them a specific context. The Valley of the Communities is part of the complex of Yad Vashem surrounded by the wider urban context of memorials and cemeteries. This is important in linking the memory of the Holocaust with the state and combining martyrdom with national heroism. Yad Layeled in Kibbutz Lohamei Hagetaot is located in an open field and adjacent to a Roman ruin, giving it a local historical context as well as a link to the agricultural landscape of the kibbutz, a myth of socialist-Zionism. Both sites have become part of the Israeli national landscape, and hence this landscape has become associated with their contents.

As noted by Azoulay, historical museums have the authority to generate new relations between individuals and the community. I have shown that both monuments reviewed here serve to integrate the visitors into the collective Jewish history, as well as to emphasize the Israeli-Zionist national answer to the circumstances of the Diaspora. In this sense, both sites serve as 'holy sites' or places of pilgrimage for the Israeli civil religion. Furthermore, since they create 'places of memory' that recall experiences and events that took place elsewhere, they serve to reproduce in Israel a sense of collective identity of the kind experienced by Israeli high school students making group pilgrimages to the death camps in Poland.

I have suggested that the Valley of the Communities at Yad Vashem is a striking example of this approach. A map of Europe has literally been transferred to the Mount of Remembrance in Jerusalem, symbolically reconstructing the entire geographic setting of the Holocaust to a local context. The site simultaneously isolates the visitors from the outside world and attempts to reconstruct real places around them. A similar technique is used at Yad Layeled in Kibbutz Lohamei Hagetaot, where the visitor is led into the earth in a symbolic decent to hell, making the outside world irrelevant for the process of remembrance.

The history of the museum complex at Lohamei Hagetaot differs from Yad Vashem in its approach to remembrance. Established in 1949 and claiming to be the first Holocaust museum in the world, it was based on the personal drive of its survivor founders to commemorate the Holocaust as well as emphasize Jewish life before, during and after the Holocaust. Statism at that period showed little interest in creating official ceremonies, buildings and patterns of collective remembrance, which were viewed as symbols of weakness and of the Diaspora. Young claims that at Yad Vashem, the Holocaust marks not so much the end of Jewish life as the end of viable life in exile,[42] an explicit view of the early state.

According to Liebman and Don Yehiya, the new civil religion expresses more sympathy with Jewish suffering and is less interested in themes of physical courage and armed rebellion. Accordingly, it is important to note that both sites commemorate the victims rather than the rebels. The Valley of the Communities deals with remembering the wider Jewish collective, the vast majority of which were killed without physical resistance. Yad Layeled is dedicated to the memory of the most vulnerable and helpless victims, the children. It is important to mention Segev's observation that Israelis have readily accepted Ka-Tzetnik's thesis of 'a different planet', which claimed that the Holocaust took place outside the boundaries of human responsibility and supported Jewish and Israeli self-perception as the ultimate victims. The two memorials were designed in a way that recalls this approach, suggesting that both clients and architects have not viewed the buildings as capable of expressing a more universal culture of memory.

Zertal points out that memory had become a central ideology of the state, an important national resource. Even statism's efforts to forget had paradoxically encouraged a project of preserving the memory of Jewish disgrace, later to be reused by the state in constant reproach towards other nations, as well as to help build up the national strength of the new Zionist society.[43] However, the idea that the Jews always carry the Holocaust with them, or the potential for another Holocaust, and that this may politically justify their collective actions, is problematic. The very shifting of the crimes from their historic place to their symbolic 'places of memory' in Israel poses a danger of banalization of the terrible events.[44] This danger is not only present in the wide use of the memory of the Holocaust in Israeli political discourse, but also in its material shape in the form of memorials and museums.

Notes

1   Quoted in Yael Zerubavel, *Recovered Roots: Collective Memory and the Making of Israeli National Tradition* (Chicago and London: University of Chicago Press, 1995), p. 4.

2   Pierre Nora, 'Between Memory and History: Les lieux de memoire', *ZMANIM – A Historical Quarterly*, 12 (1993), p. 6 (Hebrew translation).
3   Ibid., p. 15.
4   Defined by Liebman and Don Yehiya. as a symbol system that provides sacred legitimation of the social order. See: Charles S. Liebman and Eliezer Don Yehiya, *Civil Religion in Israel: Traditional Judaism and Political Culture in the Jewish State* (Berkeley and Los Angeles: University of California Press, 1983), p. 5.
5   Many researchers have addressed this issue, which will only be mentioned here briefly. See for example Tom Segev, *The Seventh Million: The Israelis and the Holocaust* (Jerusalem: Maxwell-Macmillan-Keter Publishing House and Domino Press Ltd, 1991).
6   Idith Zertal, *Death and the Nation* (Hebrew) (Or Yehuda: Dvir Publishing House, 2002), p. 90.
7   Liebman and Don Yehiya (1983), p. 23.
8   Ibid., p. 84.
9   Eliezer Don Yehiya, 'Statism, Holocaust and "Subversive Messages"', *Alpayim*, 20, (2000), pp. 88–9 (Hebrew).
10   James E. Young, *The Texture of Memory* (New Haven and London: Yale University Press, 1993), p. 210.
11   Zertal, p. 91.
12   Liebman and Don Yehiya (1983), p. 104.
13   Zertal, p. 51.
14   Liebman and Don Yehiya (1983), p. 103.
15   Young, p. 270.
16   Zertal, pp. 44–67.
17   Liebman and Don Yehiya (1983), p. 223.
18   Ibid., pp. 152–153.
19   Zertal, pp. 97–128.
20   Ibid., pp. 129–30.
21   Frances A. Yates, *The Art of Memory*, Ch. 1 (London:Penguin Books, 1966).
22   Ariella Azoulay, 'Open Doors: Historical Museums in the Israeli Public Sphere', *Theory and Criticism – An Israeli Forum*, 4, (Autumn) (1993), pp. 79–96.
23   Don Yehiya, (2000), p. 86.
24   Young, p. 225.
25   Jackie Feldman, 'In the Footsteps of the Israeli Survivor', *Theory and Criticism – An Israeli Forum*, 19, (Autumn) (2001). pp. 173–5.
26   Young, p. 220.
27   Leibovich-Dar, Sara, 'MORTAL Remains', *Ha'aretz* (daily newspaper), 20 December 2002.
28   Zertal, p. 126.
29   Ibid., p. 16.
30   Young, p. 214.
31   For more information, see Yad Vashem website (http://www.yad-vashem.org.il).
32   Young, p. 250.
33   Nora, pp. 17–19.
34   For more information, see Ghetto Fighters' House website (http://gfh.org.il).
35   Ibid. (Hebrew).
36   Nora, p. 16.

37 Israeli Center for Information, *The General Attorney Versus Adolph Eichman: Testimonies*, (1963), p. 1122 (Hebrew).
38 Tom Segev, 'Who Are You Karl Tzetinsky?', *Ha'aretz* (daily newspaper), 27 July 2001.
39 Ibid.
40 Ghetto Fighters' House website.
41 Ibid.
42 Young, pp. 215–16.
43 Zertal, p. 90–91.
44 Ibid., p. 96.

Chapter 11

Geometrical Wounds: The Work of Zvi Hecker in Israel, 1990 and 2000

Timothy Brittain-Catlin

More of a Story than a Monument

There is a character in a cartoon-strip short story by Etgar Keret and Assaf Hanuka who determines the immediate future nature of his relationships with the girls he meets by his ability to tell them how he felt when he killed someone during the Lebanese War.[1] If he tells her he felt nothing, it means that he is serious about her; if, on the other hand, he makes the story into a personal emotional drama, it means that she is to be a one-night stand. At the final frame, faced with what looks like being real love for the first time, he blurts out his story to his new girlfriend in its full horror and the reader is left wondering whether the hero cannot escape the ritual of self destruction, or whether in fact this is a happy ending in which his need to purge himself of the memories of violence and tragedy is enacted in a sort of flaccid transfiguration.

This tiny pathos-laden curiosity, six pages long, is more of a monument than a story; it is one of many demonstrations of Keret's astonishing ability to compress the moods, horrors and elations of emotional life in Israel in the 1990s into one short sequence laden with ideas compressed into symbols, or stylized text instead of dialogue. The theme that Keret here again captures so brilliantly is what has become the overwhelming characteristic of that decade, the way in which the personal and the egocentric became the determining factors of just about everything. It is the culture of the spiral, the one in which the tiny hot nucleus spins about to draw everything into a dance about its own centre.

The Spiral House

This chapter is about the way in which an Israeli sense of place is created not only out of the physical and national characteristics of its environment, but also from the voices, sometimes desperate and sometimes discordant, of its people.

First of all, however, that spiral: it was, by the way, A.W.N. Pugin in the England of the 1840s who first defined it as an original expression of architectural truth.[2] A hundred and fifty years is nothing in the lifetime of a great idea. The flats in Zvi Hecker's Spiral House in Ramat Gan were occupied in 1991 but whether the building was then finished or not is not exactly clear, for it is a structure that has gone on mutating over the decade that has passed.[3] It is a block of nine apartments in an approximately doughnut-shaped, approximately eight-floored building, and this approximation is due to the fact that the building slips down from the roadside at Tzel haGiv'ah – a romantic street name that means 'The Shadow of the Hill' – along a flight of steps that leads down at right angles to the Beit Zvi drama school (see Figure 11.1). The flats are arranged along this eastern side of the building, in an alignment that might be compared to a neat stack of bananas, pyramidal on one side, for as each floor digs down north-westwards into the hill, the banana above is slid or rotated a little bit, to a degree determined by its geometrical derivation, towards the south east. A pair of almost detached circular towers are situated to the north of the site within the curtilage of the doughnut, which is itself completed, or not, by various flying balconies and parapets, all of which are here and there supported by columns, pipes, spirals and various other bits and pieces.

What the above description does not provide however is any sense of the wildness of the construction, for every inch visible is covered in a wild assortment of materials, largely in irregular pieces of the pinky brown stone generally used for cheap cladding and not (as Peter Cook put it, when writing about the building for the *Architectural Review*) for '*serious* architecture'.[4] Some of this stone clads whole walls, but elsewhere it provides a frieze, or an edging; it provides a tasselled fringe to the bottommost whirl of the spiral stairs. There are however many materials here which are less serious still, in particular the mosaic of broken mirror-glass, some of which Hecker apparently applied himself with the slap and the dash of the gentleman artist. More extravagantly, there are some cementy creatures lurking here, most magnificently a dragon somewhere above the entrance route, and various other semi-geckos sliding about the soffites. Of course the flats themselves, in spite of their curious slithery geometry, which actually is quite well suited to the climate, are painted white, like every other flat in Israel, and their owners seem to have made some effort to conventionalize them, crazy as they are.

There is an outstanding feature of the Spiral House which lies outside its own domain and that is its location opposite Hecker's Dubiner House, designed twenty years beforehand with Eldar Sharon. This Dubiner House is the block of flats where Hecker lives. It is the Hecker of his mentor Alfred Neumann:

Figure 11.1 The Spiral House

angular, and classical in the sense of being a complete and unfaltering structure. From outside one can detect its polygonal grid but not, perhaps, the fact that it contains complex and magnificent internal public spaces of astonishing architectural control, with a full palette of variations of light and shadow extending and vanishing into different directions. Far from suffering from the usual problems of dankness and neglect that these unallocated spaces usually suffer from, these courts of Hecker's from the 1960s are airy and well-tended. Some people have put out rubbery plants in pots.

Hecker's Dubiner House brought something of Israel's heroic desert brutalism into the heart of the suburbs. It is at home there, not least because many of Israel's municipal and public structures of the 1950s and 1960s are composed of a vocabulary limited to various combinations of rough-faced overscaled concrete columns and shells, often in shades of ochre or pale brown, and when waiting at a bus stop in the sun in some suburb somewhere one can fancy, should one wish, that one is in the Mexico of Félix Candela, with strappy rubbery leaves sprouting out all over the piloti. It is this kind of work that creates the sense of place of the Israeli suburb, and it has often weathered very well. But there is always a difference between those architects who picked out and strung together their details from the magazines, and those who like Hecker were creating something altogether new. Hecker's Dubiner House entirely conquered the little street it sits over, with its scale and its panache but most of all with its discipline and its platonic conception, the benign *carceri* of the suburbs.

In fact, the immediate surroundings of the Dubiner House and its new neighbour on the other side of the road make up a picture of Israeli suburbia that has little equal anywhere, certainly for those who love the dusty green roads, the boxes on stilts, and tatty little low-lying villas, reached by jerking buses that feel as if they are made of tin, that almost scrape the boxes of nuts and seeds in the kiosks along the main streets of Ramat Gan or Givatayim.[5] Alongside 'The Shadow of the Hill' there is also of course that odd prominence itself poking up to the South, with an even odder monument on it; and down the alleyway past the Spiral House there is the straddling bulk of Beit Zvi, with its precious inmates having lessons which are intended to make them even more precious, and in fact almost grotesquely tuned into their themselves, in what for outsiders to the Israeli art of being Israeli seem like a parody of the conscious psychodrama of their daily lives. One can stumble down the public steps between the Spiral House and the school, as I did, and hear a fragment of a play being rehearsed just there, out in the open, and only the fact that the sequences are occasionally repeated reveals that this exchange of barked yet perfectly enunciated words and phrases is not a genuine conversation. The clothes that the actors are wearing are (in a country that has comparatively little native historical drama) more than likely the same ones that the characters in the play will be wearing, only to some degree more or less contrived. And it is because there is almost no difference between the form that a genuine conversation might take and that which is being spat out here (with the accompaniment of some rather stylized facial expressions) that the similarity between this and the architecture of Hecker, pushed up alongside at the time, could hardly be lost. Is the building serious or is it not? Is the *architecture parlante* reading from a script?

Given the time and expense required to create a significant building, it is surprising, and very regrettable, that more architects do not take the trouble to create a building that speaks in the coherent and symbolic manner that the Spiral House does. After all, the actor is paid to do in public a stylized version of what he does at home, and what he hears and knows about from home. Why not the architect? It should be no surprise that by far and away the most sophisticated, the most internationally recognized and most energetic and original art form in Israel is that of contemporary dance: dance is composed of the same processes that we saw with our student actors, the repetition of stylized sequences, and yet it generally succeeds in being far more disciplined than either acting or architecture. Where there is no discipline there is generally bad work, and that for the most part is what has happened to Israeli architecture. Hecker's discipline is his geometry, but it is also his alertness in identifying and representing the issues in modern life from which it draws its flavour, and giving them a comprehensible concrete form. No analysis of Hecker's architecture could be complete without investigating where his genius has trapped this fragrance – I would rather say this *nichoach*, a popular word in Hebrew literature – of modern life, and how he has left it to waft in the air to catch the nose of the passer-by.

When Peter Cook, one of the most sensitive of commentators, first met the Spiral House, he found himself appreciating its 'naughtiness': there was 'precious little naughtiness' in Israeli architecture, he thought. I do not entirely agree. There is a little naughtiness in Israeli architecture, even if it is not usually as blatant as it is in Hecker's buildings. I am not referring here to the self-conscious junkiness of much new building, usually of the speculative and commercial kind, but perhaps it would do to dispose of that particular subject as immediately as its pressing bulk demands. During the period when a very few large architectural practices dominated, architecture in Israel did have a certain discipline to it: within the overall mould of a Tel Aviv block of flats, with its restrictive building lines and heights, a talented architect might channel his originality into the most subtle of detailing, the most nifty of floor-plans, and achieve a building that had a great deal of austere beauty about it; some of the finest of these were built in the 1940s and 1950s around the Western end of Jabotinsky Street in Tel Aviv. Now of course all manner of nonsense is got up in all manner of colours; this is not 'naughtiness', but childishness, look-at-me-ness; actually, it is just a lack of discipline and of urban culture. There is a bridge erected somewhere in Jaffa a few years ago which is composed of entirely redundant archways of a sort of Arab appearance; and there is a block of flats on the Herbert Samuel promenade in Tel Aviv that is composed of various peeling planes and angled struts in ice cream colours and with a demonstrable lack of ability on the part of its designers to create a coherent composition. These things were designed by the fancier architects who appear in architects' chatter and magazines, the prima-donnas from the architecture schools, and they inevitably spawn imitations in the industrial areas of the satellite towns and along the main roads.

No, childishness is not naughtiness; naughtiness is delicious. Some of it is lying about the most sacred temples of modern Zionism, waiting to be savoured or ravished by connoisseurs, and nearly all of it was perpetrated by the great names in Israeli Modernism. One of my favourite examples can be found in the complex of the Hebrew University in Giv'at Ram. This is Ze'ev Rabina's 'Beit Belgiyah', the functional purpose of which seems somewhat unclear, but whose cultural significance is far reaching.[6] For this is a building in drag. It has a delicate stone-clad surround, and a dainty doughnut plan, but just there at the window reveals you can see bold concrete reveals: it is like looking closely down at a woman's neckline and discovering what are unmistakably a man's chest muscles underneath. And there is another agreeable piece of naughtiness nearby: the men's lavatories in the university library nearby. They are like the stage set in a Joe Orton comedy. They are stupendously grand in scale, even compared to the various odd pomposities of the rest of the building, and yet rather disgusting in detail; they are a glorious contrast to the po-faced detailing of Dora Gad's period library interiors and they are thus rather naughty and extremely funny.[7]

There is that naughtiness in Hecker, but although it is authentically Israeli it is only a small part of a tremendous repertoire: it is an element in the sophisticated conversation of a practiced raconteur. The Spiral House is as much as anything else a building of shuddering pain: even its balconies are composed of horizontal wooden boards that pierce the flesh of the soft walls like a splinter. An architecture which sings loudly of its personal pain is rooted deeply in Israeli experience: disappointment, lack of certainty, lack of money, puritanism, hard work, rebellion. That is not, perhaps, a particularly unconventional type of experience, but it is very rare that an architect can echo it and draw it into a narrative form of architecture. This Spiral House is not one of those buildings that one sees and forgets instantly, for it leaves the track of its sharp forms upon your skin, cutting and slicing as if with a pastry cutter; wounding the earth with its sharp teeth; and of course its inevitable spiral motion brings us back to Pugin's search for truth again, an architecture where every component sings with the same voice.

Voices in Israeli Culture

One of the critics to have identified an authentic Israeli voice in recent culture is Gadi Taub, whose 'A Dispirited Rebellion' was published in 1997, by which time the chaotic ugliness of the speculative architecture referred to above had established itself almost as a style. According to Taub, it is the inability of the young rebels to create an alternative, constructive worldview that makes their revolt 'dispirited'. Their revolt is merely a series of personal gestures, which are based to some degree on an ironic recycling of elements of an Israeli culture. The reemergence in the context of camp of 'Edna Lev, the High Windows, and Ilanit' (for example) has become acceptable to the new good-taste peddlers and critics of acceptable fashion because modern Israeli culture is certain enough of the quality of its more recent products.[8] The canvas from which to draw the elements of revolt is a broad one, and cannot be derived from European and American models, because:

> … we are foremost Israelis. Primarily, speakers of Hebrew. And that includes hot sun and crowdedness, and Arabic and Russian and Romanian and Moroccan, and development towns, and pebbledash villas in untended moshavim, and orange and blue signs carrying advertisements for Agfa, and Freddie Buys Everything merchants, and the ultra-orthodox and Gush Emunim, and 'Dan' and 'Egged', and the singers on the cheap cassettes sold in the markets, and meditation sessions on the hills of the Galilee, and basic training and war against another people for the same little bit of land.[9]

Nevertheless, he notes elsewhere that:

> The things that are considered the most tasteless – tracksuits, pink fur, shirts decorated with Starsky and Hutch and trousers suited for charity shops – these are the things that the new rebels want to wear.[10]

For a direct translation into architectural terms of what Taub is describing one need only see the horrible recent building for the Kalisher School of Art in south Tel Aviv. This was a new structure got up by a very Tel Aviv type of architectural practice, the principals of which teach in the smart architecture schools, and it includes a sort of tower partially faced on one side with fancy arched aluminium windows, the architectural origins of which are in the improvised stone-clad shacks that disfigure the nineteenth-century estates of Jerusalem, but also in particular characterize the new synagogues that have escaped the planning regulations south of the Mahaneh Yehudah market there.

It is conscious kitsch: it derives presumably from lashing about and trying to find comically ironic elements to draw attention to itself. Any comparison between this type of thing and the facing materials or anything else at the Spiral House can merely be a superficial one, in the exact sense of the word, since it is restricted to a similarity in the types of cladding. There is never any echo of the spatial drama of Hecker's buildings in this type of work, any more than there was with the arabesque bridge at Jaffa or the flats on the Herbert Samuel Promenade. All of these appear to be products of Taub's 'dispirited' culture, a series of detached personal statements, and they are merely visual effects, like Taub's Agfa advertisements. These kitsch buildings are suffering from what Taub calls 'postmodern distress' – which is, as he says,

> ... primarily the distress of too much freedom and too many choices, of lack of limiting norms, lack of faith, lack of uniting myths and explanations, lack of order. The distress of floating in an empty space.[11]

In other words, from a lack of discipline: it is an architecture based on nothing in particular, derived from a method of teaching that is unable or unwilling to commit itself to an ideology of any kind.

Taub's critique of the youthful rebels of the 1990s contains much of relevance to architecture: he concludes that a type of culture has developed in which stimulation has replaced feeling, observing that writers such as Keret appear to yearn to reenter the genres of the old, such as the simple detective story, but could not do so.[12] So the alternative is to make a lot of noise. The result of the 'dispirited rebellion' is, according to Taub, the extraordinary amount of violence characteristic in modern Israeli literature, a culture of decadence:

> Decadence is not only the basis for a change in order, but also a giving up of the very possibility of order. And when order ceases to be a stable rest, a basis for hope and plans, the satisfaction of the human urge is the refuge from it. A refuge which denies the possibility of stability, but nevertheless a refuge. A kind of consolation. Adhering to the refuge, more than being a subject in its own right, is the evidence of pessimism and introversion. A lack of willingness to cross the line between childhood and adolescence ... it seems that Erich Fromm's prophecy, the idea that in a crowd of society the individual becomes increasingly a 'cybernetic man' – a man who has stimulation but no feelings – is threatening at least partially to materialize.[13]

If he were an architectural critic, appalled by the violent ugliness of modern architecture perpetrated by teachers and critics as well as by the draughtsmen of the building contractors, he might well come to the same conclusion. Certainly his observation regarding the postmodernists frustration at their inability to reenter old genres carries a persuasive echo, although in fact at least one firm – Kimmel Eshkalot – has at least once most successfully reentered the genre of the classic north Tel Aviv block of flats.[14] But most of these architectural genres cannot be reentered, and some of those available to European and American architects – the neo-Classical, the neo-Gothic – have in any case never had a substantial presence here.

Taub's conclusion appears to be that contemporary writers and poets are primarily concerned with themselves and with their own knowing sense of detachment from the rest of society. It remains to be seen whether an undertaking as complex as a large building can incorporate not the disparate voices – these are the coarse shriekings we hear from the ugly buildings of the second-rate – but rather the sense of a culture that speaks with those voices: an architecture which represents the poignant cries of the rebellious and the lonely.

Kitsch and Aesthetic Violence

I have suggested that Gadi Taub's notions of kitsch and aesthetic violence are echoed in the architecture of recent years; but just as there is no attempt on Taub's part to imply that these notions are those which actually underlie modern Israeli society, it would be inadequate to claim that Zvi Hecker's architecture is composed even from a cacophony of individual voices. Hecker is not a 'new rebel'; nor is he making a simple transfer from one art form to another. Instead, it seems to me that he is making an architecture about a people that speaks about itself in individual and distinct voices.

One of the symptoms of a culture that revolves around individual self-centredness is the way in which the individual offers himself up as a victim time and time again. Although he seems to do this as a form of self-flagellation, his real purpose is to raise the intensity of an argument and to try to engage with it on emotional terms alone – preferably ones as far away from rational debate as possible. The politics of Israel has had many skillful practitioners of this art and in the mid-late 1990s Binyamin Netanyahu, for one, excelled at it. It is often astonishing how the behaviour of leaders permeates down through society, but it is possibly true that in Israel this kind of thing found particularly fertile ground. The debate of daily life, on the news, in the papers, in the post office, on the beach, revolves to an astonishing extent on the willingness of people to impress first of all their own sense of victimhood, and achieving this by making themselves as open and exposed as possible: it is of course a form of emotional blackmail. It must surely be related to the strange culture in Israel not just of voyeurism but also of wanting to be the victim of the voyeur: in part it is simply a giving of oneself up to what is sometimes a fact of life, having to live amongst too many others in too small a place. The residents of Tel Aviv publish messages in the town's weekly papers trying to make contact with those they have seen undressing almost in front of their eyes, from the kitchen window. A few years ago, the illustrator Yirmi Pinkus published a story in comic book form entitled *Observable*, which is entirely concerned with the experience of feeling that one is being watched within an urban context so intense, and so beautifully derived by Pinkus from a real address in Tel Aviv, that it has something of the quality of a stage set.[15] Furthermore, the hero is virtually always naked, or nearly naked, whenever he is seen in his own house.

The concept of voyeurism in architecture is not necessarily either abstract or metaphorical. Ya'acov Rechter's Tel Aviv Centre for the Performing Arts, provided outgoing theatregoers with the astonishing sight of the stars of the evening's performance totally and brutally in the nude, for these latter were evidently quite unaware that the large windows of their dressing rooms were placed exactly on the corner of Da Vinci Street and Shaul haMelech Boulevard.[16] A lesser architect, or perhaps quite simply a foreign one, would have placed a grand lobby or stair at this intersection, the most prominent part of the exterior of the complex, but Rechter, an Israeli through and through, decided to place here naked bodies and the genitalia of the artistes, projected by neon into the unsurprised night sky of Tel Aviv.

A conventional architect wanting to expose, as it were, the private parts of a building would choose to display the construction's own intimate details, the bits that are normally hidden, the pipes and the wires, on the outside. Plenty have done so. The fact that Rechter (whether he was aware of it or not) chose to make so bold a move, and evidently no one questioned it, illustrates to what an extent it was taken for granted by the early 1990s that no one had any right to any personal privacy any more. There is none or almost none at home, and there is certainly none in the army, and so it is simply logical that a large public building should operate on the same basis. The building is an instrument in forcing its users to play to the rules that Israelis are used to. 'Golda', as the opera house is colloquially known, is the matron that takes the naughty children in hand and punishes them.

I was reminded of 'Golda' and her cruelty to her guests on reading 'Chrysanthemums', a story about the regular nightly humiliation of adolescent boys by a boarding school matron, by the prize-winning writer Yossi Avni, who amongst his many admirers can apparently include Nissim Kalderon – the 'friend and teacher' of Taub. Avni, according to Kalderon's note on the back cover of Avni's book, 'can move in his writing from reality to fantasy whilst giving substance to both'. It is a quality well suited to architecture:

> Night after night Clara would pass through the rooms and fondle the erect penises of her students. The corridor was dark, and she would step through the rooms according in her determined order. The lads would already be waiting for her in their beds. When they heard the sound of her heels angrily striking the corridor and moving towards them, they would get into the right position and part their legs. The rustle of blankets told Clara that they were ready for her.[17]

Once one begins to look for this type of thing, an exaggerated humiliation of the individual centred about his own body, one finds they come in floods. 'Everything, absolutely everything, is coming out of the closet', as the veteran journalist Gideon Samet had memorably put it.[18] The radio editor Shy Tubali, not (unlike Avni) a writer of homosexual stories as such, won a prize from the Ministry of Science and the Arts for his short story collection *Body Language* in 1995. 'Adolescence', in some respects a conventional story, involves a boy fondling the naked body of his sleeping male babysitter; more brutally, in 'Fertilizing the Test Tube', a father, obsessively comparing and identifying himself with his adolescent son, ends by coming upon him in the night, drawing down his underpants, and pressing his fingers down towards his 'swollen testicles, and the member itself'; and having to run away from the house on being discovered.[19] And indeed at least one television drama involving scenes in showers with fathers and sons comes to mind. There is a recurring element throughout the mid-1990s of forced bodily exposure as an implement of shame and confusion. Intimacy and voyeurism, voyeurism and exposure, exposure and vulnerability; and perhaps not unexpectedly the additional elements of the male obsession with maleness and the male body carrying with it a certain glide into the realm of the homoerotic, and also into homosexuality, whether desired or not. These seem to me to be the paths that the culture of the individual inevitably must take when it is expanded and developed so far that it becomes a caricature. All that is left now is to see how Hecker succeeded in generating a new architectural form that could speak for the voices of the 1990s.

The Palmach House

Now that 12 years have passed since the inauguration of the Spiral House, the building has clearly begun to come alive. Its million faceted mirrors and stones have tarnished and weathered, and as with some kind of creeping fungus its spores have streaked across the building just as they fancy, usually but not, apparently, always in the way in which the light, or the damp, has called out to them. Its cutting edges have not blunted, and the actors on its doorstep still chatter away in dystopic fragments.

At the close of the decade over which the Spiral House ruled, the *Architectural Review* returned to Israel to record the changes that had come over Hecker's architecture.[20] The Palmach House, completed in 2000, was designed by Zvi Hecker and Rafi Segal, and is located in the non-residential university area adjoining Ramat Aviv; it was founded to commemorate the fighters of the pre-State Palmach military unit.[21] When I visited it for the first time I had come almost directly from an exhibition at the Dvir Gallery by the now internationally recognized photographer Adi Nes, whose work has very often portrayed soldiers in Caravaggio-like poses or occasionally brought them out into the world of camp: they wear make up, and engage in sometimes apparently faux-heroic activities (see Figure 11.2).

Immediately on arriving at the Palmach House I saw more crowds of soldiers, this time real ones, in purple berets and designer sunglasses, sunning themselves along the ramp of the entrance into the building. One does not have to look at Nes' photographs to see that the sexy creatures drawn long ago by Yossi Stern, those Palmach soldiers of the soft pencil line with their rounded hips and perfect backsides, have long since walked up off the page and descended into the real life of 1990s Israel. Now that the male body had become the focus of much Israeli life, a cult of male good looks and vulnerable beauty was comfortably descending into conscious self-parody. The young men here had chosen a vulnerable place to enjoy the weather and be seen, for they were poking their heads over the parapets towards the noise of the main road. The building presents a series of west-facing terraces and ramps instead of a formal elevation, and it does not have the comprehensive geometrical curtilage familiar from Hecker buildings of previous decades. It is, in fact, a building which more than anything else is distinguished by the fact that it has no distinct boundaries, and as such, as I noted at the time, it is very different from the trophy buildings of the rest of the Tel Aviv university campus.

The overall form of the Palmach House complex consists of three long structures, each one about 5m wide and from about 45 to 60m long (see Figure

Figure 11.2 Adi Nes, *Untitled*, 1996 (courtesy Dvir Gallery, Tel Aviv)

11.3). Two of these run parallel to Chaim Levanon Street, the main northeast, southwest road; the third runs from the southeast of the site towards the northwest, crossing the second block about a third of the way down from its northernmost tip. The front strip, mainly two floors high above ground, is the only complete part of the project and provides the entrance to an exhibition in the basement. The rear, intersecting, strips will eventually contain an auditorium, a library and various meeting areas, and these are located for the most part in the external 'V's formed by the intersections. The dynamic of the geometry of the plan is echoed in the vertical forms themselves, which rise

Figure 11.3 The Palmach House

and fall in counterpoint to the ramps; since the site slopes up and away from the main road, the elements of the structure further towards the rear project yet higher. Many of these are incomplete; the twin wheat sheaf symbol of the Palmach, cast in concrete, presides over what looks like an open theatre made from a Paul Nash landscape. The outstanding feature of the plan of the Palmach House is the fact that the central courtyard, between the front strip and the intersecting rear ones, has been painstakingly left exactly as was before construction began, a dusty, rocky, sparse grove of eucalyptus and pine.[22]

The building appears to have streaked or exploded through the landscape, fading away in all directions, sideways as well as up and down. The external finishes of the building, mainly strips of *kurkar*, were decided on through on-site experiment and so as at the Spiral House even the finished parts of the building altogether lack a sense of being completable.

Strangely enough, a harbinger of some of this had already appeared locally, in the form of a project at the architecture department of the Bezalel Academy of Art and Design in Jerusalem by a student called Alon bin Nun as far back as 1991.[23] This had been the year of the Gulf War, and bin Nun took as his theme the phrase used by radio announcers when about to switch on the air raid sirens: 'Following a missile attack ...' His project took the form of a building lying in modish deconstructivist shards across the diagonal of a Tel Aviv residential block, following the outline, as it were, of a detonated but still partially intact Scud missile: the building and the missile were merged with one another.

Bin Nun's point was, I think, the fragility of Tel Aviv, which had not known aerial bombardment since the Italian raids of the Second World War; his project spoke of the actual shattering of the city's rigidly orthogonal form. Unlike Hecker and Segal, he was not engaged with the fragility of the whole of the Zionist enterprise and of its soldiers. At the Palmach House, the architects' experiments with constructional techniques are explicitly derived from or recalling the practical expediencies of battle, the bunker, the sandbags, the life-or-death improvisations; there are pile-caps cropping up as windows, paths and routes going nowhere, wall planes that look like planks running up the side of a dugout. There is no sense in the finishes of the materials used that the building has been, as it were, achieved: the concrete is cast into panels and strips as if the construction (or the war) were still going on. The building commemorates the fighters of the Palmach by precisely echoing their frailties rather than their conventional triumphs, for there is a denying of heroic forms in favour of an almost pessimistic lack of boundary or mass, each part not quite being what it appears to be.

For this is the extraordinary and painful truth of the Palmach House, a building erected to commemorate a genuinely heroic period in Israel's genuinely heroic past: it portrays the army as a symbol of weakness, not of strength; it portrays the soldier as lonely, isolated, afraid, under-equipped, instead of as a hero; above all, it portrays the military bunker as an exposed, vulnerable, dirty, transient place instead of as a fortress. In so doing it reverberates with the self-doubt of the Israeli male self-image, and with the shredding of classic Zionist history, well before these things became the common language of the street. It no longer has the actors' script of the

Spiral House: it is barking commands across the platoon; deafened by the battle and by the winds. It creates a sense of place which is un-place, but an Israeli un-place.

It was thus of the Palmach House that I was reminded when I read the pained testimony of a soldier called 'David' serving in a submarine, in the compilation by the psychologist Danny Kaplan, the terrible power of which is worth experiencing here at some length:

> During my military service on submarines, playing about like gays was something well recognized and known ... it creates a massive sort of intimacy. However strange it may seem from outside, people would sometimes sleep in each other's arms in the same bed and certainly some people would feel each other up, for fun, laughing. There's a great deal of male intimacy here ... There were all kinds of tests of each other's gayness going on. First of all, you have to understand that the submarine was divided into an engine wing, which was the more masculine one, and the electronics wing, which was less so. The engine people were dirtier, covered in the grease and the filth of the machines ... so one of the soldiers from the men's wing, whose behaviour was somewhat worse, and was less of a friend of mine, and who had been in the army longer – I remember one really heavy incident with him, which he in fact started on one of the lower bunks, and he began to fondle me, like he was kind of massaging my penis. There were other people around, you're not alone, that's what stops it from being a seriously gay thing, and I, being young, had two possible reactions to choose from: one was to show him that yes, I was braver than he was, and let him get on with it until he got frightened and ran off, and the second possible reaction, which was the one I generally chose, was to try to get out of it somehow. And I remember that I said to him that time, 'stop it, that's enough', and his hand was right here, and although I was telling him to stop I was actually pressing his hand against my cock. I was doing exactly the opposite of what I was saying. He of course got it. He jumped out straight away and something in his expression which I saw as soon as he spoke seemed to show me that he understood. I remember it as a very frightening moment.[24]

This astonishing story, which is much echoed elsewhere amongst Kaplan's researches, and which also is matched by the anecdotal evidence of others, contains almost all the elements which lead to an understanding of the Palmach House. It contains the uncertainty of the balance between the heroism of the Zionist enterprise, and the delicacy of its operatives; it contains that recurring element, that is given voice by Avni, by Tubali, and in daily life, of the exposure of the individual's intimate parts and feelings both against his will but also with his complicity, and not to intimate friends but to near strangers; it contains that agreeable Ruskinian contrast (was it Charles Ashbee that brought it here?) between tenderness and brutal force; it contains a curious hovering dichotomy that tends, as it were, towards a scrunching of the gears, between pleasure and self-sacrifice; it contains the exaggerated claustrophobia that life in Israel inevitably entails; it is on the edge of a kind of ejaculation: and if relief came, it would no doubt be accompanied by waves of guilt and filth. The story, the truth, the life, the building: none of these is striving for fulfillment. The whole point is the loss, the shame, the disappointment, the sacrificing of one's privacy for nothing. 'It's crowded here … crowded'![25]

Notes

For Asaf.

Extracts are reproduced here with the kind permission of their publishers (in the case of Avni and the Friends of Natasha) or authors (Taub, Kaplan). The translations are in every case by myself.

I am indebted to my friends Esther Zandberg and Gil Klein for their generous help and advice.

1   Etgar Keret and Assaf Hanuka, 'Pa'am b'Hayim', *Streets of Rage* (Tel Aviv: Zmora-Bitan, 1997), pp. 27–33.
2   For which see, for example, Pugin's undated letter to J.H. Bloxam in the Magdalen College Oxford archive, MS 528/89.
3   Ramat Gan is a satellite town of Tel Aviv, directly across the Ayalon River to the north west.
4   Peter Cook, 'Spiral Hecker', *Architectural Review*, 188, October (1990), pp. 54–8.
5   Givatayim lies across the Ayalon River to the south east of central Tel Aviv; its architectural character is in general similar to that of Ramat Gan.
6   Opened 1967.
7   The veteran interior designer Dora Gad is mainly associated with her public and state projects, such as the Israel Museum, parts of the Knesset building, and the interiors of liners and passenger aircraft.
8   The archetypical example of this phenomenon was the reappearance in the late 1990s of the veteran songstress Shula Chen in a cameo role in the culturally aware television drama 'Florentin', after her prolonged absence from the fashionable stage. Edna Lev and Ilanit are middle-of-the-road singers of the 1970s: the former was also a tabloid gossip column celebrity, and the latter appeared for Israel more than once in the Eurovision Song Contest; the 'High Windows' was a late 1960s band closer to folk than pop.
9   Gadi Taub, *A Dispirited Rebellion: Essays on Contemporary Israeli Culture* (Bnei Braq: Hakibbutz Hameuchad Publishing House Ltd, 1997), p. 44. The original text of this extract and of all others referred to in this chapter are in Hebrew.
10   Ibid., p. 81.
11   Ibid., p. 82.
12   Ibid., p. 65.

13   Ibid., p. 77.

14   I am thinking of the block of flats at Mandelstam Street in Tel Aviv, under construction during 2000.

15   Yirmi Pinkus, *Observable*, trans. Noah Stollman (Tel Aviv: Actus Tragicus, 1998).

16   The Tel Aviv Centre for the Performing Arts opened in 1994.

17   Yossi Avni, 'Chrysanthemums', *Grove of the Dead Trees* (Tel Aviv: Zmora-Bitan, 1995), pp. 119–22. Yossi Avni is a pen name.

18   Samet was prophetically deploying the enforced self-outing of a well-known academic as a metaphor for the new maturity of Israeli society. *Ha'aretz*, 5 February 1993.

19   Shy Tubali, *Body Language* (Jerusalem: Keter Publishing House Ltd, 1996). The quotation is from p. 107.

20   Timothy Brittain-Catlin, 'Geological Formation', *Architectural Review*, 207, (2000), May, pp. 50–53.

21   Ramat Aviv is an up-market residential suburb north of central Tel Aviv. The Palmach unit operated illegally during British rule but was, in effect, tolerated because of its close ties to the Labour movement and its enmity to various other nationalist paramilitary operations. Yitschak Rabin was at one point its commander. The Palmach's greatest triumph was the securing of the road from the coast to West Jerusalem during Israel's War of Independence in 1948. The full name of the Palmach House is the 'Palmach Veterans' Memorial Centre'.

22   The landscape architects were Zvi Dekel and Shlomo Ze'evi of Tichnun Nof, Tel Aviv.

23   Hecker won the Palmach House in a competition held in 1992.

24   Danny Kaplan, *David, Jonathan and Other Soldiers: Identity, Masculinity and Sexuality in Combat Units in the Israeli Army* (Bnei Braq: Hakibbutz Hameuchad Publishing House Ltd, 1999), pp. 131–2. Kaplan's book has now been issued in an English edition: Danny Kaplan, *Brothers and Others in Arms: The Making of Love and War in Israeli Combat Units* (New York: Southern Tier Editions, Harrington Park Press, 2003). The testimony quoted here partially appears there on p. 209.

25   Michah Sheetreet, Arkadi Duchin and Yossi Alfant, 'New World Order', from *Shinuyim b'Hergelei haZricha* (*Changes in Screeching Habits*) (Or Yehudah: Hed Arzi, 1991), track 7.

PART V
PLACE/KNOWLEDGE

Chapter 12

On Belonging and Spatial Planning in Israel

Tovi Fenster

Introduction

In this chapter I will explore the discursive relationships between notions of belonging, sense of place and spatial planning in Israel. It analyses a specific case study; the conflict between Muslim organisations and communities and the local Jewish council of Nesher (north Israel) around the preservation of an old Muslim graveyard in which one of the Arab leaders Iz A-din el Kassam is buried. The local municipality wanted to construct the main road leading to the council on the edges of the graveyard and to expropriate some parts of the graveyard. This intention caused tremendous conflicts and negotiations between the two sides until a compromise has been achieved. This case study highlights interesting dilemmas as to how to define belonging as part of a sense of place; who defines which land and its symbolism belongs to whom, especially as related to the minority Palestinian citizens of Israel whose 'sense of place and belonging' contradicts or is perceived as threatening the 'sense of place and belonging' of the majority hegemonic Jews. Additionally, perhaps the most controversial issue stemms from this article is the role of spatial planning in preserving or destroying notions of belonging and sense of place.

The Tail of the Road and the Graveyard

The tomb of Iz A-din el Kassam is located in the graveyard which is adjacent to the Balad ash-Sheikh village, one of the Palestinian villages which were evicted and deserted after the 1948 war and the pronouncement of the state of Israel. Today it is the location of the Jewish Local Council of Nesher, near Haifa in the North of Israel. The graveyard became quite famous in the 1990s because of this tomb.

Iz A-din el Kassam was originally Syrian who fought against the French occupation in Syria and escaped to Palestine in 1921 where he started his combat against the British, since he claimed they helped the Zionist movement in its struggle against the Palestinians. He was killed in 1935 by the British and was buried in the Balad ash-Sheikh graveyard. After his death he became 'a symbol of radical response' (Seikaly, 1995). But it was only in the 1990s that his canonization became widely recognized and disputed among the Palestinians and Jews in Israel and in The Palestinian Authority Areas. This dispute occured when one of the sections of the Islamic Hamas movement active in the Palestinian Authority Areas adopted his name 'Iz Λ-din el Kassam Troops'. These Squads are responsible for the many killings of Israelis in terrorist attacks in Israel in the last decade and therefore his name has negative associations among many Israeli Jews who face daily threats of terror.

For the Palestinians citizens of Israel he became a symbol of political and radical resistance. One of the expressions of the importance of his memory and symbolism in the Arab-Palestinian identity formation is in street naming after his name in several Arab towns. One such example is in Umm el Fahem. This act raises yet another objections among Israeli Jews. Azaryahu and Kook (2002) for example see: 'the commemoration of Izz al-Din al-Qassam in an Israeli town as problematic' because this action is not a formally regulated decision but is perceived as an act of resistance in the eyes of some of the Jewish people in Israel. Azaryahu and Kook identify such street naming as an implied message on behalf of the Arabs of historical continuity between the struggle waged in the 1930s to that of the 1990s and also as a representation of Muslim radicalism.

If street naming after Iz A-din el Kassam causes such a dispute let alone the activities of the Islamic Movement in Israel to transform his tomb into a site of commemoration. During the 1990s the Islamic Movement took responsibility on his tomb in fencing it and maintaining it regularly. They also organized regular rituals and prayers on Fridays by the Islamic Movement believers (see Figure 12.1).

To emphasize how disputed and sensitive this issue is, I should mention here another problematic site of commemoration that of Baruch Goldstein, a Jewish-Israeli medical doctor who shoot 29 Palestinians in 1994 in Hebron. His act raised tremendous amount of condemnation among the majority of Jewish people in Israel especially around the intentions of some extremist Jewish groups to turn his grave into a site of memory and as a symbol of the Jewish fights against the Palestinians in the occupied territories. In earlier versions of this paper this latter site has been mentioned by one of the peer reviewers as another expression of the problematization of commemorating

Figure 12.1 The grave of Iz A-din el Kassam in Balad ash-Sheikh graveyard, Nesher, Haifa

sites of memory and belonging for one community which become sites of dispute for other communities. The equalization between these two cases of commemoration of disputed figures reflects in many ways the complicated and intricate side of Israeli – civil identity which still struggles for its existence and formation. The representation of Iz A-din el Kassam is perceived as problematic as the representation of Baruch Goldstein for some Jewish and Palestinians people.

No wonder that in such a bitter and complicated situation the plan of the local Jewish council of Nesher to expand the road and expropriate some of the graveyard's area caused strong objections and demonstrations amongst the Palestinian citizens of Israel especially the Islamic movement. The conflict became more aggressive when the Jewish head of the local council was life threatened mainly because of the declarations on the Jewish side that Iz A-din el Kassam is actually not buried in this tomb. An official committee was set up in early 1990s to negotiate between the two sides. The committee members included representatives of the sides involved; Muslim public figures, government representatives and the Nesher local council representatives. A compromise has been suggested to move the route of the road so that only one tomb at the corner of the graveyard would be removed. The Muslim rejected this proposition and applied to the District Court. The District Court rejected the Muslim's appeal (Berkowitz, 2000). At the end the road has been constructed as a bridge (see Figure 12.2), so that there was no need to remove any of the tombs in the graveyard. This compromise satisfied both sides to a lesser or greater extent. The local planning apparatus has honoured the disputed site of commemoration.

Clearly the tail of the road and the graveyard shows that within the process of planning there is a way to find an acceptable solution for most parties, which allows development and planning on the one hand while maintaining memory and belonging on the other hand. These resolutions take place mostly at the local/ micro level of planning. But what happens when we deal with sites of belonging and memory at a much larger scale such as the national level?

Belonging and Memory as expressed at the National Master Plan of Israel

To discuss the notion of belonging and memory at the national level of planning, I posed the question to the architect of the Israel Master Plan

Figure 12.2 The disputed bridge crossing over some graveyards, Balad ash-Sheikh graveyard, Nesher, Haifa

(TAMA/35), Shamai Assif. Needless to say, he is Jewish, a person with prosperous professional experience of planning projects in Israel. Interestingly enough, the connections between memory, belonging and spatial planning were firstly made in this official government planning document; The National Master Plan of Israel (TAMA/35) that usually articulates the hegemonic national goals and targets of the state of Israel. The architect of the plan Shamai Assif identified 'a sense of belonging to the land' as one of the three components, which consist of a 'good quality of life' (the other two are comfort and commitment to the environment, see: Fenster, forthcoming). His definition implied to the 'general public' with no particular reference to any national group. I find the fact that sense of belonging has been mentioned as target of planning in the national master plan fascinating. My immediate reaction was to look for the spatialities of belonging in the plan text. This was more difficult and it made me want to understand more about the connection between notions of belonging and planning in the eyes of the planner.

A point of clarification must be made here as to the connection between memory, belonging and spatial planning. Obviously, a sense of belonging and attachment is constructed and memorialized as part of communities collective historical identities with no direct relations to planning. However, planning can assists in legitimizing sites of memoralization and commemoration and turns them into visible and preserved spaces in which rituals that express belonging can take place. Thus, planning itself does not create belonging. Rather, it makes it visible and legitimate.

Back to the discussion with Shamai Assif. First of all I asked him a general question of his views concerning the linkages between a sense of belonging and planning:

> You identify and become connected to an arranged place that has an order that can be understood that contains and expresses certain norms and values ... One could identify or feel belong to ideas such as: green boulevards, metropolis, central transport ... these are the opposite symbols of spatial chaos occurred when the built environment is constructed as a result of pressures of private entrepreneurs and the decisions of the Israeli Land Authority. This chaos creates the opposite of sense of belonging. (Interview, March 2001)

The architect associates the expressions of a sense of belonging in planning with spatial order, that is, the 'order of space' or the 'arrangement' of space is what makes one feel belonging to a space or place. Here he refers to the meanings of a sense of belonging as a general rule in everyday life with no specific connection to specific sites representing national belonging and commemoration. The question that comes out of his narrative is what about the connection between belonging, memory and planning which is relevant not only to 'order' but to historical and or national events. So I further asked him: Belonging of whom? Or in particular what are the expressions of belonging of the Palestinians citizens of Israel in the principles of the Master Plan. In his answer he discussed the general notions of belonging and memory and then moved onwards to discuss specific 'sites of historical importance and preservations':

> There is a need to create a situation in which the Arab population feel a sense of belonging to this country, maybe there is a need to tell the Arabs that the state respects and is going to restore some of their abandoned villages and make them into memorial sites but we did not suggest it in the National Master Plan ... We live in a complicated situation of a constant threat, it is a matter of survival ... the two nations and the two communities should become more mature and more integrated with a wide agreement on the meaning of quality of life, but to suggest to restore their villages ?! ... and the second stage is (throwing us to) the sea?! ... it becomes very popular to say that we are strong enough and we can allow ourselves (to let Palestinian restore their sites of memory) but we are not so strong and we can not allow ourselves ... there is a need to respect and honor the memory of the others but between that and solutions of (restoring villages) there is a big gap. For example, the tomb of Az a dean El kasam who is a Palestinian hero ... there are situations in which a proposal to commemorate his tomb will be rejected so I told my team not to include this site (as a site of historical preservation') but we did suggested to restore other sites such as Kafar Kana or Karnei Hittin where Salah-A-Dean combat against the crusaders and this in itself is a 'revolution' that is to include such sites into the National Master Plan. I told the planning team that if we include the tomb of Iz A-din el Kassam as one of the sites for preservation, the plan would not be approved- we have to be practical. There is a plan – there is politics and there is reality. (Interview, March 2001)

A few points deserve elaboration here. First, the architect's narratives reflects some of the deep traumas and anxieties exist among the Jewish-Israelis as to their faith and destiny if they become less strong. The narrative of 'throwing the Jews to the sea' has its roots in Arab propaganda since 1948 but has been rooted and deepened into the Jewish-Israeli national identity and has been used as a mechanism of justification of a large variety of actions against the Arab-citizens of Israel in the name of Jewish survival. Here the Palestinian and the Jewish sense of belonging are presented as contradictory and clashing. It is an 'either/or' situation. The continuous holocaust memory that is transformed into today's life and serves as a mechanism to justify lack of justice and equality towards 'the enemy' – the Palestinian citizens of Israel.

Secondly, the architect's narratives expresses the dilemmas embedded in a place making process which reflects in many ways the Jewish hegemonic consensus. Here the architect becomes a practical practitioner. On the one hand, the architect wants his plan to be acceptable and approved by the authorities. On the other hand he wants to present his plan as pluralistic and as one which respects sites of memory and belonging. However, at the end of the day it is the Jewish planner who suggests the authorities which sites to include for preservations when his main consideration is the hegemonic Jewish consensus and the practical considerations of the plan's approval by the authorities. These tensions or dilemmas are also highlighted in the introduction to Forester et al.'s edited book (2001) on *Israeli Planners and Designers*. They note the tension between the democratic principles and the Jewish identity of a multiethnic and multi-religious state. These tensions echo in the Israeli planning system:

> … as professionals struggle to reconcile the duty of responding effectively to local needs with the imperatives of national security. The two dreams – the creation of a Jewish homeland and the establishment of a just society – have often been difficult to reconcile. (Forester et al., 2001, p. 3)

And the conflict between these two dreams resonates in the National Master Plan and the architect's narrative as well.

Thirdly, the decision, which sites of memory to include for preservation depends to a large extent on their historical aging. The earlier the historical period the easier it is for the Jewish to accept it. This is true both in the planning process of the National Master Plan and the process of street naming in Arab towns in Israel. The site of Hittim which commemorates the Islamic triumph victory of the Christian rule by Salah-a Dean is less controversial and more acceptable for the majority Jewish than the site of commemoration of recent period in Islamic identity such as the tomb of Iz A-din el Kassam. This is perhaps the reason why streets in Arab towns are mostly named by figures of early history of Islam (Azaryahu and Kook, 2002) because the leaders of these towns want to stay on the 'safe side' in establishing such spatial commemoration. This discourse expresses both the internal dilemma of the planner in such complicated situations but also it highlights the problematic role of the planner in creating and determining the 'boundaries of acceptable pluralism' within the context of the hegemonic Jewish belonging and memory, that is, the extent to which the planner functions as a 'technician' who merely implements policies or the extent to which s/he has the liberty, freedom and capacity to influence and change.

Expressions of belonging and commemoration at the two levels of space; the local and the national emphasize the complex situation that exists in contested spaces such as Israel and highlight the difficulties in providing spaces (in its double meanings) to Palestinian memory and belonging as part of an Israeli collective identity. The mechanism of resolution on the local level shows that sometimes reality is more flexible then ideology and the necessity to live together although separately pushes the two sides to find solutions that coincide the needs of each sides, that of belonging and memory and that of development and expansion. The relatively satisfactory ending of the *local* event reflects perhaps the fact that politics of planning do not necessarily contradict honour and memorializing of sites of belonging and attachment and that everyday life have their mechanisms in commemorating sites of belonging despite of the fact that these sites are not officially declared at the *national* level of planning.

Let us now locate these two scales of discussions; the local and the national within the theoretical and the political context of Israeli society and space.

On Belonging, Memory and the Politics of Planning and Commemoration in Israel

What is a sense of belonging? Probyn (1966, in Yuval Davis, 2003) has emphasized the affective dimensions of belonging – not just of be-ing but longing or yearning. The *Oxford Dictionary* defines belonging as composed of three meanings: first, to be a member of (club, household, grade, society, etc.); second, be resident or connected with; and third, be rightly placed, classified, or fit in a specific environment. These dimensions emphasize the membership component of belonging and its multilayered dimension (Yuval Davis, 2003). In many cases, belonging is a feeling that consists of both past and present experiences and memories and future ties and aspirations of a place and it grows with time (Crang, 1998; Fullilove, 1996).

A sense of belonging and attachment to a place is perhaps mainly associated in the literature with notions of blood, nations, sense of nationalism and identity: 'the geographies of belonging' (Mitchell, 2000). The personal sense of attachment to a place that is constructed not necessarily out of everyday life practicalities but on imagined collective memory and sentiments. Landscape or place is seen as the container of cultural belonging and it is through belonging to landscape that the connection between culture, space and people is created (Crang, 1998). One's right to belong to a space is seen as dependent on possessing the culture, which identify a specific territory as part of its content. Crang mentions that this vision of culture and space contains contradictory positions such as the fact that identity is defined by a spatially co-extensive culture or that culture is seen as the cause and not the outcome of material and symbolic practices.

How do these theoretical notions find their expressions in Israel? The state of Israel has been established as a Jewish state. As a sanctuary place for the Jews, a place for 'the gathering of the exiles' from all over the world. One of its first national goals has been to construct its collective Jewish national identity. Collective identity is usually based on symbolic myths and spatial sites of commemoration. Myths have very specific and basic function in a construction of a culture, society and nation. They are created by culture but also create cultures (Ohana and Wistrich, 1996). Myths internalize collective memory and they are usually spatial. The possibility to identify a specific geographical site in which myths took place makes them more accessible emotionally. Therefore nations make their best to commemorate and memorialize sites of myths because they serve as a strong mechanism to construct belonging and collective identity.

Sites of Jewish myths are located all over the country. Some are associated with early historical periods such as the Masada; the famous site located in Judea Desert which commemorates the myth of courage and heroism of the 'few' Jewish people in their fights against the 'many' Romans back in 63 AD. Some of the myths and sites of commemoration are more recent, relating to the times of early Zionism and early Jewish settlements in Palestine happened at the end of the nineteenth century, beginning of the twentieth century. One such important site is the battle in Tel Hai in the north of Israel which emphasizes the courageous and heroism of Joseph Trumpeldor, the leader of a few Jewish settlers that just before his death said: 'Its good to die for our land.'

The construction and commemoration of the Jewish national collective identity has become a formal project. Several government and semi-government agencies have been established to identify sites of significance from Biblical times to recent historical events. The wars against Arab countries since 1948 added myths and sites of commemoration all over Israel, memorializing army battles and soldiers killed in wars.

The construction of Israeli identity excluded in all aspects of life the myths and sites of belonging and commemoration of the Palestinian citizens of Israel.

Moreover, as it has been recently discovered, from 1965 the government of Israel took an active initiation to 'clean' the land of the ruins of the abandoned Palestinian villages (Shai, 2002). This operation took place for several reasons: the Ministry of Foreign Affairs claimed that this ruins located near main roads raise 'unnecessary questions' among tourists, other organizations thought that the ruins negatively effect the beauty of the landscape and there was even a justification for this operation among the Israeli Land Authority people that the 'cleaning' of the Palestinian villages will prevent the distress caused to the Palestinian citizens of Israel of seeing their home villages being destroyed (Shai, 2002). Even academic researches and books such as: *Myths and Memory: Transfigurations of Israeli Consciousness* published in 1996 (Ohana and Wistrich, 1996) did not include any work of Palestinian myths and memory but only those of Jewish and yet it is titled 'Israeli' instead of perhaps 'Jewish-Israeli'.

Mitchell (2000) connects such notions as who has the right to the land, who can call a land home to the work of the photographer Ingrid Pollard who challenges the stereotypical construction of identity in Britain by questioning:

> [W]hy blacks are always associated with the urban and how therefore black people will always be considered strangers in the English countryside. (Mitchell, 2000, p. 260)

Here a connection between identity construction, belonging, and spatial exclusion is made: 'when a group is excluded from ... landscapes of national identity ... they are excluded to a large degree from the nation itself' (Kinsman, 1995). These formations of inclusions and exclusions are shaped and reflect identity and nationalistic constructions. Moreover, people 'hunt down' national identity of those they consider a threat to the authenticity, purity or the mere existence of the nation. By that national identity becomes not only about belonging but also about exclusion, especially those who threaten the purified connection between 'blood and soil' (Mitchell, 2000). Here one can find a clear connection between Ingrid Polard's description and the complex situation in Israel described earlier. Identity construction and belonging in Israel both spatial and symbolic is mainly based on Jewish history, thus, excluding those who are considered as a threat – the Palestinians.

This situation in Israel is a very complicated precisely because the 20 per cent of its citizens, the Arab-Palestinians can not identify with the myths and sites of commemoration of the Israeli Jews. As already mentioned, this is so because some of those myths are based on their own associates presented as 'the enemy' 'the bad guys' especially in recent myths. These myths are founded on the conflicts between the Jewish people who arrived in mid-nineteenth century to a land that was not empty; it has been populated with some 700,000–900,000 Palestinian who lived in Palestine before 1948, some of them for many generations. The Jewish myths of those early days of settlement are formatted on the fights against the Arab-Palestinians who lived in the country, the associates of those who live in the country today. In the last years the Palestinian citizens of Israel took their own initiations and constructed some forty associations which take care and commemorate sites of memory, myth and belonging. There are also some Jewish-Arab associations that work in collaboration in putting signs of the original names of the Palestinian villages existed before 1948.

Another expression of sense of belonging that appears in the literature is the one which develops and grows with time out of everyday life activities. This aspect of belonging is discussed in de Certeau's book *The Practice of Everyday Life* (1984). For de Certeau, the corporal everyday activities in the city is part of a process of appropriation and territorialization:

> The ordinary practitioners of the city live 'down below', below the thresholds at which visibility begins. They walk – an elementary form of this experience of the city; they are walkers, *Wandersmanner*, whose bodies follow the thicks and thins of an urban 'text' they write without being able to read it. (de Certeau, 1984, p. 93)

This everyday act of walking in the city is what marks territorialization and appropriation, the meaning given to a space. What de Certeau constructs is a model of how: 'we make a sense of space through walking practices, and repeat those practices as a way of overcoming alienation' (Leach, 2002, p. 284). De Certeau actually deconstructs the process of how a sense of belonging is built up. It is a process of transformation of an unfamiliar space that by means of accumulated everyday practices becomes familiar and therefore a place of belonging. Belonging and attachment are built here on the base of accumulated knowledge, memory and intimate corporal experiences of everyday walking. It is a different type of belonging, it is a 'secular' belonging, an intimate and personal one which comes out of everyday practicalities. It is different from a collective and mythical sense of belonging, which is past and history constructed. The 'rituals', the everyday repetitions, are part of the spatial practices of belonging either everyday-intimate, or ceremonial, religious, national and collective.

To this notion of belonging we can add the discussion on memory which is perhaps the most explicit expression of a sense of belonging and a part of one's own identity. Leoni Sandercock (1998) connects between belonging and the life in the cosmopolitan city. She mentions three elements in the city, which are crucial for creating a feeling of belonging; city of memory, city of desire and city of spirit. Of the three, the first notion seems the most relevant to the notion of belonging as memory:

> Memory, both individual and collective, is deeply important to us. It locates us as part of something bigger than our individual existences, perhaps makes us seem less insignificant, sometimes gives us at least partial answers to questions like: 'Who I am? and 'Why am I like I am?' ... 'Cities', she carries on, 'are the repositories of memories, and they are one of memory's texts. We revisit the house(s) we grew up in, we show our new lover the park where, as a kid, we had our first kiss, or where students were killed by police in an anti-war demonstration ... our lives and struggles, and those of our ancestors, are written into places, houses, neighborhoods, cities, investing them into meaning and significance. (1998, pp. 207–8)

Memory in fact creates and consists of belonging. The accumulation of little events from the past, our childhood experiences which create a sense of belonging to the places where these events took place. This role of memory as part of one's own identity is another universally accepted need (Hillier, 1998).

Belonging, memory and attachment are spatially oriented. They are defined and maintained as part of an individual and collective identities. This is what motivates many of the Palestinians citizens of Israel to celebrate the day of 'El Nakba' in Israel's Day of Independence and to conduct memorial ceremonies in some of their abandoned villages. Being excluded from the national Jewish collective identity construction, the Palestinians construct their own identity and sense of belonging based on their own symbols and representations of memories. Their sense of exclusion and inability to identify with Zionist national symbols is what in a way pushed them to commemorate and internalize their own agenda of memory and commemoration. The naming of streets is a political act that expresses power and authority (Palonen, 1993). But the fact that in Israeli Arab towns streets are named after early and recent historical figures in procedures that are not formally regulated can be seen as an act of a protest, a civil protest against the lack of consideration and inclusion of Islamic history and identity in the national Israeli identity.

Last but not least let us have a brief look on the links between notions of belonging, attachment and commemoration and the politics of planning.

I wish to emphasize again that it is not argued here that the planner can commemorate or erase notions of belonging. Memories and sense of belonging are sentiments that either exist or not with no direct connection to the role of the planner. But planning do have the power to legitimize, to make sites of commemoration visible and explicit. In what follows the discussion on whether planners can play a role in the commemoration of belonging and memory is presented.

This connection between the role of the planner, memory and belonging has been challenged and criticized in recent planning literature (see for example Healy, 1997; Forester, 1999; Sandercock, 1998). The emphasis in this literature is on the negative role that planning can play, especially modernist planning in the destruction and smashing of memory and belonging more than in commemorating it. Sandercock (1998) is quite certain about the dominant role planner's play in this process. Paradigms beyond modernist planning have essential parts, she argues, in acquiring and recognizing the importance of memory, desire and the spirit of the city in creating healthy human settlements:

> Modernist planners became thieves of memory. Faustian in their eagerness to erase all traces of the past in the interest of forward momentum, of growth in the name of progress, their 'drive-by' windscreen surveys of neighborhoods that they had already decided ... to condemn to bulldozer ... Modernist planners, embracing the ideology of development as progress, have killed whole communities and destroyed individual lives by not understanding the loss and grieving that go along with losing one's home and neighborhood and friends and memories. (Sandercock, 1998, p. 208)

Sandercock makes a clear connection between planning and belonging by using the notion of memory. Planning, by reshaping space can commemorate and perpetuate memory or smash and destroy it. 'Rational planners', argues Sandercock:

> [H]ave been obsessed with controlling how and when and which people use public as well as private space. Meanwhile, ordinary people continue to find creative ways of appropriating spaces and creating places, in spite of planning, to fulfill their desire as well as their needs, to tend the spirit as well as take care of the rent. (Sandercock, 1998, pp. 213–14)

The same line of thought is presented by Hillier (1998) who analyzes planners' roles in a development process in the Swan Valley, Perth, Australia. It is an area, which is claimed and used by Aboriginal people. She contrasts the two types of knowledge involved in this project:

> [L]aypersons' knowledge, embodies tradition and cultural values; it is local and de-centered. Planners' expertise, on the other hand, is disembodied, 'evacuating' the traditional content of local contexts, and based on impersonal principles that can be set out without regard to context – a coded knowledge that professionals are at pains to protect. (Giddens, 1994, p. 85, in Hillier, 1998)

These two types of knowledge, the local which embodies memory and belonging and the 'professional', the modernist which serves as an 'evacuator' of the local illustrate the ways and means planning can serve as a vehicle for respecting and honoring sense of belonging or as a means of smashing it. The intricacies between the two types of planning depend on the extent to which the two types of knowledge intermingle (see Fenster, 2002).

Patsy Healey (1997) sheds another light on the discussion of the role of planners within the planning practice relying on two important theoretical approaches, that of Foucault and that of Giddens. The planner, argues Healey, doesn't bring power relations 'into being' as Foucault would argue but s/he can make choices to change and influence situations s/he thinks should be changed. Planners in fact have more power to change and influence the decision-makers then perhaps they take on themselves.

The architect of the Israeli master plan talked about the complicated intricacies between plan–politics–reality as three contradictory and problematic concepts, which planners have to accept as 'reality' and as barriers against radical changes. This has been his justification for not including the site of the Balad ash Sheikh graveyard where the tomb of Iz A-din el Kassam is located as a site of preservation in the national master plan. But if we take Gidden's view as presented by Healey, the planner, as much as other individuals, lives in:

> [W]ebs of relations through which structuring forces bear in on us … Through our creative efforts, we are continually making and remaking our conceptions and meanings. (Healey, 1997, pp. 47–8)

This perception takes planners as much more influential and powerful figures then it seems from the way the Israeli architect portrays his attitude. In this respect, Forester (1999) puts a greater emphasis on the role of the planner as the one who does 'much work before decision makers act' and by that both he and Healey actually view the role of the planner as much more dominant and influential then perhaps practical planners themselves do. In this context, it is interesting to look at the delicate relationships that an Arab-Israeli planner fabricates with the Mayor he works with (in Umm el Fahem). He said:

> If I come to talk to the mayor, maybe I will tell him about something he does not know. As a man that works in an official position, he does not know everything, he just knows some things. So when you go to speak with him, you have to try to listen to him, to recognize his authority, to give him respect, to let him know that you have come because you want to hear his advice and opinion. You have to let him know that you want to hear what he has to say and you have to ask him, 'Can you help us to do something? How can we do this? ... That means that you have to listen even if you know how to do what you want. (Khamaisi, 2001, p. 210)

The planner describes the tactics he uses when working with decision makers. He actually talks about involving decision makers in the planning process so that their commitment would be increased. He describes his relationships with the 'Boss', the mayor, not only as professional but also as emotional and social so that 'his heart is open'. These manoeuvres that the planner uses function to actually establish close and even personal working relationships with the decision-maker. Such relationships are established precisely so that planning issues that are not always popular could be approved. This experience of the planner is similar to what suggested earlier by Healey. Planners have the power and the ability to change perceptions and attitudes and not only to serve as technicians. This same planner describes his experience in changing the attitudes and perceptions of the state bureaucrats towards the Arabs when they met him:

> For example, I am an Arab man. When I came to meet these officials in the State agencies, they had ideas about Arabs, but where did they get those ideas? From television, and in that view an Arab was someone who threw stones, or who was very primitive. But when I met with them, I began to talk to them about the town and they changed: the blinders over their eyes and their vision began to go away. They discovered something new. They became more open. (Khamaisi, 2001, p. 210)

As much as a planner as a person can change superstitions and perceptions regarding his nationality so can planners change perceptions of decision-makers regarding emotional issues such as the inclusion of sites of commemorations of the Arabs in the National Master Plan. This is hard though. This same planner that talked about his success in changing attitudes of state officials was also a team member of the National Master Plan but could not radically change the perceptions of the team leader as to the type and nature of sites of commemorations included in the plan.

To sum, the extent to which politics of commemoration become part of the planning practice's daily agenda depends on the professional identity planners choose to adopt as agents of change. The space for manoeuvre and change in the planning practice is wide. It depends on the authority and power that the planner takes to her/himself and on the extent to which s/he manages to establish relationships of trust with decision makers. Planners can see themselves as mere technicians or as agents of change, change in perceptions and attitudes of the majority as much as the minority regarding issues that were considered as taboo before. It is probably up to the planners to choose their part and role in such processes.

Conclusions: Can the Gapes between Split Identities and Memories Can be Bridged in the Plan-making Process?

This chapter presents a dilemma emerging from the relationship examined between notions of memory, belonging and politics of planning in Israel both at the local and national scales. Belonging, attachment and commemoration became in the last decade or so a relatively fashionable field of analysis, providing overview material as well as analysis of the multiple meanings and expressions of each of these terms. This paper contributes another dimension to the existing discussion that of space and planning as mechanisms of commemorating and honoring sites of memory and belonging.

No doubt that memory and the desire to commemorate are not built up on planners actions. These sentiments exist as part of communities and individual's constructed identities and myths. But planning can encourage and promote the commemoration and honouring of sites of memory by incorporating their preservations in the planning practice.

In situations of split identities, of contradictory memories and myths in one land, such as in Israel, this general statement becomes controversial and problematic as sites of commemoration to one community are in fact sits of horror and trauma to another community. How do we bridge this gap and how do politics of planning serve their duty in such complicated situations? No doubt that spatial planning is the 'arena or the field' where these complicated situations and sentiments are becoming explicit and visual. The level in which the hegemony allows 'the other' to expose and express its spatial symbols of belonging is a matter of social, cultural and political maturity when both the hegemonic and 'the other' feel safe and comfortable within their identity contexts. However, as suggested in the paper, the spatial planning process can also be seen in a different perspective not merely as 'the field' where planners have to perform according to a set up rules. Planners can also dictate and effect rules and be perceived as agents of change and transformation in the plan making practices and their influences on decision makers is sometimes much higher than they perceive it.

Analysing the connection between memory, commemoration and the politics of planning at the local and national scales, demonstrates the importance of *the scale of the planning process* in such complicated matters. It seems that at the *local* level it is sometimes easier to bridge the gaps between two different and contradictory sets of memories and sites of belonging. The practicalities of everyday life sometimes push to compromise, an act which seems more complicated at the *national* level of planning. At this scale, the hegemonic ideology plays a much greater role in formulating planning principles, which tend to meet the goals of the majority. In both levels planners have more power to change and influence then they sometimes dare to take. Planners can sometimes play the role of conflict facilitators, transformetors or as agents of change. Either view of their role effect the way planners perceive their duties and the actual procedures they take in the plan making process. This includes their approaches to the different levels of participation, of the decisions of who to participate and in which subjects to involve people and the significance given to the local knowledge gained in such participatory processes.

The chapter poses dilemmas more than suggests solutions. Planning is usually very political especially in disputed places such as Israel. Therefore, the articulation of notions of belonging and memory of the 'other' are becoming very problematic. Moreover, the extent to which sites of commemoration of the 'other' can become part and parcel of the hegemonic planning process depends in many respects on the degree of maturity and confidence that the 'majority' hegemonic society poses. In many disputed areas in the world such an outcome is a result of a long and painful process of national reconciliation such as in South Africa. It seems that only a high degree of civil maturity and reconciliation can allow memory and belonging of the 'other' to become part of the space and landscape of the society at large which reflect its agenda and practices.

References

Azaryahu, M. and Kook, R., 'Mapping the Nation: Street Names and Arab-Palestinian Identity: Three Case Studies', *Nations and Nationalism*, 8 (2) (2002), pp. 195–213.

Berkowitz, S., *The Battles over the Holy Sites* (Jerusalem: Jerusalem Institute of Research, Israel, 2000) (in Hebrew).

Crang, M., *Cultural Geography* (London: Routledge, 1998).

de Certeau, M., *The Practices of Everyday Life* (Berkeley: University of California Press, 1984).

Fenster, T., 'Planning as Control – Cultural and Gendered Manipulation and Mis-use of Knowledge', *Hagar – International Social Science Review*, 3 (1) (2002), pp. 67–84.

Fenster, T., *The Global City and the Holy City: Narratives on Planning, Knowledge and Diversity* (London: Pearson, forthcoming).

Filkins, R. , Allen, J.C. and Cordes, S., 'Predicating Community Satisfaction among Rural Residents; An Integrative Model', *Rural Sociology*, 65, (1) (2000), pp. 72–86.

Fullivlove, M.T., ' Psychiatric Implications of Displacement: Contributions from the Psychology of Place', *The American Journal of Psychiatry*, 153 (12) (1996), pp. 1516–22

Forester, J., *The Deliberate Practitioner* (Cambridge, MA: The MIT Press, 1999).

Forester, J., Fischler, R. and Shmueli. D. (eds), *Israeli Planners and Designers* (New York: State University of New York Press, 2001).

Healey, P., *Collaborative Planning* (London: Macmillian Press, 1997).

Hillier, J., 'Representation, Identity, and the Communicative Shaping of Place', in A. Leigh and M.J. Smith (eds), *The Production of Public Space* (Oxford: Rowman & Littlefield Publishers, Inc. 1998), pp. 207–32.

Khamaisi, R., 'Planning in an Arab Municiplaity: A Profile of Rassem Khamaisi', in J. Forester, R. Fischler and D. Shmueli (eds), *Israeli Planners and Designers* (New York: State University of New York Press, 2001).

Kinsman, P., 'Landscape, Race and National Identity: The Photography of Ingrid Pollard', *Area*, 27 (1995), pp. 300–10.

Mitchell, D., *Cultural Geography* (Oxford: Blackwell, 2000).

Leach, N., 'Belonging: Towards a Theory of Identification with Space', in J. Hillier and E. Rooksby (eds), *Habitus: A Sense of Place* (Aldershot: Ashgate, 2002) pp. 281–98.

Ohana, D. and Wistrich, R.S., 'Introduction: The Presence of Myths in Judaism, Zionism and Israelism', in D. Ohana and R.S. Wistrich (eds), *Myth and Memory – Transfigurations of Israeli Consciousness* (Tel Aviv: The Van Leer Jerusalem Institute, 1996) (in Hebrew).

Pitkin, D., 'Italian Urbanscape: Intersections of Private and Public', in R. Rothengerg and G. Mcdonogh (eds), *The Cultural Meaning of Urban Space* (London: Bergin & Garvey, 1993), pp. 95–102.

Palonen, K., 'Reading Street Names Politically', in K. Palonen and T. Parvikko (eds), *Reading the Political, Exploring the Margins of Politics* (Tampere: The Finnish Political Science Association, 1993), pp. 103–21.

Sandercock, L., *Towards Cosmopolis* (London: Wiley, 1998).

Seikaly, M., *Haifa: Transformation of an Arab-Palestinian Society, 1918–1939* (London: I.B. Tauris, 1995).

Shai, A., 'The Fate of Abandoned Arab Villages in Israel on the Eve of the Six-day War and its Immediate Aftermath', *Cathedra*, 105 (2002), pp. 151–70 (in Hebrew).

Yuval Davis, N. 'Belongings: In Between the Indegene and the Diasporic', in U. Ozkirimli (ed.), *Nationalism in the 21st Century* (Basingstoke: Macmillian, 2003).

Chapter 13

Fragile Guardians: Nature Reserves and Forests Facing Arab Villages[1]

Naama Meishar

Introduction

'They don't even like nature and landscape; they have no real bond with this place. They don't feel it belongs to them', concludes Muhammad Abu al-Hayja,[2] about the attitude of the planning and nature preservation authorities towards open spaces in Israel. Abu al-Hayja is a resident and chairman of the cooperative association of the unrecognized[3] village of Ein Houd situated on Mount Carmel's western slopes. Many rural Arab communities are located next to nature reserves, national parks and forests. Nature preservation deprives these villages of their civil and planning rights, and impedes their development.[4] The borderlines between Arab communities and nature reserves and forests in Israel are the actual gaps between two types of discourse – that of nature preservation, and the discourse of planning rights for Arab villages. Following this, in this chapter I shall try to examine the gap between the discourse of Israeli planning authorities, and the Palestinian discourse of rural settlement interrelations with its natural environment.

The village of Ein Houd, on the western slopes of Mount Carmel, will serve as a case study of the friction and pain that delineate the borderlines between Palestinian villages in Israel and the open areas around them. I chose Ein Houd upon the recommendation of Mr. Hanna Sweid, Chairman of the Arab Center for Alternative Planning, who stressed the great difficulty embodied in the case of Ein Houd. I have also used the meticulous documentation offered in 'The Arab Archives' – archives of the Association of Forty, as well as the extensive information and references in Susan Slymovic's book on Ein Houd and Ein Hod (Slymovic, 1998). I shall try to describe and criticize the mode of institutive Israeli open space culture from the theoretical, physical and planning points of view. Also, I would suggest that this culture is characterized by nostalgia for an imagined and caged-in space of primal nationhood and nature. Consequently I shall argue that nature preservation in Israel does not relate to nature on the basis of commitment through biocentric ecology, nor through social ecology, but rather manifests anthropocentric-ethnocentric ecology interrelations.

In more detail, this chapter will describe the production of post-colonial Palestinian environmental culture at Ein Houd, and examine the various modes of operation used to establish it. It will look into the gap that lies between the Israeli planner on behalf of the planning authorities, and the Palestinian citizen of Israel – noting its non-egalitarian character, which will always embody hierarchy and constructed power relations that states challenges for the critical planner.

About Ein Houd

The village of Ein Houd is located in an area declared both a nature reserve, a part of the Carmel National Park, and a military fire zone.[5] The fire zone is a forest owned by the Israel Land Administration and is under the maintenance and direction of the Jewish National Fund.[6] The Palestinian village of Ein Houd used to be situated about one kilometer west of the present site. Most of its residents left it during Israel's War of Independence in 1948, and fled east to the Druse village of Daliyat Al Carmil. Here they hoped to stay until they could return to their own homes. Instead, some of them were intimidated into proceeding to the Jenin refugee camp (where families from the village of Ein Houd reside to this day), others reached various villages in the Galilee, and still others continued as far as Jordan and Iraq. One family took shelter in the shepherd huts it owned in its orchards, the site of present-day Ein Houd. But the declaration of the Absentee Property Law[7] deprived them of the land on which they settled. The absentee's lands were taken over at first by the JNF, and later by the Israel Land Administration (Golan, 1995). In 1950, Jewish artists settled in the original houses vacated by Ein Houd's residents, and founded the Artist Colony of Ein Hod.[8]

Ein Houd is still an unrecognized village. Officially, the village received government recognition through a resolution taken in December 1994. By 2000, the planning process of the village was completed, but ratification of the plan has been delayed. Without an officially authorized master-plan, recognition of the village is not yet valid. For its residents this means no official connection to water mains, public sanitation systems, electricity and communication infrastructure, no safe access road for vehicles and no obliged medical and educational services. These are the denial of fundamental civil and social rights[9] of citizens of this state. It is important to mention that one of the reasons for the status of the village is its location inside nature reserve, and its subjugation to the Israeli discourse of open spaces.

Producing Open Spaces in Israel

The discourse of official authorities planning open spaces in general and nature reserves in particular, runs along three different axes: a) the preservation of nature; b) national Israeli identity; and c) Jewish strategic control of space.

a) The Preservation of Nature Axis

This axis is based on academic-theoretical and empirical knowledge in the discipline of ecology. I shall try to show the infiltration of different types of discourse into the ecology discourse, and how it has subtly concealed non-scientific hostility towards Palestinians and their environment.

The implications of ecological research for Israel's Palestinian population have been far-reaching both economically and culturally. I shall concentrate on the developments of the ecology discourse concerning *interferences*[10] with an *ecological system* that is caused by acute *over-grazing*. Until the 1990s, it was commonly assumed that ecological habitats in Israel have suffered centuries of interferences such as fires, intensive logging and over-grazing, caused by Arabs and Bedouins. Consequently, according to this discourse, vast part of the country has not developed its *climax*. Climax is the final and stable presence of a certain natural flora, having undergone all phases of its *succession* – the changes of flora in a given area through time. Clement's succession theory of the early twentieth century is a deterministic theory[11] that dominated world ecology discourse until the early 1950s (Perevolotsky and Pollack, 2001, pp. 458–9).

The Nature Preservation/Goat-Induced Damages Law had passed in this vein in 1959. This law severely limited the number of goats legally acceptable in a private herd, as well as the size of grazing areas. The law extended inspection of official's authority and placed much stricter punitive measures for what was redefined as grazing felonies. The law went into effect only in 1977, when Minister of Agriculture at the time, Mr Ariel Sharon, decided to establish the 'Green Patrol' whose role it was to implement the law and put an end to black-goat grazing.

Until the 1970s, Ein Houd's residents owned about 1,000 goats and cows. According to Abu al-Hayja, this was their main source of income and food. It was especially significant in view of the village's inaccessibility and distance from sources of food and alternative employment. Approach to the village has been – and still is – a rough dirt road. It has not paced over since the village was unrecognized de-facto. Upon implementation of the 'Black Goat Law', Ein Houd's herd owners had to get rid of their main source of income.

Abu al-Hayja recalls that following the declaration of the property of former Ein Houd residents as a nature reserve, a part of the Carmel National Park, a forest within the charge of the JNF, and a fire zone, a major part of the environmental practices of the village has been obliterated. The terraces on the mountain slope close to where Ein Houd is situated were full of olive trees, apple, almond, fig, carob and pomegranate orchards. Until 1964, the villagers tended the terraces, pruned their trees and grazed their herds among them in the winter and spring months (in order to fertilize them and prevent thorny undergrowth in the summer). In 1964, the village houses were fenced in by barbed wire and the JNF planted the orchards over with cypress trees. Over the years, the terraces that remained untended were ruined.

Thus, in a gradual process of limitations, marginalization and discipline of movement and employment, the life tissue of an Arab rural community has deteriorated, and all evidence of its interrelations with its environment has been erased. Abu al-Hayja claims that the nature preservation authorities cannot contain human beings and their natural environmental interrelations. 'Nature, for these authorities, is a sterile entity'. For Abu al-Hayja, 'nature gives back to you what you give it'. Agricultural exchange with the land, too, is a give-and-take relationship. Now the enormous know-how that enabled the community to live and prosper in its environment is disappearing. The daily bond – of intimacy between the plants, the picking hand, the cooking pot and the family's mouths to be fed – is gone.

Cessation of grazing and tending of Palestinian lands in 1948 has caused numerous *dunam*s to turn fallow. Jewish farmers were not interested in tending these agricultural lands in a state of uncertainty, in view of the possible return of Palestinian refugees. In such an eventuality, refugees would reap the fruits of Jewish labor invested in the lands they had left during the war. This was a paradise for the researcher studying climax formations. The negative disturbances were removed, and now the researcher would meet pure, stable and original climax formations. In this context the ecologist Ze'ev Naveh (Naveh, 1959, p. 3) argues that:

> After centuries of *wild* grazing by goats and logging for firewood, woods have
> gone to *waste* ... The Arab shepherd let his herd grow regardless of the effect
> of over-grazing on the quality of future pasture. (Emphasis is mine, NM)

Naveh was looking for the most effective formation of natural pasture for
the modern Jewish settlement in the fallow of deserted fields. Naveh's book
Natural Pasture in Israel indicates an intense planning effort to feed a rapidly
growing Israeli population, with maximal usage of resources, and minimal
interfering with the ecological systems while instituting an autarchic Israeli
economy. Naveh was one of many researchers who pointed out the disturbances
of Palestinian grazing upon local climaxes. (Perevolotsky and Pollak, 2001,
pp. 276–7). It is interesting to note that two decades later, on 15 June 1978,
in a letter to the *Jerusalem Post*, Naveh and two other senior desert ecologists
proclaimed that the 'Black Goat Law' and its implementations damaged the
Negev's ecological systems.

At this point I address the claim made by Nurit Kliot (Kliot, 2000, p. 218)
about the anthropocentric basis of nature preservation in Israel. Kliot argues
that until the early 80's, nature preservation in Israel differed in character
from nature preservation norms in the west, despite Israeli pretense of an
enlightened and rational approach, which differed completely from the
Palestinian approach. Kliot cites Azaria Alon, former chairman of The Society
for the Protection of Nature in Israel that holds an established authoritative
position in the discourse of nature preservation:

> He writes that Israel is an island of nature preservation compared to its
> neighboring countries, who do not consider this subject important at all. (Kliot,
> 2000, p. 224)

Philosophy of nature preservation in the west is based upon a moral,
independent commitment towards all species of nature. Nature preservation in
Israel, however, is based upon the importance of the environment and nature
values from a human's point of view. The central considerations for nature
preservation were retaining open spaces for the sake of the citizens, their
bond with the land and the shaping of their ideology. Alternately, preserving
nature grasped as a tourist-economic element or as scientific resource (as a
genetic and biological pool). Less central reasons for nature preservation were
the universal moral and biocentric ones. In view of the constructed hostility
towards Palestinians as points of departure in scientific studies of grazing, I
argue that the basis for nature preservation in Israel is ethnocentric as well,
and not just anthropocentric, as claimed by Kliot.

b) The National Axis

I shall try to show that by means of western cognition and scientific practices, nature researchers-guardians tried to constitute their own positive, native identity, versus the Other, negative native Palestinian identity. Climax as narrative blends with various Israeli narratives that have legitimized the injustices inflicted upon the Palestinians.

A narrative such as 'Fatherland', for instance (Benvenisti, 1997, pp. 7–29) justified Jewish settlement in Israel while driving its Palestinian residents out of its borders, or within the borders, out of their homes, taking their land into Jewish ownership. A similarity may be found between the use of the term 'our forefathers Biblical time' to legitimize acts that wrought disaster upon the Palestinians, and the use of the term 'original climax formation' of Israeli space, that should be restored and protected from Palestinians, both in the ecological theory and in its practice. This restoration made possible by the destruction of the interrelations of Palestinians and environment. In both cases, the Palestinian Other disrupts the realization of the two favored entities – sovereign Jewish life in Israel, and bringing nature to its climax phase.

Palestinians only confirmed and revived those ancient times (Benvenisti, 1997), by controlled grazing of their herds, for example, as a means for ecological study and production of landscape conducive to recreation and tourism in the forests and reserves. Such controlled grazing is performed with no linkage, representation or fulfillment of the needs and character of a Palestinian society's environmental culture. Priority is given to the climax, to the national-Jewish landscape and to the comfort of vacationers.

The ecological dominant metaphor of the Zionist narrative is that the appearance of Zionism in this space constitutes the climax formation of the space itself. Similarly, modern Jewish culture is at its climax phase, having overcome turmoil and crises, and brought back to its deterministic state of the original species, namely the Jew as a Hebrew upon the land of his forefathers. Zionism created a new Jew, who possesses a new body (Gluzman, 1997, pp. 145–62), and maintains a new, healthy relationship with the land that is his by right of the Scriptures, on the one hand, and by rational, international resolutions on the other. He belongs to a strong, independent and sovereign community. This entity and the space it occupies have to be continuously protected from physical and symbolic disturbances, such as the presence of a Palestinian national entity in the very same original space. This entity maintained interrelations with the land – relations that were defined as a disturbance within the ecology discourse and therefore needed to be undermined.

In the mid-1970s, ecological theories of non-equilibrium emerged and took hold of ecology discourse (Perevolotsky and Pollak, 2000, p. 197). These theories stress ecological disturbances as a dominant factor that prevents balance. The ecology discourse in Israel changed its borders, and shook free of the climax theory and the national narratives that had been attached to it. Less ethnocentrically biased research took hold in the 1990s. Perhaps as a sequel to changes in the perception of self, Israeliness and relations with the Palestinians. All the previous arguments about the influences of grazing reached a turning point. I shall not try to prove here the conditions that I have found conducive to change.

According to Perevolotsky and others, to what extent can grazing develop into over-grazing and completely destroy the Mediterranean forest that had known thousands of years of grazing and adaptations? If grazing is removed from the system, the growth of trees will increase considerably and with it, the dangers of destructive forest fires. It has even been assumed that the whole country is actually undergrazed, and that the encouragement of grazing is an ecological must for the keeping of species spectrum (Perevolotsky and Pollak, 2000, pp. 651–2). Perevolotsky reviews ecological studies (Perevolotsky and Pollak, 2000, pp. 647–52) showing that areas that underwent combined grazing and logging did not fail the criterion of wide species spectrum compared to areas that reached so-called climax. The wide spectrum of species is a central variable to assessing the value and sensitivity of an ecosystem (Perevolotsky and Pollak, 2000, pp. 491–4).

Tom Selwyn (Selwyn, 1998, p. 129) illuminates the construction of Jewish Israeli identity through nature and landscape versus Palestinian identity. He also questions the common expression 'protection/defense of nature' and the name of the well-known Society for the Protection (Defense) of Nature in Israel (Hachevra Lehaganat Hateva), whose title in Hebrew rings amazingly similar to the term 'Israel Defense Forces' (Tzva Hagana l'Israel). From whom is nature to be protected? How have the enemies of the state and the enemies of nature and landscape – the Arabs – been constructed together as the evil Other. I shall propose that the Hebrew expression 'protection (defense) of nature' could also have been termed 'preservation of nature'. Did the choice of 'protection (defense)' resulted from the fact that areas that had become 'nature' after 1948 were not worthy of preservation as they contained traces of Palestinian culture.

Nature had to emerge out of the destruction that Palestinian entity had wrought upon its own environment, and out of the very same Palestinian space, ruined by the Israeli army during the War of Independence in 1948.

Nature was supposed to conceal the evidence of this double destruction by a stratum that reproduces pure primordial ecosystem. As growth led to climax, the need arose to protect this original nature from the Other – the Palestinians remaining in Israel.

This protection was established not only through legislation and supervision of open spaces, but also through conscientiousness raising programs in the school system and the youth movements, towards producing Israeli landscapes. These frameworks mediated space through learning experiences and extensive hiking. The closeness to nature was institutionalized as a national objective in order to master a new body of knowledge, through major physical exertion of effort. In yearly school hikes, the individual had to exert him/herself to reach the finishing line and become part of the normative group (Selwyn, 1998, p. 12; Almog, 1997, pp. 252–88).

c)   The Axis of Jewish Strategic Control of Space

Muhammad Abu al-Hayja recounts things he was told by an inspector of the Nature Reserve Authority:

> If you built in fear, you will destroy nature less. You must not be given permits, you should be kept crawling, close to the ground.

The Axis of Jewish control of space is the subtlest of the three. In practice, on the way to climax, nature preservation became a means to mark borders of Arab localities, and a limiting and supervising procedure. The eye that observes Arab communities by various means, limits the basic civil and planning rights (Fenster, 2001, p. 5) of the rural Arab sector. Inspectors of the Nature Reserve Authority, the Green Patrolmen, hikers, tourists and aerial photography guard the internal borders between Jewish and Palestinian citizens of the State of Israel. These guards, applying explicit or implicit force, replace the military rule that was forced upon Arab communities inside Israel from 1948 until 1966.

I shall return now to Ein Houd.

The Master-plan[12]

Master-plan for the regulation and changes of borders of the Carmel Park, and of granting permanent status and recognition to the village of Ein Houd, was deposited by the District Committee for Planning and Building of the Ministry of the Interior in the Haifa District on 29 December 1998. Initiators of the plan are the Planning Administration in the Ministry of the Interior, and the Israel Land Administration. The village of Ein Houd is now located, as I have already indicated, on land owned by the Israel Land Administration, and defined as Carmel National Park, a nature reserve and a fire-zone.

The Israel Planning and Building Law is a legacy of British Mandate law. From the colonizer's point of view, local authorities run by natives were not capable of managing colonial interests with any rational efficiency. Therefore, most of the planning force and authority at every level – spatial and governmental both – were exercised by the Mandate government.[13] This legacy weakens social rights, and weakens Israeli civil society today. It enables the government, through the planning committees of the Ministry of the Interior and the Israel Land Administration, to follow closely and control planning and building processes in the country, and thus a major political dimension introduced to every planning action. Official and practical recognition of Ein Houd is regulated by ratifying its master-plan, on the part of the District Committee for Planning and Building in the Haifa District of the Ministry of the Interior. This committee has linked recognition and planning of Ein Houd with the regulation of *administrative* disorders of land misuses in the Carmel Park area. The misuses are:

1   pasture for the cattle of Moshav Nir Etzyon (exists de-facto) – *151 dunam*;
2   farm structures of Nir Etzyon – cowsheds (exist) – *74 dunam*;
3   hotel and recreation facilities in Nir Etzyon (exist) – *68 dunam*;
4   educational facility Yemin Ord (Jewish religious boarding school) (exists) – *231 dunam*;
5   village of Ein Houd (exists) – *80 dunam*;
6   approach road to Ein Houd – *20 dunam*.

Clearly, the areas thus deducted from Carmel National Park, now managed by the Israel Nature and Parks Authority, are areas that do not function – de-facto – exclusively as park areas, and are already misused lands. It is also clear that the change of function of these areas in Jewish communities and institutions is strictly procedural. Exceptions in the use of these lands are not questioned and do not involve the suffering of any of the bodies responsible for them. In contrast, as described above, enormous damage and suffering is inflicted upon Ein Houd residents, in their status as present-absentees.[14]

While working on the master-plan of the village itself, the planners recommended that Ein Houd occupy an area of *170 dunam*. During the planning process the area was reduced to *80 dunam*.[15] I should mention here that the areas allotted hotel and recreation purposes in Moshav Nir Etzyon amount to 68 *dunam,* and the areas allotted the Moshav's cowsheds amount to 151 *dunam,* larger than the entire area allotted Ein Houd. The population-density set for Ein Houd is six residential units per *dunam,* versus two units per *dunam* in rural Jewish settlements in Israel. The programmatic needs of Ein Houd in living areas for the long run, as well as areas for guesthouses and employment, have been rejected on grounds of maximal protection of sensitive open spaces.[16]

An additional argument for denying the allocation of area for guesthouse and other employment areas was the comparison of the Ein Houd community's character with that of Jewish settlements that are 'bedroom' communities that do not contain any guestrooms accommodation or employment areas. This comparison naturally ignores the different cultural codes of Jewish settlements and Arab villages, a vast difference in levels of income, population mobility, and ethnic barriers to employment outside the village and the inherent contradiction between Palestinian tradition and the employment of women outside their home and village.

The process of planning and approving anyhow, reached an un-surprizing turning point.

The Open Spaces Exchange Affair

In addition to the difference in significance and urgency between ratification of the Ein Houd master-plan and the implementation of other administrative changes in this part of the Carmel area, the plan was burdened with yet another factor. As compensation for deducting a total of 544 *dunam* from the Carmel National Park, 7,034 *dunam* of adjacent woodland (declared in the National master-plan for Forests and Forestation 22) – owned by Israel Land Administration, managed by the JNF, and actually a fire-zone – would now transfer to the responsibility of Israel Nature and National Parks Authority.

On 28 April 2002, the JNF appealed to the Haifa District Court against the District Committee for Planning and Building to repeal the committee's resolution to deposit the local master-plan. The JNF is surprised to discover that the District Committee includes in its plan the transfer of JNF areas to the Nature Reserves and National Parks Authority, instead of separately promoting the official issue of the plan – a master-plan for the village of Ein Houd. I shall note here that the JNF's main role is to ensure that state lands remain in Jewish hands. I shall propose that the JNF's refraining from giving up the forest, is a direct result of perceiving its role in this way. Clearly, because of its military uses, this forest cannot serve its original function, as recreational wooded area that lessens the pressure of tourists in the nature reserves.

If so, what function of this area is the JNF giving up, as it does not own but only manage this forest? Giving up the forest to the Israel Nature and Parks Authority would pave the way for ratifying the Ein Houd master-plan and completing its recognition processes. Thus lands belonging to the Israel Land Administration would be sold or passed over to non-Jewish hands. Furthermore, as soon as the Nature Reserves and National Parks Authority gives up areas for Ein Houd, and receives areas managed by the JNF, there is no stopping its repeating this process in the future, when the village needs further areas for future expansion. As long as the JNF is around, it may pose reservations and obstacles in implementing the plan, just as it objects and defers the approval of the plan at present, in the appeal process. Thus another significant delay is assured in the plan ratification, within the winding corridors of the Israeli justice system, in a legal confrontation that is – supposedly – not political. Also, I shall mention that the suggestion to destroy Palestinian villages during the War of Independence came from Mr. Yosef Weitz, director of the Purchase Department of the JNF (Golan, 1995, pp. 403–44). Although the government stopped him, military commanders on the different front-lines, had already internalized and adopted the demolition acts he had outlined (Falah, 1999, pp. 99–100).

Abu al-Hayja intends to file legal suit on behalf of the Ein Houd Cooperative Association against the District Committee for Planning and Building concerning the delay in ratifying the master-plan. An essential part of this delay results from the legal confrontation between the JNF and the District Committee for Planning and Building. Some of the following inspections might not be used in the appeal to court, but may help to understand the legitimating system that enables the District Committee to handle thus Ein Houd's master-plan.

Naturalizing[17] Planning Discourse

Several times in the planning process, it has been suggested to the chairman of District Committee for Planning and Building to separate the recognition of Ein Houd from regulating the other misuses of lands in the Carmel Park. But the chairman argued that objects of this discussion might not be separated as they all belong *to the same geographical unit*.[18] Thus, using this 'professional' term, he turns the planning process under his charge into a naturalized, obvious procedure. He is not even convinced when a conflict arises between the JNF and the Israel Nature and Parks Authority, a conflict that means years of deferments of the ratification of the plan. It seems that the committee chairman was not able to distinguish units of environmental justice.[19] He could not discern the different ethnic units that share the very 'same geographic space', nor perceive the suffering and injustice inflicted only upon the one ethnic group during the past 53 years in the area under his committee's charge.

The notion of 'one geographic unit' is an example of the biased use made of supposedly scientific geographic discourse in an attempt to reduce this discourse of its present scope and contents. Cultural geography, now offers wider contexts for space, and maintains an active exchange with sociology, anthropology, theology and philosophy (Soja, 1989; Cosgrove, 1985). Furthermore, in the committee's discussion, the following expressions are sometimes used: 'humane solution',[20] 'acute human problem',[21] or 'the program was carried out sensitively'.[22] These expressions block any criticism or doubt regarding planning. Such terminology reveals the committee's attitude to the village residents. It is not the attitude to a group of citizens entitled to equal rights, immediately. No attempt is made to recognize the loss and disaster their community had undergone in Israel's War of Independence in 1948, and in their many years of unique suffering as residents of an unrecognized village and their right to compensation. On the contrary, as Mr Mordechai Ben-Porat said at the meeting of the National Parks Council:

> I adamantly object to the argument that the village existed there prior to the founding of the state. This was not the government's motive to grant recognition. The resolution was made out of *humane considerations*.[23]

This is an attitude adopted toward a weak ethnic group that needs *favoring*, that requires *charity*. All this without losing control, 'so that matters should not get out of hand, there must be a set of regulations that regards the area'[24] or 'limits should be placed and we must be strict about this'.[25] Taking care not to overstep the boundaries of traditional discourse on the priorities of environment and climax, and adhering to the traditional procedure of curbing the planning rights of Arab communities in Israel.

To what physical and spiritual extent can different kinds of borders to be layout on Ein Houd's community from outside? I will try to follow the production of borders of actual open space and the visions concerning it, from within the village community.

From Exotization to Postcolonial Insight – Primary Impressions from the Garden

I seemed to identify the Palestinians' interrelation culture with their environment through Abu al-Hayja's descriptions of the grazing and terraced orchard-tending practices. But some factors kept creeping in to the foreground that I kept trying to push aside as marginal. Such that are not compatible with that Other, original Palestinian place. I shall focus here on the flora I have identified in Muhammad Abu al-Hayja's garden.

Even from afar, Abu al-Hayja's yard stood out with some mature and wide-canopied Indian rosewood (Dalbergia siso) and Small-leaved Fig (Ficus oblique) trees. Other species marking the edges of his yard were Monterey cypress (Cupressus macrocarpa 'goldcrest') and olive trees (Olea europaea). The British imported the Indian rosewood to Israel from India in 1925. Abu al-Hayja received three saplings from his uncle who worked as a gardener for the Haifa municipality. He planted an Indian rosewood tree in his garden in order 'to have shade, fast'. Later he planted a small-leaved fig tree next to the Indian rosewood, to replace it when the time comes. He explained to me that the Indian rosewood is not desirable; its pods cause much waste. For me the Indian rosewood is home ground. Such trees grow in my northern Tel Aviv neighborhood, and were a part of my kibbutz childhood landscape. Strange to find this species in a garden at Ein Houd, where I expected to see various fruit-bearing local endemic species.

By planting these trees, Abu al-Hayja possessed the Indian rosewood, which is an example of representations through plants of the British imperialist power. The introducing and acclimatizing of flora from the colonies to the continent or to other colonies were a powerful means of representing domination of space. It actually changed landscapes as a part of the subjugation process of the consciousness of peoples and ethnic groups (Ephraim, 1993; Mitchel, 1994, pp. 53–5). Abu al-Hayja used the British tactic inherited by Zionism, and used introduced pioneer plants – rapidly growing trees that lend an immediate effect of shade and greenery at new residential colonial areas – to change his environment fast, and to prove domination. He created thus a physical *third space*. In the term 'third space', of which Homi Bhabha and others (in various ways) made extensive use (Bhabha, 1994, pp. 19–39; Lavie and Sweedenberg,

1996, pp. 67–86), I have found a fruitful tool of interpretation. Third space refers to a formation of expression, which lends meaning and a frame of reference through ambivalence. Ambivalent expression that is conscious of its lack of uniformity and cohesiveness of cultural symbols, and lacks the possibility of possessing, reproducing and translating these symbols.

This formation of expression undermines representation as an act of reflection, contrast and comparison, through which cultural knowledge is usually perceived as a whole. It undermines the authenticity and continuity of identities, and the perception of historicity of nations. It undermines the teleology of Western time. Third space is, therefore, a split and hybridized space by which we may abandon the politics of polarity and appear in it as the others or different of ourselves.

As Abu al-Hayja's trees were taken from him brutally, he didn't hesitate to take possession of the acclimatized species, alongside the use-bearing endemic ones. The acclimatized species in his garden, reflected to me, with wonder and irony, the ambivalent image of my own landscape, both of national-colonial occupation and of a private cherished childhood. My own essentialist search of the Other and his relations with his environment, now seems to me degrading, captive and of limited fruition. The space of Muhammad Abu al-Hayja's garden was not an authentic native space, his planting was not just mimicry that turned into hybridity. It was a third space, in which he had broken and decentralized the faked homogeneity (Lavie and Swedenburg, 95, p. 74) of the hegemonic Zionist landscape image, through reflecting and possessing. This third space contained *agency*.

Arbitrary Closure, Agency and Differential Nostalgia, or Some Second Thoughts of Hybridity

The hybrid third space model seemed to shake after another meeting with Muhammad Abu al-Hayja, (and the same happened to the confidence I had in my ability to interprate and translate hybrid third spaces). It was not efficient enough for describing identities and agencies for change, or the narratives, visions and actions of the enfeebled community concerning its space and environments. I have found that the notions of *arbitrary closure* and *identity politics*, that yield agencies for change, must be part of the theoretical tools for the interpretative work of environmental culture.

Arbitrary closure is a term coined by Stewart Hall (Hall, 1993, pp. 134–8). It represents a move in which a subject that has become racinated, learns and consolidates his/her racinated identity at moment in which he/she is placed as the Other from the *outside*. He/she enters a state of minimal self (a post-traumatic state of conducting minimal social interrelations), and fixes the borders of his arbitrary closure – the community within which he/she feels secure and worthwhile – in order to enable him/herself definitions and targets of agency, for changing his/her status and situation, as well as changing hegemonies.

Abu al-Hayja was summoned to a plenary meeting of the National Parks Council that took place on 1 February 1996, to discuss the official recognition of Ein Houd. Some of the participants tried to prevent him from staying and speaking, claiming that he had only been invited to the meeting as a guest. Abu al-Hayja, having thought he could realize his rights within the structure of Israeli democracy, experienced his otherness here, in a new manner. He exited the hall and burst into tears. On this occasion he first experienced racism so bluntly: 'In one second your whole life suddenly seems one big lie'. After this incident he realized he is subjugated twice over. His own narrative is placed outside the hegemonic narrative that has been exposed as racist, and furthermore, the place and framework for expressing his narrative are not at all self-evident. At such moments of acute racination, arbitrary closure is defined.

Arbitrary closures are important for beginning the discourse and struggle over representing identities. They are not committing or final. In identity politics, closures open and close again according to changing definitions of means and objectives. Abu al-Hayja alternately works with the establishment (architects, advisers and officials of the planning system), and against the establishment (as part of the Association of Forty or with other organizations).

Abu al-Hayja shares with me his vision in which the surroundings of his village would turn from a nature reserve into rural Palestinian farmland. It would contain terrace horticulture with fruitful orchards that would contribute to the community's economy both in its use value, and its exchanged value. All this would be achieved through the use of *advanced agricultural technologies* of hothouse cultivation. The profits would go to the Cooperative Association of Ein Houd, which is, the first Arab community in Israel that is a 'moshav' – an Israeli entity of cooperative association farming village. This would offer momentum for cultural, educational and social progress of the community.

I shall try to trace some evidential framework for a translation of Abu al-Hayja's vision. According to Slymovic, the village was known during the British Mandate years, of its honey, carobs and olive oil, and was permitted to market its farm produce in Haifa's markets (Slymovic, 1998, p. 128). This description attests to a society that produced more than it consumed, and was well integrated in the western economy of the British Mandate in Palestine. Hirsch's book of 1939 on agriculture in Palestine describes Palestinian agriculture and the relations between the Palestinian farmers and the British government:

> During the past years, general improvement has been noted in Arab farming in the country, thanks in part to the considerable efforts that the [British, NM] government has exerted in this direction, by founding numerous experimental stations and fields at Arab centers, and by the comprehensive guidance offered by the Agriculture Department to Arab farmers. It is no exaggeration to mention that 95 percent of the government's actions for improving the farmer's conditions in Palestine are aimed at the Arab fellahin; it allocates its lands, distributes saplings for orchard planting, gives out choice seed, etc. Indeed in the past years the Arabs have planted numerous fruit-tree orchards under the guidance of Agriculture Department officials, and in two-three years' time, these orchards shall begin to bear fruit. (Hirsch, 1939, p. 19)

But when Abu al-Hayja tried to start his own bee-keeping endeavor for honey in the past two years, he realized that:

> The West has stolen from me and is not returning to me… I do not have a Western identity, only a Western way of life. I am, undoubtedly, an Arab, I do not waver but am being forced to lead a Western life. If I don't get out of here [the village, NM], I shall die! Literally! For they have turned this into a ghetto, a prison. We were farmers, and they wrenched us out of that, forcibly, by taking our land, by making it impossible to graze, by fencing us in. The years have gone by, and I want to go back but this is not what I was thirty years ago. My farmer's know-how is gone. When I was a child, we planted vegetables, we gave them water, we fertilized them, we knew when to plant, prune, how to raise goats, care for them, grow their young, milk and shear them, raise chickens and have hens lay fertilized eggs. How do I get back to all of that? I am caught in the Internet. The Internet wrenched me out of my knowledge, in order to re-teach me. The researches I read on the web have been made about me, about what use to be my home!

Palestinian citizens of Israel are the ultimate Other of hegemonic Israeli identity, but also of Palestinian identity. Their citizenship in a Jewish state breaks up their national narrative (Lavie, 1992, pp. 55–96). Still, Palestinian citizens of Israel have absorbed into their own culture, at times forcibly, parts of Israeli culture.

In order to cope with the ongoing process of absorption while at the same time having his community marginalized by Israeli authorities, Abu al-Hayja developed a vision that is simultaneously contains and avoids several types of *nostalgia. Essentialist nostalgia* locates culture beyond place and time and seals it as a closed entity (Lavie, 1996, p. 69). This is the kind of nostalgia that often characterizes arbitrary closures in identity politics, as well as characterizing the way hegemonies grasp these closures. Essentialist nostalgia provokes criticism because of its fixed, separatist nature. To a certain extent, essentialist nostalgia is present in his vision, as a form of *imperialistic nostalgia* (Rosaldo, 1989, p. 68), in which former colonizers yearn for native cultures as they had been in imperialist times, and their critics look for pre-colonialist indigenous cultures. An example of such imperialistic nostalgia is what does the JNF at the Sataf site, near Jerusalem. There, only Jews have leased terraced land that is cultivated in semi ancient technics. The terraces dated as ancient Israelite horticulture, while the ruins of the Palestinian village of Sataf appear among them, unnoticed.

But Abu al-Hayja, in his vision, offered *not only* traditional terrace horticulture and thus both evaded and contained essentialist-imperial nostalgias. He legitimately introduced the modernization of hothouses, which is typical modern Israeli farming culture, in order to progress the condition of the village. He cannot afford to embrace fragmentizing hybridity proposed by Bhabha to individuals, as he must fight hybridity's forced and differential character. Hybridity as well as third spaces contain hierarchies and power relations within them.

A translation more complexed to the village's vision is what I shall coined '*differential nostalgia*', following Chela Sandval`s *differential consciousness*. Differential consciousness offers groups of subordinate to maneuver tactically between five locations of consciousness: equal rights, revolutionary, supremacist, separatist and differential. The addition of the differential serves as a factor of mobilization and transformation between the other four locations (Sandoval, 2000, pp. 53–8). Abu al-Hayja works with nostalgia for pre-colonial Palestinian mountainous environmental culture (a culture the comprehensive knowledge of which would be the result of historical research alone), nostalgia for hybrid Palestinian environmental culture influenced by the modern British Mandate, and yearning for possessing Israeli farming culture and its advantages, not available to Palestinian citizens of Israel because they have been deprived of most of their lands, deprived of their right to possess modern goods and to become hybridized freely for their benefits.

How can Abu al-Hayja transmit his differential nostalgic vision to Israeli planning authorities? It seems far outside the borders of Israeli discourse of open spaces, I have already interpreted.

Discourses in *Differend*

In view of the gap between the village residents' discourse and that of the planning institutions, I turn to the term *differend* coined by Jean François Lyotard (Lyotard, 1996, pp. 139–50; Ophir 1996, pp. 149–90). The differend at this junction is the unbridgeable gap between the planning discourse of planners, institutions and committees, and the discourse of identities, rights and ethnographies of suffering, of Palestinian citizens of Israel. Abu al-Hayja cannot attest about the damage inflicted upon his community while in the context of the planning discourse genre. The District Committee, which is to decide upon the present interim phase of planning, is not the venue for voicing the narrative of the Arab village in particular and the narrative of the Palestinian citizens of Israel in general. The committee is limited to an old fashioned professional-planning discourse.

The committee is a government entity, an executive authority of those governments that try to produce, coerce and regiment the ethnic landscapes and narratives of racinated groups. The District Committee is drowned in Israeli national discourse that produces and maintains the myth of a daily struggle of survival on the land that is the property of the Jewish people. And as I have shown earlier, the discourse of nature preservation, as well, hardly manages to disentangle itself from this national morass. How can it include Abu al-Hayja's vision of *Palestinian-Modernized-Terraced-Agriculture-Environmental-Reserve*?

These are the conditions for *differend*, in which the discourse genre of the jurisdiction authority is the discours of the perpetrator of damage – the legislator and the planner. In this case the community of Ein Houd cannot testify, using its own discourse genre, about the damage and suffering inflicted upon it. Injustice is done to the community of Ein Houd, and it becomes a victim. The victim, nevertheless, continues to develop its visions, using differential consciousness and differential nostalgia in order to maintain its existence and agencies.

Conclusions

The case of Ein Houd exemplifies the effectiveness of the arguments of protection-preservation of the environment against the rightful demands of the residents of a Palestinian-Israeli village to have an acknowledged exist on their fathers land, have the state acknowledge their basic and legitimate civil and planning rights, and maintain an environmental culture linked to their identity in order to lead a prosperous and dignified life. The Israeli hegemony's main purpose in the village's long-term planning is to limit, forcibly maintain and perpetuate the community in conditions of constructed deprivation legislated by the master-plan. The planning authorities produce the Palestinian village as an estimated, panopticious and (forced) heterotopian space (Foucault, 2003).

The density forced upon Arab villages inside and adjacent to nature reserves does not match the concept of rural settlement in an area of high environmental sensitivity in general, and certainly not the density standard in Jewish settlements adjacent to such areas. The planning policy of the periphery, in which the woods and nature reserves are produced and preserved, folds the Arab localities in upon themselves with a sense of frustration, that might evolve into an arbitrary closure that is strictly essentialist and fundamentalist. This might be an arbitrary closure that would lead to the construction of physical and spiritual life-tissue outside the law, inside and outside the settlement borders. At worst, this would deteriorate into an arbitrary closure that forfeits surviving.

As I tried to show in this chapter, lands in Israel, including forests and nature reserves – that constitute the ground for the planning practices as well as its raw-material – are first and foremost the bearers of strategic and ethnocentric conceptions in the labyrinths of statutory and metaphorical ownerships. The government's planning establishment retains the most decisive say on planning and producing space. It places barriers and constantly produces the enemy behind a green curtain of forests and nature reserves. This curtain – the fragile guardians – enjoys a network of public relations, professional and aggressive legal advice, and obtains the virtual character of a fragile natural entity that needs protection. This is a constructed production of essentialist nostalgia for nature that is innocent, fragile and vulnerable.

The discourse of one whose civil and planning rights have been denied in order to protect this 'nature' would describe the very same nature as a produced, aggressive and callous nightmare that only inflicts damage and injustice.

Israeli space is racinated and conflicted. It needs the actions of agencies of the weakened subjugated communities as well as a more moral aware planners actions. These actions must be taken from a state of consciousness, concerning the impossible neutrality and the constant hierarchy inherent to representations, and translations in the planning discourse. Actions of mediative entities such as that of Bimkom – Planners for Planning Rights, are crucial for conducting a critical planning activity in an attempt to overcome the differend between the weakened and the planning authorities. These critical planning activities are a must for various communities in Israel, Palestinian as well as weakened Jewish ones, who are on the brink of disaster in 2003 Israel, on the verge of imploding into a strategy of separatism, entirely outside the law.

Notes

1   The core of this chapter was first presented as a lecture at the Technion, Haifa, in the discussion 'Ethics and Landscape', conducted at the 2002 graduation ceremony of the Landscape Architecture department. My thanks go to Muhammad Abu al-Hayja, Chairman of the Ein Houd cooperative association and Chairman of the Association of Forty, who shared with me his experiences, ideas and vision, and commented on my ideas with precision and patience. I am also thankful for the translation of Tal Haran.

2   In this chapter I quote Muhammad Abu al-Hayja from our meetings at Ein Houd on 18 July 2002, 31 September 2002 and 27 January 2003.

3   An unrecognized village is an Arab settlement that, usually, was already in existence prior to the founding of the State of Israel, and whose lands have been taken over by the Israel Land Administration under the Absentee Property Law because its residents were absent from their homes in the years 1948–50. The village does not appear on official maps. The state denies the residents their planning and civil rights in order to make them evacuate their village.

4   A partial list of these localities according to Muhammad Abu al-Hayja and Hana Sweid: Ras el Ein and Ussfiye's Ruhna neighborhood that are both unrecognized localities, and the recognized Arab settlement Daliyat il Carmil, Sakhneen, Arrabe, Beit Jann, Eilaboun, Kufr Kanna, Tur'an and Buiena Nujidat.

5   Fire zone: an area used by the Israeli armed forces for live-ammunition practice and maneuvers.

6   Jewish National Fund – JNF – Land Development Administration. A non-governmental organization controlled by the World Zionist Organization. It was founded in 1901 following the 5th Zionist Congress resolution, in order to purchase lands in Palestine from Palestinians, turn them into the property of the Jewish people that would never be sold but only leased. Forestation is a JNF's main concern, especially as a practice of retaining hold of untended land, by the actual planting of forests, their maintenance and inspection, and by turning them into recreation areas.

7   Absentee Property Law was passed by the Knesset in March 1950. By defining the term Absentee, vast expropriations of Palestinian-owned lands were made possible by the state.

8   This information is based on the testimony of Muhammad Abu al-Hayja and on Slymovic's book (Slymovic, 1998). A different Jewish-Israeli version claims that the Abu al-Hayja family resettled only in 1950.

9   Social rights are founded on the assumption that every resident has the right to fulfill the basic needs for respectable human existence, including work, wages, working conditions, education, schooling, health and social welfare. It is founded on the recognition of the value of man, the sanctity of his life and liberty.

10 An occasion that destroys organisms in an ecological environment and opens space for a renewed settlement of other organisms.

11 Climate has dominant influence over climax. Climax is predictable, and every stage of the flora is in a state of constant change of species, towards the same climax.

12 The background is given for the period from 1998 until the present, stressing matters related to 'green' organizations. The source for this review is Abu al-Hayja's archive that contains correspondences about the various aspects of the masterplan planning. Additional background: Slyomovic, pp. 126–9.

13 I was acquainted with this aspect of The Israel Planning and Building Law in a conversation I held with Architect Daniella Posek, Tel Aviv City Architect.

14 Since the legislation of the Absentee Property Law, that created the unrecognized villages, Palestinian citizens of Israel sarcastically call residents of unrecognized villages 'present-absentees'.

15 Summary of the Council of the Israel Nature and Parks Authority of 10 July 2000, item no. 4.

16 Letter from the District Committee for Planning and Building: resolution of the Sub-Committee for Objections of the District Committee 19 December 2001, item 3.

17 Naturalizing is a term coined by Roland Barthes, referring to turning thing or phenomenon into a natural thing, an obvious phenomenon, a myth. See: Barthes, 1998.

18 Letter/summary of the meeting of the sub-committee for objections of the District Committee for Planning and Building of 19 December 2001, Chapter 5, item no. 4.

19 Environmental justice relates to the ecological conflicts involving weakened groups that enjoy less nature resources, suffer more from environmental risks, and pay a crucial price for the practicing of nature preservation by wealthy groups.

20 Letter from the director of the Carmel National Park to the director general of the National Parks Authority, from 10 November 1994, p. 1.

21 Minutes of the meeting of the council of National Park Authority, of 20 November 1994, item No. 1.

22 Letter from the District Committee for Planning and Building, The Subcommittee for Appeals, 19 December 2001, items 2/4 and 5.1. See also minutes of the Subcommittee of Planning of the District Committee for Planning and Building, 17 January 2000, access road to Ein Houd, Mr Y. Shacham's statement.

23 Minutes of the Israel Nature and Parks Authority Council no. 82, of 1 February 1996, meeting.

24 Minutes of the meeting of the Israel Nature and Parks Authority Council no. 82, of 1 February 1996, architect Meir Peleg's statement.

25 Minutes of the meeting of the Israel Nature and Parks Authority Council no. 82, of 1 February 1996, Mr Yoav Sagi's statement.

References

Almog, Oz, *The Sabra – A Profile* (Tel Aviv: Am Oved Publishers Ltd, 1997).

Barthes, Roland, *Mythologies* (Tel Aviv: Babel Publishers, 1998).

Benvenishti, Meron, *Sacred Landscape: The Buried History of the Holy Land Since 1948* (Berkeley: University of California Press, 2000).

Benvenishti, M., 'The Hebrew Map', *Theory and Criticism*, Vol. 11 (Tel Aviv: Van Leer Jerusalem Institute and Hakibbutz Hameuchad Publishing, 1997), pp. 7–30.

Bhabha, K. Homi, *The Location of Culture* (London and New York: Routledge, 1994).

Cohen, Ephraim Shaul, *The Politics of Planting: Israeli-Palestinian Competition for Control of Land in the Jerusalem Periphery* (Chicago: The University of Chicago Press, 1993).

Cosgrove, Denis E., *Social Formation and Symbolic Landscape* (London: Croom Helm, 1985).

Falah, G., 'The Transformation and De-signification of Palestine's Cultural Landscape', in I.A. Lughod, R. Heacock and K. Nashef (eds), *The Landscape of Palestine: Equivocal Poetry* (Birzeit: Birzeit University Publications, 1999).

Fenster, T., 'Foreword', in *Planning Rights in Israel: The Relations Between Community and the Establishment* (Jerusalem: Bimkom, 2002).

Foucault, Michel, *Heterotopia* (Tel Aviv: Reseling Publishing, 2003).

Gluzman, M., 'Longing for Heterosexuality: Zionism and Sexuality In Hertzel's *Altneuland*', *Theory and Criticism*, Vol. 11 (Tel Aviv: Van Leer Jerusalem Institute and Hakibbutz Hameuchad Publishing, 1997), pp. 145–62.

Golan, A., 'The Transfer to Jewish Control of Abandoned Arab Lands during the War of Independence', in S.I. Troen and N. Lucas (eds), *Israel: The First Decade of Independence* (Albany: State University of New York Press, 1995).

Hall, S., 'Minimal Selves', in A. Gray and J. McGuigan (eds), *Studing Culture: An Introductory Reader* (London: Edward Arnold, 1993), pp. 134–8.

Hirsch, Max, *The Practice of Hebrew Agriculture in the Land of Israel* (Tel Aviv, 1939).

Kliot, N., 'Nature Preservation in Israel', in G. Barkay and E. Schiller (eds), *Landscape of Israel – Azaria Alon's Jubilee Volume* (Jerusalem: Ariel Publishing House and The Society for the Protection of Nature in Israel, 2000), pp. 217–28.

Lavie, S., 'Blow-Ups in the Borderzones: Third World Israeli Author's Gropings for Home', in S. Lavie and T. Swedenburg (eds), *Displacement, Diaspora and Geographies of Identity* (Durham, NC: Duke University Press, 1992).

Lavie, S. and Swedenburg, T. (1997) 'Between and Among the Boundaries of Culture', *Theory and Criticism*, Vol. 7 (Tel Aviv: Van Leer Jerusalem Institute and Hakibbutz Hameuchad Publishing, 1997), pp. 67–86.

Lyotard, J.F., 'Le Differend', *Theory and Criticism*, Vol. 8 (Tel Aviv: Van Leer Jerusalem Institute and Hakibbutz Hameuchad Publishing, 1996), pp. 139–50.

Mitchel, W.J.T., *Landscape and Power* (Chicago and London: The University of Chicago Press, 1994).

Naveh, Zeev, *Natural Grazing in Israel* (Merchavia: Sifriyat Poalim, Maanit and Hakibbutz Haartzi Publishers, 1959).

Ophir, A., 'Damage, Suffering and Differend: Toward a Postmodern Ethics', *Iyyun*, Vol. MH (Jerusalem: S.H. Bergman Center for Philosophical Studies, 1996).

Perevolotsky, Avi and Pollak, Gad, *Ecology – Theory and The Israeli Experience* (Jerusalem: Carta, 2001).

Rinat, Z., 'The Battle Over Ruling the Forests and Nature Reserves', *Haaretz*, 8 August 2000 (2000).

Rosaldo, Renato, *Culture and Truth: The Remaking of Social Analysis* (Boston, MA: Beacon Press, 1989).

Sandoval, Chela, *Methodology of the Oppressed* (Minneapolis and London: University of Minesota Press, 2000).

Selwyn, T., 'Landscape of Liberation and Imprisonment: Towards an Anthropology of the Israeli Landscape', in E. Hirsch and M. Ohanlon (eds), *The Anthropology of Landscape: Perspectives on Place and Space* (Oxford: Clarendon Press, 1995).

Slyomovic, Susan, *The Object of Memory: Arab and Jew Narrate the Palestinian Village* (Philadelphia: University of Pennsylvania Press, 1998).

Soja, Edward W., *Postmodern Geographies: The Reassertion of Space in Critical Social Theory* (London: Verso, 1989).

Tibi, Rakefet, 'The Process of Landscape Design and Evaluation for Preservation: Case Study of Meron Mountain', research thesis for Masters degree, Geography Department, The Hebrew University in Jerusalem (2000).

Yiftachel, O. and Kedar, S., 'Landed Power: The Making of the Israeli Land Regime', *Theory and Criticism*, Vol. 16 (Tel Aviv: Van Leer Jerusalem Institute and Hakibbutz Hameuchad Publishing, 2000), pp. 67–100.

EPILOGUE

A Moment of Change? Transformations in Israeli Architectural Consciousness Following the 'Israeli Pavilion' Exhibition

Shelly Cohen

The Berlin Affair as a Manifestation of a Change in the Political Consciousness of the Architectural Community in Israel

The 'Israeli Pavilion' exhibition[1] was presented in the Architects' House Gallery in Jaffa, in July–August 2002. The exhibition presented Israeli proposals for two international architectural exhibitions: The Architecture Biennale in Venice, which opened in September 2002, and proposals for the International Union of Architects (UIA), which was conducted in Berlin in July 2002. The exhibition took place at a moment of change. A moment in which new directions were formed in the Israeli architecture by a new generation of architects, curators and researchers, who identify in the Israeli architecture clear trails of Israeli politics. The works in the exhibition proposed a critical reading of the local architecture.

The candidates for curators of the Israeli pavilion at the Venice Biennale were summoned by a subcommittee appointed by the Department of Art in the Culture Administration of the Israeli Ministry of Science, Culture and Sport. The candidates for curators of the Israeli pavilion at the International Union of Architects in Berlin were summoned by a steering committee appointed by the United Architects' Association in Israel. The architect Zvi Efrat and his team were chosen, and he curated the Israeli pavilion in the Venice Biennale. The architects Rafy Segal and Eyal Weizman were chosen as curators of the Israeli exhibition at the international architects' congress in Berlin. However, the committee eventually rejected Segal and Weizmans' final work – a catalogue and a collection of articles that appeared under the headline 'Civilian Occupation, the Politics of Architecture' (Segal and Weizman, 2002) – due to its political position. The works of an exhibition that had already been presented in the Architects' House Gallery were finally sent to Berlin, presenting projects that were built in Israel in recent years.[2] The affair raised questions regarding the extent to which Israeli architecture and the professional union representing Israeli architects are political, and regarding the freedom to express a political, harsh and extreme position in an international exhibition. The Berlin case reflected both the change of consciousness in Israeli architecture and the resistance that this change has awakened in the architectural community. At the end of the day, canceling the exhibition did not silence its political messages, awakening instead a storm within the Israeli architects' community, which developed into a dispute in the printed media[3] (Zandberg, 2002, 2003). The story was also widely discussed in the international media (Ruding, 2002), following which the curators, architects Segal and Weizman, were invited to lecture and present their work in various places in the world.

In this chapter, I shall present the 'Israeli Pavilion' exhibition and the occurrences surrounding the Berlin Congress, as a test case for the coming of age of the architectural discourse in Israel. In the first sections of the article, I shall discuss the visual and textual representation of Israeli architecture, presenting the difficulties that exhibiters encounter when they are required to translate the problems of local architecture into universal language and to embrace an international agenda. Further on, I shall present the Israeli 'sense of place', as it is presented in the works of the 'Israeli Pavilion' exhibition, as a site of construction and destruction.[4] I shall point out the relation between architecture in Israel and the politics of this state, observing the new local discourse that is replacing the regional discourse in Israeli architecture and attempt to outline the theoretical context of its transition into a political discourse. On the basis of the findings of this survey, I shall point out the gallery or the museum as a site for establishing the Israeli critical image. Finally, I shall discuss the critical possibilities that this exhibition is opening up for architectural theory and practice.

The Representation of Israeli Architecture in International Exhibitions

The proposals for the Israeli Pavilion in Venice and Berlin were required to address the subject of architecture, not politics. The subject of the eighth Venice Biennale was: NEXT. The subject of the 21st convention of the Architects' Union (the UIA) was 'Resource Architecture'. The Israeli committee even set a specific subject for the Berlin exhibition: Modernism in Israel. Just as the term 'politics' is customarily attributed in Israel to anything and everything, so it is possible to attribute anything in Israeli architecture to 'modernism'. However, only few of the works that were proposed for the Berlin exhibition directly addressed modernism. Of these works, only the curator Zofia Dekel, in her work 'Israeli Modernism – Between Tel-Aviv and Jerusalem', actually referred to modernism as an architectural style. The architects Yair Avigdor, Yosy Klein and Eran Neuman, in their proposal 'Modernism under Dispute – Breaches in Israeli Architecture', referred to modernism as a central practice in the formation of Israeli nationality.

On the other hand, many of the respondents, in their attempt to formulate architectural vision for the place in which they live, opened their proposal with a description of what is taking place outside the architectural firm, writing about what is referred to in Israel as 'the situation'. In most of the works, the political situation made its impact: the bright shades that are usually reserved for futuristic and technology-abundant architecture were stained this time by reality, in earth colors. As formulated by the culture researcher Sigal Bar-Nir and the landscape architect Yael Moria:

> It is not easy at this moment to speak of the future of architecture in Israel, or of the future in general, when we are in the midst of a cycle of violence and fear. When the rate of unemployment is rising, the social gaps are wider than ever, and architecture in its essence is an inseparable part of the social and political life. It would be pure escapism to address technological or formal aspects of the future.

This quote directly calls the circumstances of the discourse to mind – the request for proposals for *international* exhibitions, in which Israeli architecture is mediated to foreign eyes, accustomed to identifying Israel with hostile confrontation situations. The circumstances of the work led the candidates to internalize the look from the outside inwards, with a twofold effect: on the one hand, they served as catalyst for a political reading of Israeli architecture. On the other hand, in some of the works they led to self-censure and moderated criticism. In the open debate conducted at the 'Israeli Pavilion' exhibition, several of the presenters confessed to having doubts about the extent to which a critical stance, contradictive to the policy of government, might be accepted by the establishment to represent the state. And in fact, the Berlin case indicates that these fears were founded in reality. One cause for the objection was the apprehension, that a radical criticism against Israel would not be understood in Europe, in which the complexity of the Israeli-Palestinian conflict is not fully comprehended. Even some of those who did support the contents of the chosen project contended that these contents should first be opened to a preliminary discussion in Israel.

The architect Hillel Schocken, curator of the Israeli pavilion at the former Venice Biennale – the seventh Biennale – exposes how the look from outside rearranges the local architectural agenda:

> ... when nominated curator of the Israeli pavilion at the seventh international Architecture Exhibition in Venice I found myself facing a dilemma: What can one show that be of significance to the world of architecture and planning and is specially Israeli? ... I looked back at the second half of the 20th century, the period of the rebirth of the Jewish homeland and its consolidation into an energetic and thriving modern state. I searched for aspects of the built enviorment that accompanied this process which it would be meaningful to present. Should I present the heroic kibbutzim movement and its special social and environmental impact? Should I show the unique contribution of the international style ... to the Israeli urban environment? Should I show the influence of the Arab-Israeli conflict on prevailing policy of spreading the opulation throughout the country, and its impact on our rural settlements, development towns and big cities? I resolve to do none of the these. (Schocken, 2000, pp. 20–21)

In addition to the optimism regarding the Israeli-Palestinian conflict, which characterized the time of the Oslo agreements, this text also reveals how the task of representing Israel to the world minimizes the range of exhibition topics and excludes the contemporary issues that are crucial for the Israeli architecture.

Underlying these conflicts are the feelings of inferiority of an architecture that has embraced a universal western identity as part of the Zionist cultural project (Chinsky, 1993, p. 120; Chinsky, 2002). When it is intended for show in Europe, in Venice or in Berlin, the Israeli architecture examines itself by European standards. Reverberations of this can be found in the rare honesty of the 'Arkod Architects' team, in their proposal for the Venice Biennale:

> Thus, when we attempted to relate to the Biennale curator's general subject, we were faced with many questions: What are projects of an international interest? Is it their location that determines their international interest? Is it their size? Their design? The architect who designed them? Does 'international' necessarily mean the west? Does 'international' refer to the architects' community? Why is it important for this project to be of international interest? Or perhaps an international interest is related to the project's contribution to humanity on the level of creating residence textures for poor populations, for refugees, foreign workers, immigrants ...

The Image of the Israeli Place as a Site of Construction and Destruction

A look at the proposals for the Venice Biennale reveals that the place image that guides the presenters is of a place under construction, in a process of formation. In 'Closure', their proposal for the Venice Biennale, the architects Rafy Segal and Eyal Weizman speak of 'rapid processes of change in the landscape and in the built environment'. Two proposals in this exhibition (that of Arkod Architects and their team, and the proposal that I myself submitted in cooperation with the landscape architects Naama Meishar, Amy Tsruya and Zofit Tuvi), share the title 'Under Construction':

> The exhibition 'Under Construction' will point at constant change as the principal characteristic of the Israeli built environment. It is based on the recognition that Israel is undergoing a continuous process of construction and reconstruction, as a direct result of the ever-urgent national effort to maintain a quantitative and spatial demographic advantage. The thesis of this exhibition relates between the historic parallelism of Zionism to the project of 'building the land of Israel', and the contemporary reality in Israel, in which a wave of patriotism in the media is currently accompanied by a momentum of building.[5]

According to the architect Zvi Efrat's proposal, the borders of Israel are blurred. The name of his work, stemming from this fact, was borrowed from the field of psychology – 'borderline disorder'. The architect and artist Bilo Blich, the culture researcher Sigal Bar-Nir and the landscape architect Yael Moria find that the intensive building processes of the Israeli space are accompanied by processes of erasure and destruction. Bilo Blich, in his work 'Erasures – The familiar landscape and the foreign city', presents the act of erasing as an omnipresent, intense and rapid process. Despite the desire to rebuild, this process always leaves behind traces of the past: the remnants of a terrace or a mosque, citrus wood and ornamentations. Sigal Bar-Nir and Yael Moria are opposed to the destruction, and seek 'Tikun' (correction) as an alternative to the act of destruction. This is also how they named their work. They return to the Jewish concept:

> Referring to the simple act, carried out at a regular time: reading, prayer or a meal. The power of such an act is in amending the state of affairs in the world. 'Tikun', repairing, in its everyday sense, is intended to return the object to a state of functioning; from shoe repairing to the repairing, or renovating, of a building.

In the 'Israeli Pavilion' exhibition, Bar-Nir and Moria showed a presentation that was comprised of images of destruction – destruction of buildings and of a built environment, stemming from political reasons or as the result of development acts.

Is it Possible to Separate between Architecture and Politics in Israel?

Thus far, I have discussed the self-awareness of Israeli architecture and the extent to which it is political, as represented in international architecture exhibitions. Now, I wish to discuss Israeli architecture itself, and the question whether it is possible to separate it from Israeli politics. What does my use of the term 'political' refer to?

Politics is the theory and practice of government. Originally, '*polity*' meant partnership in the Ancient Greek city-state. In modern society and modern theory, this term discerns between various spheres of action. The political sphere determines the power relations and the authority boundaries in all the other spheres, maintaining the separation between the political center of power and other areas, in which the discussion is free of power relations.

This perception stems from the Habermas idea of separation between society and the state as a condition for a bourgeois public sphere. In other words, the desire to separate between the political and the non-political, as an attempt to maintain the independence of civilian and professional spheres, was not invented by the architectural discourse. However, even Habermas showed how this separation has become undermined by the competition between the various power centers. Foucault clarified that power relations are not found exclusively in the political sphere. They take place outside this sphere, too, and the political involves the representation and undermining of power relations. The political interpretation of architecture in the 'Israeli Pavilion' exhibition is not political in this wide sense of the term, according to which anything that manifests power relations is political – but rather in the narrower sense, according to which architecture in Israel is political because it is inherently related to the political sphere, and serves the policy of the state. Thus, the use of the adjective 'political' does not refer to the internal power relations between the municipal authorities, the planners and the market powers, or to the power relations within the professional community. In the 'Israeli Pavilion' exhibition, this interpretation focuses on the preoccupation with the Israeli-Palestinian conflict. Also, it occupies itself with social issues in Israel.

The Israeli-Palestinian conflict, just like other social conflicts and various topics that Israel has to cope with – such as territory, borders and the land regime, has a significant spatial dimension. Israeli building and architecture are the result of a governmental ideology and policy, no less than they are affected by international developments and professional styles. This was well formulated by the architect and theoretician Sharon Rotbard, in the catalogue edited by the architects Rafy Segal and Eyal Weizman for the Berlin exhibition:

> The most significant aspect of Israeli architecture, at once most evident yet so well concealed, is its political dimension. In Israel, just like war, architecture is a continuation of politics through other means. Every act of architecture executed by Jews in Israel is in itself an act of Zionism, whether intentional or not. The political dimension of 'building the land of Israel' is a fundamental, albeit often latent, component of every building in Israel, and the political facts it creates are often more dominant and conclusive than any stylistic, aesthetic, experimental or sensual impact they may have.

What, then, are the conditions in which an architectural act is considered political? Is any architectural act – even the choice of tiling for an apartment – considered political? The catalogue that was edited by the architects Rafy Segal and Eyal Weizman for the architects' congress in Berlin – 'A Civilian Occupation, The Politics of Israeli Architecture' – is the most direct work in the 'Israeli Pavilion' exhibition, referring to the politic nature of Israeli architecture and to the Israeli-Palestinian conflict. By focusing on the occupied territories, in which most new Israeli settlements have been built in the last decades,[6] their work provides an indirect answer to this question. The catalogue presents landscape as a battlefield, in which a struggle is carried on for power and political control. The architect Uri Zrubavel, director of the United Architects' Association in Israel, attacked Weizman and Segal's geographical focusing:[7]

> Since the establishment of the state of Israel, its population has been multiplied by ten – from 600,000 to 6,000,000 residents. The need to provide for such an accelerated increase, together with the desire to settle along the borders of the state, stemming among other reasons from security motives, led to the building of new civilian settlements as well as military settlements along the borders. In the past two decades, many settlements were built within the Israeli territory, inside and outside the 'green line', including new suburbs in existing towns and cities. Settlements such as Reut, Maccabim and Kochav Yair were

built, as well as the new city of Modi'in, and many existing settlements were reinforced, in various areas of Israel ... The [rejected – SC] exhibition ignored Israeli architecture as a whole, addressing only one aspect, i.e. the West Bank settlements and the Israeli presence there, a topic that in any case is greatly disputed among Israelis.

A comparison between Rotbard's words and those of Zrubavel indicates that the perception of the architectural act as a political act is no less charged than the actual setting of settlement boundaries. Admittedly, the wider answer, contending that any architectural act in Israel is political, reinforces the change in the professional discourse – from the perception of architecture as a 'pure' professional practice to a sweeping political awareness, but it also neutralizes the effectiveness of architectural criticism. When any architectural act is perceived as political, the single act is seemingly exempt of responsibility. On the other hand, the narrower answer excludes crucial questions from the professional discussion, and serves as an excuse to avoid a critical approach in architecture. Thus, Segal and Weizman's focus on the occupied territories is legitimate and important, outlining a way towards architectural criticism.

The catalogue 'A Civilian Occupation, The Politics of Israeli Architecture' is comprised of a collection of articles, photographs, maps, blueprints and other effective visual materials. The map of the West Bank, as it is presented in the catalogue, exposes the gap between the built Jewish areas in the settlements (1.7 per cent of the West Bank territory) and the judgment borders and areas intended for future Jewish building (41.9 per cent), showing the actual results – fragmentation and lack of a Palestinian territorial continuity.

The catalogue, which was printed on newspaper, was intended as the principal exhibit in the rejected exhibition. The red silhouette on its cover, in the shape of the occupied territories, 'stained' the headline 'Civilian Occupation'. The catalogue's sharp design, by the graphic designer David Tartakover, was perceived by those who objected to the catalogue as a direct visual expression of the project's radical nature (see Figure 14.1).

In the article written by the geographer Professor Oren Yiftachel, which appears in this catalogue and addresses the subject of 'Settlements as a Reflex-action', the contention is that 800 settlements of various types have been built in Israel to this day – the largest number of settlements-per-person in the world. Yiftachel argues that building such a large number of settlements is damaging to the relations between Jews and Arabs and to Israel's security, and that it exacerbates the social gaps in Israel (since the settlements that are built are peripheral, with high unemployment and poverty rates, and their existence

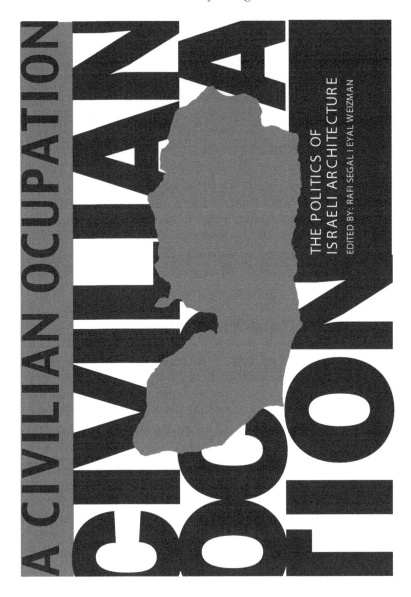

Figure 14.1 R. Segal and E. Weitzman, *A Civilian Occupation: The Politics of Israeli Architecture* (2002), cover of the catalog. Designed by Tartakover Design.

only serves to further weaken the periphery), as well as wasting public funds on decentralization that has already been recognized by national planners as a geographical and urban mistake.

'The Mountain', an article written by the editors, architects Eyal Weizman and Rafy Segal, analyses the settlement of the Gush Emunim movement in the late 1970s and early 1980s, on mountaintops in the West Bank, bringing many Jewish settlers to areas with no Jewish affinity. Israel established its ownership over the untilled lands on the mountaintops, relying on the Ottoman law. Even though, the transfer of civilian population into the occupied territories is considered a war crime by the Fourth Geneva Convention (Levi, 2002). Weizman and Segal describe the 'Community Settlment',[8] the new type of settlements, as an adaptation of concentric morphology to radical ideology and secluded social organization, or in their words – 'the built array of claustrophobia'. They regard these settlements as optic means of gaining domination and supervisionsurveillance and control over the Arab towns and villages.

In an interview that appears in the catalogue, architect Thomas Leitersdorf speaks of the way in which architecture and urban planning served political purposes in the planning of the new West Bank town of Ma'ale Edummim. The town was planned very quickly in the first place, against international political pressure, with the intention of reinforcing the Jewish presence in the occupied territories and of enabling the domination of roads in these territories. The town was planned in an area with difficult weather conditions, at the edge of the desert between Jerusalem and Jericho, within the territory of the state according to public consensus. It is interesting to read the planner's words on the way in which the planning principles had to measure up to commercial requirements, even though the project was planned for political reasons. For instance, the principle of a dense population in Ma'ale Edummim was contradicted to the tendency to disperse small settlements on every mountain and hilltop, and enabled a higher service level and the maintaining of open spaces. This principle was commercially justified when the residence units were sold and attracted a large population to the new town.

The recognition of architecture's political power – or perhaps its weakness and its subjugation to the government – can be regarded as the result of the maturation of a group of individual architects, who ceased to perceive themselves as individual creators and began to observe the society in which they were working. However, the Berlin Exhibition affair indicated that the professional union itself, whose role is to elevate itself above the personal-economic interests of individual architects and to struggle for the benefit of architects and society as a whole, objects to this line of thinking. In the discussions of the Berlin case that were conducted by the Isreli Association of the United Architects', the repeated argument was that the Architect's Association should maintain a non-political professional nature. The director of the United Architects' Association argued that 'use has been made in this case of the Association and its cultural and material assets in order to send across significant non-professional political messages'. This objection reflected a certain resistance to taking a stance, which is not shared by all the Association members. When the power of the state decreases, the professional organizations of the civilian society grow stronger. The question is whether these organizations duplicate the government's attitudes, or become agents of change (Shenhav, 2000, p. 6). In my opinion, in this case, the Architects' Association missed an opportunity to use its status in order to promote political change.

A New Local Discourse is Replacing the Regional Discourse in Israeli Architecture

In the professional discussion, recognition of the political nature of architecture, at least as a dominant trend, is a new recognition. To this day, the professional and cultural discourse of architecture has rarely addressed political issues. As an example of the absence of politics from the professional discussion, I shall note another, earlier exhibition that was hosted by the Israeli pavilion in the Sixth Venice Biennale of Architecture. The exhibition 'Visible and Beyond', presented Israeli architecture as accepted in the hegemoneous architectural discourse. It included prominent projects and buildings, focusing on 'universal' issues of style. It reflected the *regional* discourse in Israeli architecture, in addition to some manifestations of an historical postmodern style. The architects David Guggenheim and Omry Eytan curated the exhibition. In the text for the exhibition, architect Omry Eytan characterizes such architecture as combining east and west, old and new, and manifesting a 'Mediterranean or middle-eastern' identity. He argues that such architecture reflects cultural pluralism and historical stratification, by bridging between the influences of ancient local cultures and modern technologies.

The regional discourse in Israeli architecture was, and still is, an alternative to the multifaceted Israeli modernism. The history of Israeli architecture can be read as existing between these two approaches – locality and modernism. Whilst modernism concentrated in applying edvanced universal technology and architecture, the regional architecture searched for an authentic language, suitabile to climatic and particular cultural elements of the locality. In practice the dichotomy between the two paradigms is not a real one. International style has had different expresions at different countries and regions. The 'tour' proposed by Zofia Dekel in the 'Israeli Pavilion' exhibition, between Tel-Aviv and Jerusalem examines the various versions of modernism in these two cities. The appeal to the vernacular has been influenced by modernism itself. Le Corbusier, for instance, one of the founding fathers of modernism, regarded the vernacular in the Mediterranean countries as a source of inspiration.

The regional Israeli architecture has usually been influenced and mediated by Western architectural styles (such as the eclectic style in the 1920s). A late combination between regionalism and modernism in western architecture appeared in the 'critical regionalism' formulated in the 1980s by the historians of architecture Kenneth Frampton, Alexander Tzonis and Lian Lefaivre. 'Critical regionalism' proposed a criticism of modernism, a reaction against its undesirable implications – mediocrity and the abolishment of local creativity – while embracing the spirit of modern progress. Its goal was to promote a vital local culture, which combines modernity with a return to origins. It encouraged reference to the environmental context of architecture and a modern technological interpretation of local elements, calling architecture to transcend stylistic and formalistic characteristics. Thus, it distinguished itself from the vernacular – which refers to climate, culture, myths and local arts – and from regional historical precedents (Frampton, 1980; Tzonis and Lefaivre, 1996).

Visually, the regional theory was translated in the Sixth Venice Biennale of Architecture into a formal vocabulary and a limited and repetitive building material repertoire that is considered authentic – eastern arches, inner yards, means of shading, building in stone, etc. The exhibition presented mainly preservation and restoration projects in the old cities of Nazareth, Acre and Jerusalem.

In the 'Israeli Pavilion' exhibition, shutters are the sole vernacular element. Through these shutters, (named Tri'sol after a 1960s brand name of Israeli PVC product) truths about Israeli architecture were exposed and camouflaged simultaneously. But shutters, an inexpensive and efficient means of protecting against the strong Israeli sun, are not a typical characteristic of the vocabulary of forms in regional architecture. They are neither representative nor oriental. On the contrary, shutters were commonly used as a deprecated means of closing in balconies, an illegal addition to residence apartments. Through his choice of this element, Zvi Efrat marks Israeli modernism as regional and

Israeli western architecture as no less improvising, lowly and popular than conventional regional models.

It is interesting to compare the Israeli exhibition in the Sixth Venice Biennale with the proposal for the Berlin convention, 'Architecture of Insecurity, or Brief Thoughts of an Old and Beautiful City on the Meditterenean' presented in the 'Israeli Pavilion' exhibition. The curator Eytan Hillel and the architect Yael Ben-Aroya also addressed the subject of architecture in mixed-population cities[9] in Israel, focusing on Acre as a test case. However, their awareness of post-colonial theories enabled them to read stylistic effects as symbols of political power relations. Their work focused on the way in which the establishment is present in the public space through a series of nondescript modernist buildings, or through the use of early symbols of power such as the British police building.

Thus, a critical view of Israeli regionalism exposes the Orientalism in the regional discourse and stress the 'relationship of power, of domination, of varying degrees of a complex hegmony' (Said, 1979, p. 5), that underlies the regional atitude towards the orient. In the Sixth Venice Biennale, the depiction of the rehabilitation process of old Arabic sites, avoid any referrance to the 1948 war, and to the everlasting Jewish effort to gain domination over the land of Israel. The focus on the preserved architecture presented the eastern architecture as a fixed essence, severed from its historical and political context (Said, 1979, pp. 60, 97). The architectural practice of the regional school narrows political and cultural questions down to a mere formal and stylistic preoccupation. More severely, it is possible to argue against the Mediterranean school of regionalism architecture that it possesses indigenous Palestinian forms as means of establishing Western dominance of space.

The embracing of a Mediterranean identity requires additional deconstruction: the Mediterranean identity provides Israeli architecture with a sense of belonging to a large and appreciated geographic unit, while avoiding both the threatening eastern identity and the East-European spirit of the Diaspora, from which Zionism has distinguished itself. This identity is a manifestation of the dialectics between acceptance and denial of the Israeli place. For the planning architect, 'Mediterraneanism' is a conceptual hammock in which to rest from the colonial hurling between Europeanism and Assianism, between the position of conqueror and the position of the conquered who bows to the West.

Thus, through an open political debate, the 'Israeli Pavilion' exhibition formulates a new local discourse as an alternative to the regionalism discourse in Israeli architecture.[10] The new discourse gives 'local' a wider interpretation, directed at seeking for the source of architectural phenomena in the planning procedures in Israel and in political and economic issues. For instance, the work 'Area D', by architects Yehoshua Gutman and Rinat Berkowitz, binds together the implications of Israeli politics with the architecture of consumer society in Israel, characterizing several Israeli phenomena as spatial mutations that combine the Israeli aspiration to normalcy with the political state of emergency. Thus, the new local discourse can be read as meeting some of the goals of 'critical regionalism' more fully: it is neither formal nor material, and it is critical. (In the chapter that describes the critical strategy of the proposed works for the Venice Biennale, I shall employ the concept of criticism as it is used in 'critical regionalism').

The Gallery as a Site for Establishing a Critical Image for the Israeli Place

Architectural discourse in Israel did not turn into a political discourse spontaneously, by itself. This change was nurtured by various sources, by the escalation in political events and by developments in the theoretical discourse. I shall attempt to outline a primary, non-chronological sketch of the theoretical context that has broken the ground in recent years for this trend.[11] As follows, a number of the participants in the present exhibition paved different ways, for political interpretation of Israeli architecture.

The 'Israeli Project' exhibition,[12] curated by the architect Zvi Efrat, the curator Meira Yagid and their team, addressed the physical planning of Israel in the 1950s. The exhibition was crucial in its importance for formulating the relation between the Zionist project and the building and architectural enterprise in the first decades of the State of Israel. This exhibition also presented some of the implications of Jewish settlement for the Palestinian population of Israel. The culture researcher Sigal Bar-Nir and the landscape architect Yael Moria had written about the 'conquering of the wilderness' as one of the leading myths in landscape architecture in Israel, in an article for the 'Point of View' exhibition (Gaon and Paz, 1996). Later on, in an exhibition titled 'Shaping the Memory',[13] Bar-Nir addressed the subject of the Israeli landscape as a site for shaping the Israeli identity. In the exhibition 'Pastoralia',[14] the landscape architect Naama Meishar pointed at the political implications of the Israeli forestation policy, showing how power and domination relations are involved in shaping the seemingly innocent and natural Israeli nature. In many exhibitions throughout the years, the architect and artist Bilo Blich has examined alternative channels of observing conventional architecture. In the previous decade, Blich worked with groups of young architects, and curated exhibitions at the Ami Shteinitz Gallery and at the gallery of the Camera Obscura School.[15]

Influence over the local discourse can also be attributed to the international discourse in architecture, and to the title of the previous Venice Biennale – 'Less Aesthetics, More Ethics' (Fuksas, 2000, p. 10), which has influenced the local architectural consciousness.[16] Geography, a field that is close to architecture, has preceded architecture in developing an awareness of the political power of planning. Outside the field of architecture, and within the public discourse in Israel, political criticism is published on a regular basis. Discussion of the Israeli identity and culture intensified towards the celebration of Israel's jubilee, and continued after it. The post-Zionist discourse began in radical academic faculties,[17] but continued to penetrate into the popular and journalistic writing.[18] The extent of its influence over the internal architectural discourse can only be imagined.

This brief review indicates that galleries and museums can be regarded as the site that establishes a critical image for the Israeli space.[19] Most of the architectural practice is submerged in the material and political world, while the exhibition space enables a pause in which a critical and moral discussion of Israeli architecture can emerges.

Architectural Criticism or Critical Architecture

Is architectural criticism possible at all, beyond the space of galleries and museums, within architectural firms? What are the relations between critical theory and architectural practice? In order to answer this question, we must clarify the meaning of the term 'criticism'.

I wish to embrace the definition of *criticism* used by the historians of architecture Alexander Tzonis and Lian Lefaivre (1996, p. 488). According to Tzonis and Lefaivre, criticism has two components: it challenges the existing social order, and activates a reflective self-observance, through self-examination and self-assessment. They derive the later characteristic from the philosophy of Kant and from the Frankfurt School. In the 'Israeli Pavilion' exhibition, the arrows of criticism are aimed both outwards and inwards. Within the field of architecture, criticism challenges the architectural agenda and self-consciousness, undermining the accepted boundaries in this field and arguing, for instance, that the political, which has heretofore been considered irrelevant for the field of architecture, is imbedded in its essence. The arrows of criticism are also directed outwards, touching on life in Israel, as social or political criticism.

In the gallery forum conducted as part of the 'Israeli Pavilion' exhibition, there were those who argued that the critical position is disconnected from the professional practice, or that it obliges the planning work to cease, and thus it is the privilege of theoreticians. This emphasized the distance between theory and practice.[20] Critical theory, on its part, objects to its detachment from architectural practice, not merely because it fears for its status but also because of a sense of political and moral commitment and a desire to affect reality.

Thus, the act of constructing a pavilion for an international exhibition takes place at a crossroad between the two fields in which architecture acts: as a high cultural act, it performs in a limited architectural-cultural field and in the elitist exhibition space, in this case – international architecture exhibitions. However, beyond the private case of the Biennale, as a building act, architecture as a rule takes place in the public space. It would be more correct to say that it *builds* the physical public space, and thus its success in formulating criticism through architectural means has a larger visibility potential than that of artistic objects. While artistic objects have aspired in recent decades to expand the boundaries of their influence and to 'get out' into the street, architecture *is* the street. In addition to the desire to influence, a resistance to theory and preference of the architectural object, underlying the search for criticism through architecture– for critical architecture that will replace architectural criticism. This is a manifestation of an influence of the field of art, in which there is a clear preference of the artistic object over any other object in the field (Azulay, 1999, p. 63).

How, then, can architectural practice implement architectural criticism? The larger the project, the wider the influence of criticism. Political awareness in the primary planning stages encourages a critical reading of municipal and regional building plans, which define the contours of planning. The proposal for the Venice Biennale that was submitted by the architects Perla Kaufman and Fara Goldman, called 'A Way with Land', shows the planning horizon of such awareness in relation to urban planning. In this proposal, the architects point at the Negev area as the future site of planning in Israel, in the hope that this planning will be carried out through sensitivity to social and political problems.

A social commitment was inspired in modern architecture, by international conventions (CIAM) – held in the first half of the twentieth century in Europe

– which encourged the planning of minimal living conditions through the standardization and rationalization of planning. But in the Israeli economy, which has gone through privatization processes in the past few decades, there are only a small number of projects of this type, and these are usually offered in the private market, with limited social emphases. A more active approach is proposed by the 'Bimkom' (the Hebrew word for alternative), an NGO which works to reinforce the relations between human rights and planning systems in Israel. This stems from an assumption that 'spatial planning is fundamentally political' (Fenster, 2002, pp. 5–9), and from a desire to correct the political and social wrongs that are caused by planning. However, even the activity of this association pertains more to the conditions that precede the actual planning, to issues that pertain to land rights and land destination, rather than to the critical possibilities of the planning language, of the architectural design itself. Not so the proposals for the Venice Biennale that are shown in the 'Israeli Pavilion' exhibition. These proposals attempt, as described by the architect Zvi Efrat, 'to talk about architecture through architecture', and to develop a real critical architectural language.

The demand to implement criticism through planning leads to a question regarding the ability of architecture – which is by nature physical, material, formal and nonverbal – to express critical contents, objection, and to declare a social or political agenda. Can a built wall, a dome or the design of a building's doorways possess moral contents of themselves? After all, architectural shapes, building elements and materials can serve both moral and non-moral goals to the same extent. What is the meaningful unit in architecture? This is not the place in which to fully investigate these important questions.

Critical Architecture in the Proposals for the Venice Biennale

The fact that the Israeli pavilion in the Venice Exhibition Grounds (the 'Gardinni') is undergoing renovation enabled some of the proposals for the Biennale, to design alternative exhibition conditions, outside the building, through the use of critical architectural language, in addition to and in combination with their curator thesis. In doing so, they created a sub-group of works within the 'Israeli Pavilion' exhibition, enveloping the building with a 'second skin' – a screen wall that is a common and trendy architectural element in international architecture. It delimited a narrow pathway that was also intended to serve as a semi-closed exhibition path, attached to the external wall of the renovated Israeli pavilion.

The architect Zvi Efrat built a multidisciplinary installation 'Borderline Disorder' at the Biennale – which combied the work of architects and artists .An enlarged cartographic image of the Israeli-Palestinian terrain, was printed on the shutters, in such a way that brings to mind a military camouflage net. Behind the shutters, a horizontal panoramic photograph of the horison of the Israeli architecture was presented, along with a computer animation that depicted the the historical expansion processes of the Israeli space and a video that depicted a fragment of a demonstration (see Figure 14.2). Sounds from a sound installation, depicting and manipulating the Israeli daily audio experience, accompanied the path to the pavilion.

Arkod Architects, in their proposal, proposed to wrap the pavilion in three-dimensional scaffolding, as in a building site, and to turn it into a space for exhibiting installations. The proposal I myself submitted, in collaboration with the landscape Architects Naama Meishar, Ami Tsruya and Zofit Tuvi, also proposed to wrap the building in a semi-opaque 'scaffolding of images' that would hide the renovation process in the Israeli pavilion and present images and texts that document the current building process in Israel (see Figure 14.3). The architects Rafi Segal and Eyal Weizman, in their proposal

Figure 14.2 Z. Efrat, 'Borderline Disorder' (2002), detail of the Israeli installation at the 8th Architecture Biennale

Figure 14.3 S. Cohen, Z. Tuvi, A. Tsruya and N. Meishar, 'Under Construction' (2002), detail of the Israeli installation at the 8th Architecture Biennale

'Closure', sought to provide the Israeli pavilion with an image of 'architecture under siege'. They proposed to surround the Israeli pavilion with two soil and stone dikes, climbing gradually up to a height of approximately three meters and creating a kind of fortification or blockage array, which would protect the building but would also close it off and prevent free approach.

Most of these works used a combined strategy: they turned the architectural addition into a reflective component, formulating criticism of Israeli architecture. More specifically, they made use of the double wall, as a metaphor for the exposure and concealment of moral truths. The building materials for this wall were taken from a world that is not architectural or representative – soil dikes, scaffolding or window shutters – and thus they are charged with military meanings, meanings taken from the world of advertising and art, or from the early stages of building. Alexander Tzonis and Lian Lefaivre had written about the displacement of architectural components and their change of context as a poetic mean, that turns the familiar into something foreign, which distorts the immediate and natural automatic perception of the building. Unconventional and 'low' building materials has a similar critical role. It suspends the purposeful use of the building and paves the way for the creation of a critical space. Then, the visitor to the building is invited to decode the meanings that this architecture imports from other fields, such as photography or geography – fields that can be interpreted and made to speak, from which it is possible to extract sayings about architectural. Use and context are enlisted together for the purpose of reconstructing the Israel sense of place in the Venice Biennale.

The exhibition context, in which architecture seeks to become surprising and thought-provoking, allowed the critical message of the work built in the Venice Biennale to be accepted. Buildings are not accompanied by written explanations, but exhibitions are not mute. The attempts made by the proposals for the Venice Biennale, to formulate criticism through architectural means, were not based exclusively on architecture. They were accompanied by clear and strong texts, from which I have quoted here. In the Architects' House Gallery, the formal and contextual similarity between the various proposals, which I have noted here, caused the different works to echo each other and to reinforce their critical content.

Notes

1   As curator of the Architects' House Gallery, as submitter of one of the proposals for the Venice Biennale, and as participant in meetings of the steering committee appointed by the United Architects' Association for the Berlin exhibition, I appealed to the heads of the committees that chose the candidates, and through them to the various candidates, in a request to present their proposals in a combined exhibition. Seven out of the nine candidates for the Venice Biennale, and five out of the six candidates for the Berlin exhibition, consented to exhibit their work, which was then presented in the 'Israeli Pavilion' exhibition. The works that were presented in the exhibition were usually faithful to the primary idea that had been presented to the choosing committees.

2   The works that were sent were taken from the exhibition 'Space 2001 – The Israeli Architecture Biennale', which was presented in January 2002 in the Architects' House Gallery, and curated by the Exhibition Committee of the United Architects' Association in Israel.

3   Esther Zandberg, architecture critic of the *Haaretz* newspaper, brought this affair to the knowledge of the general public. Zandberg criticized the decision that was made by the United Architects' Association in Israel, to reject Segal and Weizman's work.

4   The complete list of works and presenters in the exhibition (the names of the team leaders are emphasized in heavy print. In the text itself I shall refer only to their names):
    Proposals for the Israeli Pavilion at the Venice Biennale
    A.   'Borderline Disorder' – Curator: Zvi Efrat, Production: Michael Gov, Arad Turgeman, Installation Design: Efrat-kowalsky Architects, Zvi Efrat, Meira Kowalsky, Keren Avni, Engineers: Leonid Berzon, Ya'akov Achbert, Cartograhic

image: Eyal Weizman, Panorama: Daniel Bower, Sound work: Yossi Mar-Haim, Video: Avi Mugrabi,Graphic Design: Yotam Bezalel, Computer Animation: Yehoshua Gutman, Donny Valer, Racheli Rotem, Matan Sapir, Malkit Shoshan, Vitala Tauz, Rinat Berkovitch, Helena Gibel, Yulia Umaneski, Tehila Megiar, Ronit Markovitch, Tamar Ziv, Tamar Makover, and a production team.

B.   'Tikun' – Sigal Bar-Nir, Yael Moria, Asaf Galay, Ido Nissenbaum, Rebecca Sternberg.

C.   'Erasures – The familiar landscape and the foreign city' – Bilo Blich.

D.   'Under Construction' – Shelly Cohen, Naama Meishar, Kav Landscape architecture – Zofit Tuvi and Ami Tsruya, Raffy Tsruya.

E.   'Closure' – Rafi Segal, Eyal Weizman.

F.   'Under Construction' – Orit Siman-Tov Pinchas, Doron Pinchas, Arkod Architects, Najud Mazrib, Anat Frenkel, Jonathan Shaked, Stephen Mati, Arie Rotenberg, MeirGal, Orit Shershevski Mor.

G.   'A Way with Land' – Perla Kaufman, Fara Goldman, Pazit Shauli, Peach Visual System Design Ltd.

Proposals for the International Architects' Union Convention in Berlin

I.   'Modernism in Dispute – Breaches in Israeli Architecture' – Yair Avigdor, Yossi Klein, Eran Neuman.

II.   'Area D.' – Yehoshua Gutman, Rinat Berkowitz.

III. 'Israeli Modernism between Tel-Aviv and Jerusalem' – Zofiya Dekel, Lilach Dekel, Nizan Ram.

IV.   'Architecture of Insecurity or Brief Thoughts of an Old and Beautiful City on the Meditterenean' – Eytan Hillel, Yael Ben Aroya.

V.   'Civilian Occupation – The Politics of Israeli Architecture' – Rafi Segal, Eyal Weizman, Zvi Efrat, Daniel Bower, Meiron Binvenishti, Geoge Dufin, Nadav Harel, Oren Yiftachel, Miloten Labudovitch, Gideon Levi, Ilan Postach, Micky Keratsman, Sharon Rotbard, Efrat Shvili, Eran Tamir-Tawil, Pavel Wolberg.

5   From the proposal I myself submitted in collaboration with the landscape architects Naama Meishar, Ami Tsruya and Zofit Tuvi.

6   Some of the articles in the catalogue address the history of Israeli building from its very beginning and before the occupation of the territories at the 1967 war.

7   In the 'Israeli Pavilion' exhibition, alongside the competing works, the original version of the proposal for Berlin was presented. This version was different from the final catalogue (more moderate, according to the heads of the exhibition's steering committee), and was accompanied by the response of the head of the Architects' Association, from which I quote here, and by the response of the steering committee head.

8   The community Settlment is a new and urban form of settlement, which began in the 1980s and 1990s in West Bank settlements and in suburbs close to and beyond the Green Line. Legally, community Settlments are collaborative associations, and this enables them to be selective in accepting new members.

9   Mixed cities are cities in which Palestinian refugees continued to live, within the borders of Israel, after the 1948 war.

10 The series of exhibitions that I have been curating in the Architects' House Gallery received the name 'Local', after this new local discourse.

11 The social discussion of Israeli architecture have many more precedents than the political one. I shall note here only the preoccupation with social polarization in Israel, which has focused on the topic of public housing in Israel.

12   The exhibition 'The Israeli Project – Construction and Architecture 1948–1973' was presented in October 2000, at the Helena Rubinstein Pavilion, Tel-Aviv Museum.

13   The exhibition was presented at the 'Askola' school gallery in 2000.

14   The exhibition 'Pastoralia', part of the 'Local' series, was presented at the Architects' House Gallery in May–June 2001.

15   The exhibition 'Evacuation-Construction' was presented in the Camera Obscura Gallery in March 2001.

16   See for example the annual convention of the United Architects' Association that was held in Ma'alot-Tarshicha and addressed the subject of involving the public in planning.

17   The journal *Theory and Criticism* (The Van Leer Jerusalem Institute and Hakibbutz Hameuchad Publishing House, Tel Aviv), led the preoccupation with critical aspects of the Israeli culture.

18   For example, the series of books *The Israelis*, edited by Gideon Samet and published by Keter Press.

19   The exhibition space is the natural existence space for critical architecture, but of-course not every architecture that is presented in it is necessarily critical, as I have demonstrated in relation to previous Israeli exhibitions presented in the Venice Biennale.

20   Azulay argues that the critical position embraces an external apperance in order to enable criticisim, even though it is posioned within the artistic field. (Azulay, 1996, p. 66).

References

Azulay, A., *Training for the Art Critique of Museal Economy* (Tel-Aviv: Hakibbutz Hameuchad, 1999).

Chinski, S., 'Silence of the Fish, The Local versus the Universal in the Israeli culture', *Theory and Criticism – An Israeli Forum*, 4 (Autumn) (1993), pp. 57–86 .

Chinski, S., 'Eyes Wide Shut: The Acquired Albino Syndrome of the Israeli Art Field', *Theory and Criticism – An Israeli Forum*, 20 (Spring) (2002), pp. 105–22.

Efrat, Z., 'Foreword', in Z. Efrat (ed.), *Borderline Disorder: The Israeli Pavilion The 8th International Architecture Exhibition* (La Biennale di Venezia, 2002), pp. 24–5.

Efrat, Z., 'Borderline Disorder', in Z. Efrat (ed.), *Borderline Disorder: The Israeli Pavilion The 8th International Architecture Exhibition* (La Biennale di Venezia, 2002), p. 31.

Fenster, T., 'Opening Words', *Planning Rights in Israel: The Relations between Community and Establishment Relations* (Jerusalem, 2002).

Frampton, K., *Modern Architecture: A Critical History* (London: Thames and Hudson, 1980), pp. 313–27.

Fukasas, M., 'Less Aesthetics, More Ethics', in M. Fukasas (ed.), *Less Aesthetics, More Ethics, 7th International Architecture Exhibition* (Marseilio: La Biennale di Venezia, 2000).

Gaon, G. and Paz, A. (eds), *Point of View: Four Approaches to Landscape architecture in Israel* (Tel-Aviv: The Genia Shreiber University Gallery of Art, 1996).

Gugghenheim, D. and Eytan, O., in H. Hollien (ed.), *Sensing the Future, 6th International Architecture Exhibition* (Electa: La Biennale di Venezia, 1996), pp. 392–5.

Levy, G., 'The Lowest Point in Israel', in R. Segal and E. Weizman (eds), *A Civilian Occupation: The Politics of Israeli Architecture* (London and New York: Verso, 2002), pp. 81–2.

Rotbard, S., 'Homa Umigdal', in R. Segal and E. Weizman (eds), *A Civil Occupation: The Politics of Israeli Architecture* (London and New York: Verso, 2002).

Ruding, A., 'Are Politics built into Architecture?', *New York Times*, 10 August 2002.

Said, E.D., *Orientalism* (New York: Vintage Books, 1978).

Schocken, H., *Intimate Anonymity: The Isreli Pavilion. The 7th International Architecture Exhibition* (La Biennale di Venezia, 2000).

Segal, R. and Weizman, E., 'The Mountain', in R. Segal and E. Weizman (eds), *A Civilian Occupation: The Politics of Israeli Architecture* (London and New York: Verso, 2002), pp. 42–6.

Shenhav, Y., 'Space, Land, Home: On the Normalization of a "New Discourse"', *Theory and Criticism – An Israeli Forum,* 16 (Spring) (2000), pp. 3–13.

Tamir-Tawil, E., 'To Start a City From Scratch', in R. Segal and E. Weizman (eds), *A Civilian Occupation: The Politics of Israeli Architecture* (London and New York: Verso, 2002), pp. 81–2.

Tzonis, A. and. Lefaivre, L., 'Why Critical Regionalism Today?', in K. Nesbit (ed.), *Theorising a New Agenda for Architecture* (NewYork: Princeton Architectural Press, 1996), pp.484–92.

Zandberg, A., 'The Urge to Create a New World', *Haaretz*, Gallery Section, 1 May 2002.

Zandberg, A., 'The Drawing Table as a Battlefield', *Haaretz*, Gallery Section, 19 July 2002.

Zandberg, A., 'There is No Such Thing as Architecture Neto', *Haaretz*, Gallery Section, 25 July 2002.

Zandberg, A., 'Blurred Boundaries Through the Window Shutters', *Haaretz*, Gallery Section, 4 September 2002.

Zandberg, A., 'Israel is Not Just the Army', *Haaretz*, Gallery Section, 3 October 2002.

Zandberg, A., 'Civilian Occupation', *Haaretz*, Gallery Section, 23 January 2003.

Zandberg, E., 'The Exhibition that was Canceled', *Hagar: International Social Science Review* (Beer Sheva: Ben-gurion University, forthcoming).

Index

Abu al-Hayja, Muhammad 303, 306, 310, 313, 315–20, 322, 323
Acre 188, 340, 341
Ahad Ha'am 17, 22, 23, 32, 43, 46
Andromeda Hill 211, 212, 220
anti-Semitism 22, 23, 25, 32
Assif, Shamai 289
Averbuch, Genia 60, 63
Avigdor, Yair 331, 349
Avni, Yossi 274, 279, 280, 281, 348

Baerwald, Alexander 56
Balad ash-Sheikh 285–8
Balfour Declaration (1917) 22, 27
Bar, Michael 180–83
Bar-Nir, Sigal 331, 333, 334, 342, 349
Bauhaus (see also International Style) 17–19, 24, 27–8, 30, 33, 45, 47, 57, 74, 78, 91, 175, 196, 214, 217
Beer Sheva 87,108
Beit Jalla 136, 138, 149, 151, 152, 153, 160
Ben-Aroya, Yael 341
Ben-Gurion, David 48, 76–8, 82, 88, 91, 107, 143, 168, 196, 249
Berkowitz, Rinat 288, 342, 349
Berlin Congress 12, 330
Bethlehem 149, 150, 151, 160
bin Nun, Alon 278
Blich, Bilo 333, 343, 349
British Mandate (1948) 10, 19, 27, 43, 59, 78–9, 81, 166–7, 174, 185, 311, 317–19
Brutzkus, Eliezer 82, 85
Buber, Martin 17, 31–4, 43, 49, 77, 245

capitalism 5, 10, 193, 194
Carmel National Park 106, 107, 304, 306, 310–13, 323

Children's Memorial Museum 248, 257, 258
Christaller, Walter 83, 84
colonialism 7, 10, 36, 193, 194
Cook, Peter 167, 265, 268
cultural identity 146, 231, 237
cultural Zionism 17, 22, 32, 36, 41, 43

Dekel, Zofia 281, 331, 340, 349
demographic engineering 123, 168, 170, 182
de Certeau, Michel 139, 156, 295
Diaspora 3, 25, 26, 32, 195, 204, 227, 239, 250, 260, 261, 341
Dinur, Yehiel 253, 258, 259, 261
Dizengoff, Meir 59, 62, 194, 197, 217
Dome of the Rock 227, 228, 231, 233, 240, 241, 243
Don Yehiya 249, 250, 252, 261, 262
Dubiner House 265, 267

ecology discourse 305, 308, 309
Efrat, Zvi 8, 9, 76, 124, 125, 329, 333, 340, 342, 345, 346, 348, 349
Einstein, Albert 120, 230
Ein Houd 11, 303, 304, 306, 310–15, 317, 320–23
El-Hanani, Arieh 60, 61, 71
Elhanani, Aba 96, 102
el Kassam, Iz A-din 11, 285, 286, 287, 290, 291, 298
Eretz Israel/Eretz Yisrael 23, 26, 30, 52, 55, 62, 65 ,69, 73
Eytan, Omry 339, 349

Fascism 23, 32, 37
Foucault, Michel 6, 298, 321, 334
Frampton, Kenneth 340

Fuksas, Massimiliano 205–6, 212–14, 221, 343

Galilee 92, 124, 125, 126, 144, 147, 270, 304
garden cities 87, 89, 91, 98, 100, 104, 108, 111, 115, 171, 177, 178, 195, 197
Gaza 112, 174, 214
Geddes, Patrick 101, 113, 195, 231–5, 243–4
gesamtkunstwerk 41, 67
Ghetto Fighters' House Museum 248, 257
Gilad, Shlomo 106
Gilo 9, 136–9, 143–5, 147–55, 159–60
globalization 121, 139, 200, 205
Goldman, Fara 344, 349
Goldstein, Baruch 286, 287
Green Line 87, 153, 159, 349
Guggenheim, David 339
Gutman, Yehoshua 342, 349

Habermas, Jürgen 334
Haifa 47, 48, 57, 79, 90, 92, 93, 101, 104–6, 133, 177, 188, 205, 285, 287–8, 311–12, 315, 318, 322
Hall, Stewart 316
Hanuka, Assaf 264, 280
Harpaz, Yoseph 238, 239, 244, 245
Hebrew University 10, 23, 37, 39, 43, 51, 67, 94–5, 171, 174, 227–44, 269
Hebron 174, 286
Hecker, Zvi 11, 264–81
Herzl, Theodor 17, 22, 23, 43, 46, 76, 194, 214
Hillel, Eytan 332, 341, 349
Holliday, Clifford 167, 172–4, 177, 187
Holocaust 10, 11, 81, 168, 247–54, 256–261
Holocaust Martyrs' and Heroes' Remembrance Authority 11, 247, 249, 253

Holy Temple 227, 231, 234–5, 240–41
Howard, Ebenezer 84, 91, 195

'Israeli Pavilion' exhibition 12, 329–49
identity politics 316, 317, 319
International Style (IS) (*see also* Bauhaus) 17, 18, 21, 27, 42–3, 45, 47, 52, 54–7, 63–4, 66, 72, 194, 196–99, 203, 208, 214–5, 221
Intifada 9, 136, 151, 153, 156, 204,2 05
Islamic Hamas 286
Israel 3, 8–12, 18, 21–3, 25–6, 2 8, 30–31, 43–5, 72–3, 76–80, 82–5, 88–94, 96–8, 100–5, 107–8, 110–14, 122–9, 132–3, 142–5, 147, 152, 154, 158–9, 161, 165, 167–70, 183–4, 188–9, 194, 197, 201, 205, 207–9, 212–15, 219, 227, 239–41, 243, 247–54, 257–61, 264–5, 267–8, 272–3, 275, 278, 280–81, 285–96, 300, 301, 303–5, 307–15, 317–23, 329–36, 338, 341–46, 348–9
Israeli-Palestinian conflict 331, 332, 334, 335

Jaffa 10, 100, 167, 175, 177, 183, 192–5, 197–16, 218–21, 234, 268, 271, 329
Jericho 237, 338
Jerusalem 9, 10, 20, 31, 62, 73, 79, 94–5, 114, 133, 136, 145–6, 148–9, 151–3, 159, 167, 173, 202–3, 205, 214, 228–32, 234–7, 239–41, 243, 248, 253, 256, 260, 270, 278, 281, 319, 331, 338, 340, 349
Jewish immigration 9, 21, 79, 124, 168, 169, 170, 178, 189, 201
Jewish National Fund (JNF) 67, 104, 158, 207, 304, 306, 312–14, 319, 322
Jewish settlement in Palestine 52, 55, 62, 64
Judaea 34, 44, 241
Judaism 22, 32

Judaization 125–6, 128–9, 132, 166, 179, 182, 185, 187, 220

Ka-Tzetnik 135633 *see* Dinur, Yahiel
Kalderon, Nissim 274
Kaplan, Danny 279–81
Karmi, Ram 238–40, 245, 248, 257–9
Kauffmann, Richard 56, 60, 65, 66, 70, 73, 108
Kaufman, Perla 344, 349
Keret, Etgar 264, 271, 280
kibbutz/kibbutzim 26, 27, 28, 48, 80, 86, 96, 108, 111, 179, 196, 257, 260, 315, 332
Kikar Hamedina 99, 100, 101, 103, 114
Kimmerling, Baruch 242, 245
Kiryat Gat 107, 126
Kiryat Shmona 85, 86
kitsch 271, 272
Klein, Yosy 113, 280, 331, 349
Kliot, Nurit 307
Kook, Peter 206, 286, 291
Krakauer, Leopold 56
Kuzinski, Dov 56

Labour Zionism 23, 27, 42–4, 47–8
Lebensraum 84
Lefaivre, Lian 340, 343, 347
Lefebvre, Henri 5, 6, 139, 156, 165, 182, 215
Leitersdorf, Thomas 338
Levant Fair 8, 52–74
Le Corbusier 17, 19, 24, 30, 55, 57, 94, 96, 196, 197, 215, 340
Liebman, Charles S. 249, 250, 261, 262
Lod 10, 165–88, 210
Loewe, Heinrich 229
Lohamei Hagetaot 257, 260
Lurie, Harry 60
Lydda 166–7, 170–76, 178, 180, 185, 188, 189
Lyotard, Jean François 320

Ma'ale Edummim 338
Mandate Palestine 7, 17, 18, 21, 23, 29, 42, 194
Mandate period 10, 17, 21, 46, 166, 175, 197, 198, 206
Mandel, Sa'adia 184–6
Marcuse, Herbert 3, 6, 141
Masada 293
Mears, Frank 232, 233–5, 244
Meishar, Naama 11, 303, 333, 342, 346, 347, 349
Mendelsohn, Erich 7, 17–51, 56, 113–14
Mizrahim 120, 124–33, 169, 179, 188
modernism 7, 8, 17, 18, 22, 24–5, 34, 45–7, 52, 53–7, 63–4, 70–72, 96, 101, 112, 146–7, 196, 269, 330–31, 339–40, 349
Moria, Yael 331, 333, 334, 342, 349
Mount Carmel 90, 92, 93, 104, 303
Mount of Remembrance (Har Hazikaron) 248, 253, 256, 260
Mount Scopus 10, 149, 227–8, 230–31, 235–43
Mumford, Lewis 41, 244

nationalism 6, 7, 10, 25, 31–3, 41, 48, 165, 187, 193–4, 231, 243, 292
nationalist 10, 194, 227, 237, 281
National Master Plan 288–91, 299, 300
National Outline Plan 112
national socialism 91
Naveh, Ze'ev 306, 307
Nazareth 188, 340
Nazis/Nazism 22, 36, 49, 50, 84, 196, 251, 258, 259
Negev 9, 41, 80, 90, 92, 107–8, 110–11, 124–7, 307, 344
Nes, Adi 275–6
Nesher 11, 285, 287–8
Netanyahu, Binyamin 272
Neufeld, Yoseph 23, 24, 28, 47, 59, 60
Neuman, Eran 331, 349
Neumann, Alfred 265

New Towns 77, 82, 85, 86, 87
Niemeyer, Oscar 8, 89–14
Nitzan, Yaakov 68
Nora, Pierre 247, 252, 256, 257

Old City 95, 145, 217, 229–30, 232, 234, 237–44
orientalism 7, 25, 31, 33, 45, 49, 56, 62, 341

Pacovsky, Richard 60
Palestine 7, 8, 10, 17–19, 21–3, 25, 27, 29–38, 41–3, 45–55, 57–70, 72–4, 91–2, 114, 123, 152, 166, 170, 172–5, 177, 188, 192–7, 199–202, 204, 210, 214, 219, 221, 227, 229, 232, 234–5, 243, 286, 293, 295, 318, 322
Palestinian immigration 9, 21, 79, 124, 168–70, 189, 201
Palestinian Intifada 9, 136, 151
Palestinian refugees 169, 178, 306, 349
Palmach House 275–79, 281
Peres Center for Peace 206–7, 212–14, 221
Pinkus, Yirmi 273
Pinsker, Leon 22
places of memory 247–9, 251, 252, 256, 257, 259–61
Polcheck, Otto 167, 175, 176–8, 180, 187
political Zionism 17, 22, 23, 43, 46
population dispersal 83, 92, 105, 124, 143
Posener, Julius 25, 26, 28, 29, 30, 35, 42
postmodernism 140, 147, 208, 211, 215, 219, 271, 339

Rabin, Yizhak 237, 281
Ratner, Yohanan 19, 56
Raviv-Verobeichic, Moshe 69, 74
Rechter, Zeev 20, 23, 24, 56, 245, 273
Redstone, Louis 60

Reznick, David 238–40, 243, 245
Rotbard, Sharon 335, 336, 349
Rupin, Arthur 51, 73, 228

Said, Edward 6, 33, 166, 175, 187, 341
Samet, Gideon 274, 281
Schocken, Hillel 20, 31, 42, 332
Segal, Rafi 275, 278, 329–30, 333, 335–8, 346, 348–9
Segev, Tom 258, 259, 261, 262
sense of belonging 123, 167, 289–90, 292, 295–98, 341
settler societies 4, 120, 122, 132, 142, 184
Sharet, Moshe 62
Sharon, Arieh 19, 23–5, 27–9, 44, 60, 78–9, 82, 88, 91–2, 96, 217
Sharon, Ariel 305
Sharon Master Plan 7785, 87, 91, 92, 112, 124, 184–6, 244, 290
Sheinkin, Menachem 228
Six Day War (1967) 9, 10, 136, 145, 159, 204, 235, 237, 241, 245, 349
socialism 8, 17, 19, 21, 23, 26, 27, 29, 31, 41, 42, 43, 45, 73, 87, 90, 97, 111, 260
Sokolow, Naum 228
Spiral House 264–69, 271, 275, 278–9

Taub, Gadi 270, 271, 272, 274, 280
Technion 57, 67, 72, 205, 322
Tel Aviv 7, 8, 10, 17–20, 23–4, 27–8, 30, 42–3, 46, 51–3, 54–61, 64–8, 72–4, 79, 88, 90, 95, 98–101, 103–4, 113–14, 133, 192–220, 268, 270, 272–3, 275–6, 278, 280–81, 315, 331, 340, 349
Temple Mount 230, 231, 238, 240, 241, 242, 243, 244
traditionalism 62
Tsruya, Amy 333, 346, 347, 349
Tuvi, Zofit 333, 346, 347, 349
Tzonis, Alexander 340, 343, 347

Ussishkin, Menachem 228, 229, 230, 244

Valley of the Communities 248, 254–6, 260–61
van der Rohe, Mies 19, 55
Venice Biennale 329–30, 332, 333, 339–48, 350

War of 1948 78
War of Independence (1948) 281, 304, 309, 313–14
Weizman, Eyal 329–30, 333, 335–6, 338, 346, 348–9
Weizmann, Chaim 23, 31, 38, 39, 42, 228, 243
Weltsch, Willie 60, 61, 64
West Bank 112, 127, 144, 147, 153, 158, 159, 336, 338, 349
World War I 21, 34, 41, 195
World War II 77, 82, 249, 253, 278
World Zionist Organization 19, 23, 31, 38, 322

Yad Layeled 248, 257–61

Yad Vashem 247–9, 253–4, 256, 260–61
Yahalom, Lippa 248, 254, 255
Yashar, Yitzhak 87
Yiftachel, Oren 3, 123, 125, 130, 131, 169, 336, 349
Yishuv 17, 19, 23, 25–7, 29, 30–31, 42–3, 45–6, 54–61, 65, 67–9

Zionism 3–4, 7–10, 17–19, 21–9, 31–4, 36–8, 41–7, 51–2, 54, 59, 61, 65–74, 77–8, 80, 82, 84–5, 89–91, 94, 100, 104–5, 107, 113, 124–5, 132–3, 145, 153–4, 166, 179, 185–6, 188–9, 192–8, 200, 202–3, 206, 208, 211, 215, 217, 227–9, 232, 234–5, 242, 243, 249–50, 260–61, 269, 278–9, 286, 293, 296, 308, 315–16, 332–3, 335, 341–2
Zionist Enterprise (HaMiph'al HaZioni) 77–8
Zionist ideology 8, 24, 48, 90, 192, 195, 229
Zionist narrative 45, 48, 185, 308
Zionist Project *see* Zionist Enterprise
Zur, Dan 248, 254, 255